# Nanotechnology in Environmental Science and Healthcare

Acharya Balkrishna · Naveen Thakur ·
Vedpriya Arya · Muralikrishnan Dhanasekaran ·
Ashwani Kumar
Editors

# Nanotechnology in Environmental Science and Healthcare

Springer

*Editors*
Acharya Balkrishna
Patanjali Research Foundation
Haridwar, Uttarakhand, India

Vedpriya Arya
Patanjali Research Foundation
Haridwar, Uttarakhand, India

Ashwani Kumar
Patanjali Research Foundation
Haridwar, Uttarakhand, India

Naveen Thakur
Faculty of Allied and Health Care Sciences
Desh Bhagat University
Mandi Gobindgarh, Punjab, India

Muralikrishnan Dhanasekaran
Graduate School
Auburn University
Auburn, AL, USA

ISBN 978-981-95-6713-3 ISBN 978-981-95-6714-0 (eBook)
https://doi.org/10.1007/978-981-95-6714-0



This Springer imprint is published by the registered company Springer Nature Singapore Pte Ltd.
The registered company address is: 152 Beach Road, #21-01/04 Gateway East, Singapore 189721, Singapore

# Preface

Nanotechnology, with its vast potential in environmental remediation, medicine, and public health, offers immense promise. This book delves into the application of nanotechnology in environmental and public health management, encompassing the latest advancements, challenges, and practical suggestions. Contributions from internationally renowned academicians ensure a comprehensive, up-to-date perspective that equips readers with deep insights and future readiness. This book goes beyond mere compilation by offering newer interpretations of the existing literature.

The coverage areas include nanomaterials' implications for environmental remediation, antimicrobial resistance, water purification, and controlled agrochemical release, helping readers grasp novel technologies, risks, and solutions for pollutants like microplastics, heavy metals, and dyes. Metal nanoparticles for disease diagnosis, nanomaterial-based drug delivery systems, nano-approaches for neurodegenerative diseases like Alzheimer's, cancer nanomedicine trends, and critical perspectives on nanotoxicology also constitute a major portion. Tackling challenges such as agrochemical misuse, delayed diagnosis, blood-brain barrier hurdles, and limitations in cancer therapy, the book presents recent advances, practical solutions, and a future roadmap. While nanotechnology's compelling features attract researchers, its environmental and clinical translation faces hurdles, particularly those related to the toxicity of nanomaterials. Thus, this book also addresses the long-term environmental and public health risks of nanomaterials.

This book serves as a valuable resource for researchers, academicians, and students (undergraduate and graduate) in nanotechnology, physics, chemistry, engineering, biology, biotechnology, and pharmaceutical sciences. Its broad scope makes it ideal for the pharmaceutical and biomedical industries, academic institutions, research organizations, public policymakers, and environmental regulators.

The editors express heartfelt thanks to revered Swami Ramdev for his unwavering support and inspiration.

We sincerely appreciate all authors for their quality contributions to this book. Finally, we thank Springer Nature Publishing for their instrumental support in bringing this book to fruition.

Haridwar, India — Acharya Balkrishna
Fatehgarh Sahib, India — Naveen Thakur
Haridwar, India — Vedpriya Arya
Auburn, USA — Muralikrishnan Dhanasekaran
Haridwar, India — Ashwani Kumar

# Contents

# Chapter 1
# Nanobiosensors for Environment Quality Monitoring: Recent Developments, Challenges, and Perspectives

**Lacy Loveleen, Riya Singh, Nidhi Gupta, and Surendra Nimesh**

**Abstract** The increasing environmental pollution has led to the development of sensors for decades. The contamination of any biosphere elements not only has eventually led to the deterioration of the quality of life and environment, but also leaves behind a vulnerable future for generations to come. The development of the sensor, followed by nanobiosensors, has been a boon for sensing and controlling environmental pollution. Still, since there are significant challenges that prevent their use in real time, the chapter provides information about types, components, and applications of nanobiosensors to ensure further adequate studies and their development for real-life use.

## 1.1 Introduction

For the sustainability of life on Earth, the three most essential components of the biosphere, air, water, and soil, are crucial, on which organisms are dependent. In the current scenario, about 0.05% of the total water, 25% of fertile land, and a very small proportion of breathable air are available, which may degrade rapidly over time due to many factors, such as continuous deforestation, lack of crop rotation, and pollution of air, land, and water bodies by the release of toxic chemicals produced in factories and households (Hargrove et al. 2023). Utilization of these contaminated resources gives rise to various physiological diseases. For instance, air pollution not only affects respiratory tract diseases (such as common inflammation and infection, chronic obstructive pulmonary disease, and cancer), but also contributes to heart diseases (including stroke and coronary artery disease) and skin diseases (such as eczema, psoriasis, and acne). Thus, the need of the time is to

L. Loveleen · R. Singh · S. Nimesh (✉)
Department of Biotechnology, School of Life Sciences, Central University of Rajasthan, Ajmer, Rajasthan, India
e-mail: surendranimesh@curaj.ac.in

N. Gupta
Department of Biotechnology, IIS (Deemed to be University), Jaipur, Rajasthan, India



A. Balkrishna et al. (eds.), *Nanotechnology in Environmental Science and Healthcare*,
https://doi.org/10.1007/978-981-95-6714-0_1

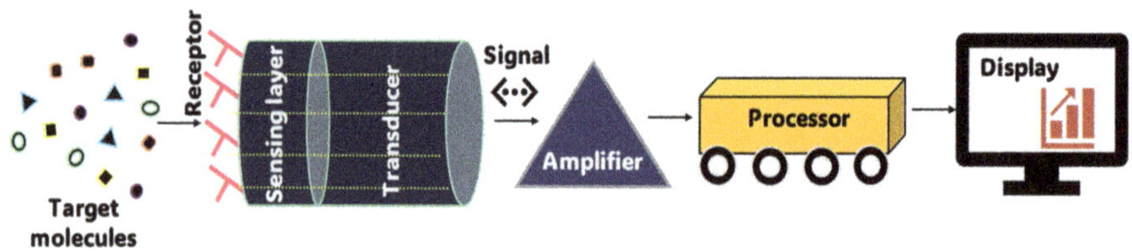


**Fig. 1.1** Brief structural description of biosensors

improve and maintain the quality of the environment. Environmental quality monitoring is a process that enables the observation and analysis of environmental quality through data collection and statistical analysis. The initiation step involves the development of sensors, wireless communication systems, remote management software, and audits that are portable, sensitive, accurate, and reusable (Huang et al. 2023). Although numerous sensors are available for monitoring environmental quality, they inadvertently contribute to the significant amount of anthropogenic substances that further pollute the environment. Some of the environmental sensors are connected to individuals to track objects that are implanted or embedded into the organism. These sensors can provide information about the location, position, and movement of the individual. Thus, we need biosensors that can naturally detect pollutants present in the surrounding environment. The parameters typically used for screening include biochemical oxygen demand (BOD), chemical oxygen demand (COD), total organic carbon (TOC), and total dissolved solids (TDS) (Omer2019).

Traditional chromatographic techniques have been used as analytical methods for detecting environmental pollutants. However, the high cost of reagents, time consumption, and utilization led to the development of biosensors, which are more sensitive, cheaper, faster, easier to operate, and portable. Biosensors comprise three components: a biological sensing element, a signal processing system, and a transducer or a physicochemical detector (Fig. 1.1).

The sensing elements are usually biological molecules, such as nucleic acids, enzymes, tissues, cells, their receptors, organelles, and microorganisms. The transducer facilitates the conversion of biological interaction signals into electrical signals that can be easily quantified. Sometimes, the sensors can be used for quantitative analysis, such as colorimetric detection of changes resulting from biological interactions. Biosensors are mainly classified into two categories based on their transduction principle and recognition element (Tetyana et al. 2021).

Optical fiber evanescent wave sensors have been widely explored for the detection of water, pesticides, hydrocarbons, and other gaseous pollutants, which are examples of optical biosensors. Another typical example of optical biosensors is the application of surface plasmon resonance, as seen in UV–Vis spectroscopy (Chevalier et al. 2018). Amperometric and impedance biosensors are electrochemical biosensors used to detect organic pollutants, heavy metals, pesticides, BOD, cyanides, and some biological toxins (Woldu 2022). The detection of air, water, and soil contamination using quartz crystal microbalances, which involves observing the change in mass and resonance frequency with a very small sample, is an example of piezoelectric biosensors (Marco et al. 1996; Etorki et al. 2007). Furthermore, based on the recognition element, biosensors are classified into immunosensors, enzymatic biosensors, genosensors, and aptasensors. Although immunosensors and enzymatic sensors have been explored more for detecting environmental pollution, aptasensors have gained prominence over time. They are thermally stable, easily modifiable sensors that can be used to target various functional groups and rehybridize.

## 1.2 Nanobiosensors

Biosensors, as the name suggests, are sensors that utilize biological components to detect or sense any biological or chemical compound present in a sample or the environment. It measures biological or chemical reactions by converting them into electrical signals. Biosensors constitute three major components, which are a bioreceptor, a transducer, and a signal processing system (Bahadır and Sezgintürk 2017). The bioreceptor, being the most crucial part of biosensors, is used to detect the target compound (also known as the analyte) and is attached to an electrode. The transducer helps to convert the chemical or biological signal into an electrical signal, and the signal processing system generates the output result, which is used for further analysis. Biosensors have applications in multiple facets of life, including health and diagnostics, environmental quality monitoring, drug discovery, food pathogen discovery, and toxin identification.

Despite being a highly useful technology, biosensors face several challenges, including low stability, low durability, slow response times, and short lifetimes. Recent advancements are being used to overcome these challenges, providing better solutions to the limitations encountered. One such development is the use of nanotechnology-based sensors, known as "nanobiosensors."

Nanobiosensors are a modification of biosensors that utilize materials and technology at the nanoscale to detect the target analyte. They have at least one component with measurements in nanometers, which is directly involved in the detection process (Huang et al. 2021). Nanobiosensors are tiny and compact structures with applications in various fields, such as drug discovery, diagnostics, food quality monitoring, environmental pollution monitoring, and many more (Balkrishna et al. 2024). Nanoscale results in the alteration of physical attributes of the bulk material; thus, the use of nanotechnology for the formulation of nanobiosensors imbues them with enhanced physical properties, such as better stability, high sensitivity, longer lifetime, and cost efficiency, compared to conventional biosensors (Salouti and Derakhshan 2020).

Nanobiosensors, like biosensors, consist of three major components: a receptor, a transducer, and a detector system. The receptor is a bioreceptor comprising a biomolecule for interaction with the analyte. The biomolecule can be a nucleic acid, protein, enzyme, immune component, or a whole cell-based entity. The transducer is used for converting biological signals into an electrical output and comes in various

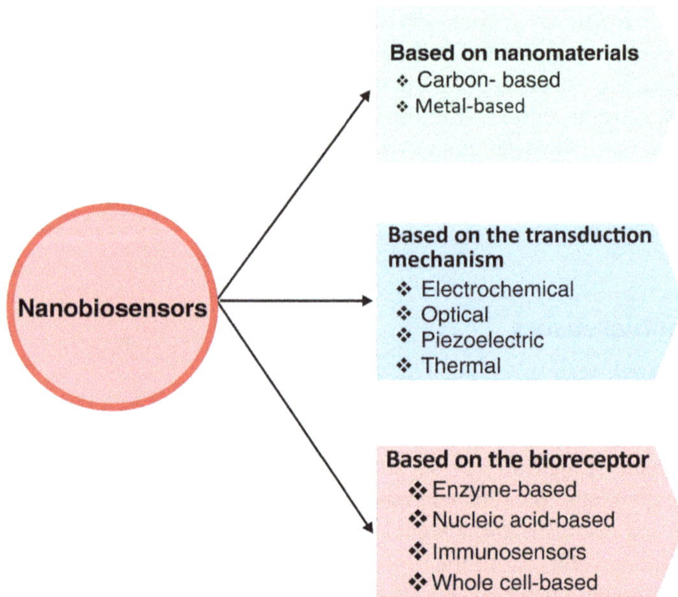


**Fig. 1.2** Different types of nanobiosensors

types, including electrochemical, optical, magnetic, and piezoelectric, which form the basis of nanobiosensor classification (Mahmood et al. 2024). The classification of nanobiosensors is based on three criteria: (i) based on the nanomaterial used, (ii) based on the transducer, and (iii) based on the bioreceptor. The various types of nanobiosensors for environmental monitoring are depicted in Fig. 1.2.

### *1.2.1 Based on the Nanomaterials*

- **Carbon-Based Nanobiosensors**

Carbon-based or carbon-derived nanomaterials are frequently utilized in the design of nanobiosensors due to their versatile nature. Properties, such as high conductivity, mechanical strength, flexibility, and high electron mobility, make carbon-based nanomaterials a highly useful component in the formulation of nanobiosensors (Chung et al. 2018). The carbon structure is responsible for the non-toxicity toward the environment, making them an eco-friendly and sustainable option along with making them a cost-effective alternative as well (Jaleh et al. 2022). Carbon-based nanomaterials, including graphene, fullerenes, carbon nanotubes, and carbon nanofibers, are utilized in various applications, such as drug delivery, medical applications, energy storage, solar cells, pollution monitoring, and many others. Originating from graphite, graphene is a carbon allotrope known for its planar honeycomb structure.

It consists of a single layer of atoms and is popular for its exceptionally high tensile strength, transparency, thin structure, and electrical conductivity. The one-layered thickness of graphene exposes nearly all the carbon atoms to environmental signals, which play a crucial role in sensing applications (Afsahi et al. 2018).

Graphene and its derivatives, mainly graphene oxide, have applications in environmental quality monitoring. The biodegradability, stability, larger surface area, and enhanced conductivity make them efficient sensor materials. They have been used to detect the presence of heavy metals, such as lead ($Pb^{2+}$), using $Pb^{2+}$ DNAzyme as a bioreceptor (Wen et al. 2013). A graphene oxide–based nanobiosensor was reported for the detection of phenolic compounds in pesticide monitoring (Zhang et al. 2012). Nanorods made from graphene or graphene oxide are also used for sensing applications.

Fullerene, another allotrope of carbon, is also used for the architecture of nanobiosensors. Fullerenes are either hollow spheres, ellipsoids, tubes, or other shapes. The unique structure imparts extraordinary properties to fullerene, such as high conductivity, high ductility, a chemically inert nature, and high tensile strength, which result in various applications of fullerene. Carbon nanotubes are one such form of fullerenes that have great application in the field of nanobiosensing. Carbon nanotubes have a diameter in the range of nanometers, while their length can be in millimeters. Their linear structure and side wall modifications facilitate their interactions with other materials. Carbon nanotubes are widely reported for use in gas pollution detection due to their high sensitivity and rapid response time. Several formulations of carbon nanotube nanobiosensors are reported for sensing air pollutants. A metallic–carbon nanotube nanobiosensor was reported to detect the presence of $H_2O_2$ pollution (Liu et al. 2019a, b). Similarly, a modified carbon nanotube nanobiosensor was developed for detecting H2S contamination in the environment (Lawrence et al. 2004).

Other carbon-based nanobiosensors like carbon nanowires, carbon nanospheres, graphene quantum dots, etc., are also known for their applicability in nanobiosensing for environmental quality monitoring. Their enhanced unique properties enable their use in various applications in day-to-day life.

- **Metal-Based Nanomaterials for Biosensing Applications**

Nanomaterials derived from metals are known as metal-based nanomaterials. Copper, zinc, aluminum, nickel, and iron, along with noble metals like gold, silver, and platinum, are the most common metals used for the synthesis of metallic nanoparticles. Metal-based nanomaterials offer significant advantages, making them well suited for synthesizing nanobiosensors. Their compact nature, high electrical conductivity, large surface area, and ease of architecture make them an exceptional option for this purpose. Metal nanomaterials can be used individually or in combination with other nanomaterials to enhance sensitivity, sensing/ response time, and precision of recognizing the analyte. Noble metals, such as gold, silver, and platinum, are primarily involved in the formation of metal-based nanobiosensors. They are inert metals with a longer lifespan compared to other metals that are prone to corrosion through oxidation.

Gold nanoparticles, in combination with biological materials, such as DNA, enzymes, antibodies, and proteins, are widely utilized in biosensing applications across various fields. Gold nanoparticles are widely used due to their inert, non-toxic, and highly biocompatible nature (Rafique et al. 2017; Balkrishna et al. 2023). Similarly, silver and platinum nanoparticles are associated with the synthesis of nanobiosensors for similar properties. Along with metals, metal oxides are also involved in the synthesis of nanobiosensors. Oxides of copper, zinc, nickel, tin, iron, cobalt, and manganese are very popular metal oxides used for this purpose. They have been used extensively due to their remarkable physical and chemical properties, which are essential for sensing applications. Among all these metal oxides, oxides of copper, zinc, and iron are used the most for the synthesis of electrochemical nanobiosensors due to their excellent magnetic properties and electron mobility (Shi et al. 2014).

Quantum dots–based nanomaterials, nanowires, nanorods, and dendrimers are some other types of nanostructures involved in the formation of nanobiosensors. They are either metallic, carbon-based, or a combination of both.

### *1.2.2 Based on the Transduction Mechanism*

- **Electrochemical Sensors**

This type of sensor is based on an electrochemical reaction on the transducer surface between the bioreceptor and the target analyte, which is then presented in the form of an electrochemical signal. The signal is produced according to the concentration of the analyte present in the environment. It is the most widely used type of biosensor as it does not damage the sample and is non-toxic to the environment. The electrical signal generated can take various forms, including current, voltage, capacitance, and impedance. Based on the signals produced, electrochemical sensors are categorized into four types: potentiometric, voltammetric, impedance spectroscopic, and amperometric sensors (Wang et al. 2014).

- **Optical Sensors**

Optical nanobiosensors are sophisticated analytical tools that detect and measure biological molecules with high sensitivity and specificity using light-based technologies and nanomaterials. They are beneficial in areas, such as health care, environmental monitoring, food safety, and drug development, as they integrate nanotechnology, optics, and biology to enable real-time monitoring of molecular interactions. The analysis is based on changes in optical properties, such as frequency, wavelength, absorbance, fluorescence, and refractive index, which produce signals in combination or individually (Long et al. 2013).

- **Piezoelectric Sensors**

A special type of biosensor, piezoelectric nanobiosensors utilize nanoscale piezoelectric materials to detect infections, biological substances, or biochemical alterations. These devices utilize the piezoelectric phenomenon, which occurs when mechanical stress is applied to certain materials, causing them to generate an electric charge. They are useful for research applications, environmental monitoring, and medical diagnostics due to their excellent sensitivity and specificity. Signals are produced when changes in mass, resonance frequency, and viscosity are detected. The resonance occurs because the magnetic field is altered (Muenchen et al. 2016).

- **Thermal Sensors**

One type of biosensor that utilizes thermal characteristics at the nanoscale to detect biological interactions, such as the presence of pathogens, biomolecules, or specific cellular markers, is known as a thermal nanobiosensor. These sensors work by monitoring variations in thermal characteristics resulting from metabolic reactions or interactions, such as temperature, thermal conductivity, or heat generation. Thermal biosensors are not widely used in environmental quality monitoring due to their low sensitivity and signal productivity. They are used in combination with other biological molecules, such as enzymes and nucleic acids, to enhance their sensitivity; however, they are not very successful when used individually (Ramanathan et al. 1999).

### *1.2.3 Based on the Bioreceptor*

- **Enzyme-Based Sensors**

Due to their high sensitivity, specificity, and adaptability to various environmental conditions, enzyme-based nanobiosensors hold great promise as tools for monitoring environmental quality. By catalyzing particular processes, these biosensors use enzymes that have been immobilized on nanomaterials to detect environmental contaminants. In this context, enzymes are often globular proteins composed of linear chains of amino acids that can fold into three-dimensional structures, providing the system with the specificity it needs. By expanding surface area, enhancing signal transduction, and permitting downsizing, nanotechnology improves sensor performance. Enzyme-based nanobiosensors are often incorporated with an electrochemical transduction mechanism (Zhao et al. 2017).

- **Nucleic Acid–Based Sensors**

Nucleic acid–based nanobiosensors are an advanced technique for monitoring environmental quality, utilizing DNA's sensitivity and specificity to detect and measure pollutants in the environment. These tiny devices utilize a combination of nanotechnology and biological recognition elements, such as DNA sequences, to develop highly sensitive tools for identifying pollutants, including heavy metals, pesticides, pathogens, and organic compounds. The detection process relies on the hybridization of the DNA sequence with its complementary sequence, the target (Kavita 2017).

Aptamers are molecules made of single-stranded DNA or RNA that can bind to specific molecules with high specificity and affinity. They are commonly known as "chemical antibodies" because they can identify and bind to targets, such as proteins, small molecules, and even entire cells. Aptamers are generating significant interest in the development of nanobiosensors due to their unique properties. Aptamers can form various structural patterns, such as hairpin loops, stems, and inner loops, through van der Waals forces, hydrogen bonding, and electrostatic interactions (Zuo et al. 2018).

- **Immunosensors**

Immunosensors are conventional nanobiosensors based on the basic antigen–antibody interaction. Antibodies are used as detection biomolecules, and the target analyte acts as the antigen. There are two types of immuno-nanobiosensors, namely labeled and non-labeled. Labeled nanobiosensors are more specific and sensitive in their detection mechanism, whereas non-labeled nanobiosensors measure and quantify the signal based on antigen–antibody interactions (Pérez-Fernández and de la Escosura-Muñiz 2022).

- **Whole Cell–Based Sensors**

Whole cell–based sensors are derived from live cells or organelles and measure pollutants based on the interaction of the live cells with the target analyte, generating an electric signal directly proportional to the amount of contaminants present. Whole cell–based nanobiosensors have gained popularity in recent times due to their reduced sensitivity to environmental fluctuations and precision signaling in open environment conditions. The entire cell or organelle is immobilized on the transducer (by entrapment, adsorption, cross-linking, etc.) to facilitate the detection procedure (Turdean 2011). Microbial nanobiosensors are also used to detect the target molecule in the sample. Microorganisms can detect changes in cell behavior and metabolic pathways. Whole cell–based nanobiosensors employ genetically modified microbes for the detection process, which contain the desired gene to detect the target organism or compound (Strosnider 2003).

## 1.3 Land Quality Monitoring

Land pollution refers to the degradation of soil quality, often caused by harmful human activities. It has severe effects on the quality of the ecosystem and can cause hazardous health effects. Various sources contribute to land pollution, including deforestation, waste disposal, mining activities, urbanization, and construction. Land pollution leads to a loss of soil fertility, the decline of wildlife, deteriorating human health, and multiple climate-related hazards. Land, being an essential part of the ecosystem, needs to be assessed for maintaining the quality and fertility of the soil. Proper quality of soil ensures sustainable agriculture and the use of land resources. This, in turn, results in the maintenance of ecosystems, food quality, adequate water resource management, and human health.

Manual quality monitoring or primitive methods help maintain soil quality, but are not very efficient. Sensors have proven to be beneficial in quality monitoring as they are mechanical and have fewer chances of errors. Among the sensors popular for this purpose, the nanobiosensor is the most beneficial option because of its eco-friendly behavior and cost-effectiveness. Nanobiosensors have proven to be an excellent option for monitoring soil quality based on parameters, such as pH, humidity, constituents, pesticide load, and microbial load (Mandal et al. 2020).

### *1.3.1 Assessment of Soil Quality*

Nanobiosensors have proven to be highly useful in assessing soil quality by detecting changes in pH, temperature, texture, soil constituents, such as heavy metals, and soil fertility. There are various nutrients and minerals present in the soil that maintain its fertility, and the change in the balance of these minerals results in the degradation of soil quality. Lipid-based nanobiosensors have been developed to detect the presence of nitrate and its forms in the soil. High nitrate levels in the soil indicate the presence of insects that feed on soil, and the detection of elevated nitrate signals proves useful in maintaining the population of these insects (Siontorou and Georgopoulos 2016). Soil moisture is also an essential factor that determines soil fertility and affects crop farming. Biosensors with integrated capacitive sensors have been developed for the rapid detection of soil moisture content (Surya et al. 2020). Nanobiosensors have also been reported to detect heavy metals and toxic chemicals in the soil. The presence of urea, arsenic, mercury, zinc, and iron has been detected using nanobiosensors. Metallic, polymeric, and metallic–polymeric combination biosensors using electrochemical transduction mechanisms have been reported for the same purpose (Numan et al. 2021). Nanomaterials popular for detecting heavy metals include carbon-based nanomaterials, polymers, and DNA-based metallic nanomaterials. Bismuth nanoparticles have been reported for the successful detection of heavy metals, including lead and cadmium (Lee et al. 2007). Modified DNA-based titanium oxide nanotubes hybridized with carbon were reported for the detection of lead in the environment (Liu et al. 2010). Multiple reports demonstrate the effectiveness of nanobiosensors in evaluating the quality of land and present them as a superior and more efficient alternative.

### 1.3.2 Assessment of Pathogens in Soil

Soil and plant pathogens in the land are responsible for considerable losses in the farming industry. Monitoring of microbial load present in the soil is necessary to maintain its fertility and quality, as crop sustainability is essential to provide feedstock for the growing population. Microbes, such as *Pseudomonas syringae, Pantoea stewartii, Ralstonia solanacearum, Pseudocercospora fijiensis, Botrytis cinerea*, and *Leptosphaeria maculans,* are more prevalent in destroying crops (Dyussembayev et al. 2021). In 2013, DNA-gold nanoprobes were formulated for the colorimetric detection of *Pseudomonas syringae.* The change of color from red to purple indicated the presence of the pathogen (Vaseghi et al. 2013). Another sensor was developed using a different technology for the detection of multiple pathogens in soil. The manufactured nanobiosensor employed the surface-enhanced Raman scattering method for the simultaneous detection of multiple plant pathogens (Lau et al. 2016). The bioreceptor mainly found in nanobiosensors utilized for pathogen detection is DNA. DNA plays a crucial role in recognizing pathogens through gene-to-gene interactions and in monitoring pathogen–host interactions, behavioral patterns, and genetic transfer. Optical nanobiosensors incorporating antibodies have proven to be significantly faster for detecting pathogens. It was primarily used for detecting E. coli, but later proved beneficial for the rapid detection of various bacteria and fungi as well (Khiyami et al. 2014). Various nanobiosensors have been reported for the abovementioned purpose, particularly those involving gold and DNA-based nanomaterials. They are fast, precise, and more efficient compared to the primitive techniques and can detect a plethora of pathogens found in soil and plants.

### 1.3.3 Assessment of Pesticides in Soil

Pesticides are chemical compounds used by farmers to protect their crops from unwanted pests and pathogens. Pesticides and fertilizers gained importance in modern agricultural practices as they are an easy and cost-effective method for pest control. The regular use of pesticides, although beneficial for crop yield, poses multiple health hazards when accumulated over time. Its overuse is not only harmful to human health, but it also affects the fertility and the quality of the soil. Therefore, a need has arisen to monitor the amount of pesticides used in the field and to assess the toxicity caused by the elevated levels of pesticides present. Nanobiosensors are the answer to the abovementioned problem. Various detection techniques, such as electrochemical sensing, optical sensing, and enzyme-based sensing, have been demonstrated to check the optimal levels of pesticides present in the soil.

Gold nanoparticle–based nanobiosensors have been successfully used for detecting pesticides, such as chlorpyrifos and carbofuran (Sun et al. 2014). Electrochemical-based sensors have also been utilized for pesticide detection by incorporating graphite and copper oxide nanomaterials for this purpose (Gholivand et al. 2018). Fast and early detection is always preferred over a time-consuming detection process. For this purpose, a single-step analysis biosensor was developed and synthesized using carbon dots for the detection of kaempferol (He et al. 2020). Similarly, multiple reports and research articles demonstrate the use of various types of nanobiosensors for detecting pesticides in soil. These nanosensors are a versatile, quick, eco-friendly, and sustainable form of technology that can serve multiple purposes simultaneously.

Nanobiosensors have therefore proven to be highly effective in determining and maintaining soil quality. It provides an exceptional alternative to the existing technologies for monitoring land quality and productivity.

## 1.4 Water Quality Monitoring

Over time, due to the development of industries and technologies, the amount of waste released into water bodies has increased significantly. Since water is essential for life on Earth, the quality and quantity of drinking water must be carefully managed. There are several methods for monitoring water quality, but the inability to retrieve these data contributes to water waste. Additionally, their low sensitivity increases the likelihood of using nanomaterial-based biosensors for contaminant detection. These nanobiosensors usually use nucleic acids as their detectors, with promising and improved results in terms that they are comparatively more stable, sensitive, rapid, accurate, and portable.

Most contaminants that affect the water system are characterized into two groups: microorganisms and micropollutants (Soukarié et al. 2020). The latter includes organic and inorganic pollutants, such as ions and pesticides. The sensing of these substances has been categorized into two groups: first, through (de) hybridization of DNA, and second, through nucleic acid probes, which include aptamers and DNAzymes. So, for the first category, the ability of nucleic acids to bind to a complementary strand of DNA present in the environment is explored. Single-stranded DNA strands are synthesized for the detection of probable contamination. To this, the complementary DNA strands present in the water bodies form H-bonding, indicating their presence. Additionally, the change in conformation is measured with respect to the consecutive recognition events. Another alternative is to add a fluorophore to the single-stranded oligonucleotide, and the complementary sequence may carry the quencher and vice versa.

The second method elaborates on the use of aptamers. These are single-stranded oligonucleotides that range in length from 20 to 100 nucleotides and act as probes that can specifically bind to their target site. Through the systematic evolution of ligands by exponential enrichment (SELEX), a randomized DNA library with more than 1014 sequences, flanked by specific primers, is incubated with the immobilized target. The bound sequences are eluted and amplified in PCR. The obtained results are further subjected to sequencing (Sharma et al. 2017). Aptamers are capable of detecting small molecules, ions, pathogens, and even eukaryotic cells. Likewise, attempts have been made to manufacture RNA-based aptamers; however, due to their low stability and fragile nature, they have not yet been utilized for monitoring purposes. Nevertheless, there are protein-based nanobiosensors, also known as immunosensors, based on antibody-mediated detection.

The in vitro selection of oligonucleotides that possess catalytic properties and employ metal ions as cofactors is known as DNAzymes. These oligonucleotides consist of a substrate strand and an enzyme strand, each containing an adenosine at a cleavage site for RNA-cleaving DNAzymes, which confers catalytic activity. Initially, for toxic metal ion detection, such as lead and mercury, DNAzymes that had LOD in the order of nanomolar were used, which was eventually reduced to the picomolar range after using gold and silver nanomaterials. Another matter of concern is the persistent use of toxic chemicals and pesticides in agriculture and other industries. The toxic release gets involved in the food chain and further magnifies through biomagnification processes. The breakdown of these toxins takes decades, yet cannot be removed entirely from the environment. A dozen aptamers and DNAzymes have been developed to monitor pesticides in water bodies, with aptamers against atrazine being the most effective example (McGhee et al. 2017; Rangel et al. 2018; Abraham et al. 2018).

Most pathogenic diseases, such as typhoid, cholera, and diarrhea, are caused by contaminated food and water resources. One of the major lethal toxins released by cyanobacteria that has gained attention is microcystin-LR, which is toxic to hepatocytes and has been detected using aptamers. This led to the development of more sensitive nanomaterial-based aptamers, commonly referred to as aptanano sensors. The conserved 16S rRNA sequences from bacteria are used as sensors, wrapped along with carbon nanotubes, for detecting the presence of pathogens and their genetic material in water bodies (Ozkan-Ariksoysal et al. 2017; Liu et al. 2019a, b).

Some of the other ongoing developments include the use of DNA origami and single-stranded DNA-mediated water quality monitoring. The DNA origami is a small DNA nanostructure over which the recognition element can be grafted and used for broad potential applications (Goodman et al. 2005). Some of the 2D and 3D DNA origami structures that have been experimentally used for the detection of genetic materials in water resources have been constructed lately. Similarly, a DNA origami device, known as a nanoactuator, is designed in a way that is more advanced and can detect small and large molecules. Single DNA sensing is even more sensitive than simple DNA origami, as some of these methods can be used as biochips that simultaneously acquire data from over 500 molecules, making them more sensitive (Ke et al. 2016).

## 1.5 Air Quality Monitoring

The respiratory process of every organism present on Earth is dependent on gaseous exchange. Thus, the air that we respire must help in good bodily functioning rather than cause various respiratory-related diseases. Air pollution can therefore be briefly described as the increase in organic and inorganic substances in the air, thereby decreasing air quality. So, the detection of poisonous gases before they spread around requires their early detection using biosensors. The traditional methods involve heavy and complex instrumentation, which consumes large amounts of electricity and incurs significant maintenance costs. There is a growing need for the development of sensitive, accurate, handy, and affordable tools.

Nanotechnology has, in a way, developed methods to measure, monitor, and manage air pollution, thereby improving air quality. Instances provide evidence that titanium dioxide and silver-based nanomaterials can aid in degrading toluene and other airborne pathogens present in vapors (Mohamed and Awad 2020). Similarly, Cu-Cd/$TiO_2$-Pt nanospheres were designed to photodegrade isopropanol from the gaseous phase (Mohamed and Do 2020). Nanotechnology has also developed the ability to sense poisonous gases using nanoplatforms.

Carbon and graphene nanotubes offer excellent mechanical strength, thermal and electrical conductivity, and a large surface area. These structures are effective in detecting nitrogen dioxide, ammonia, methane, carbon monoxide, sulfur dioxide, and hydrogen sulfide. Quantum dots, on the other hand, are inorganic nanoparticles that possess narrow fluorescence emission bands, thus rendering them excellent optical transducers. These can be modulated in conjunction with Cd, Se, Zn, Te, and graphene (Lesiak et al. 2019).

Nanobiosensors for air quality monitoring are primarily based on enzymes, DNA, antibodies, and other biomolecules that bind to receptors. Transducers are categorized into electrochemical, optical, piezoelectric, and colorimetric sensors. The DNA-based sensors are based on either affinity between the molecules or DNAzymes, as discussed earlier in this chapter. Similarly, the immunobiosensors are based on antigen–antibody interaction. Another major category of nanobiosensors is cell-based biosensors. These kinds of biosensors are sensitive and cost-effective compared to enzyme-based biosensors, which are expensive and can even function without the availability of enzymes (Long et al. 2013).

Apart from detecting gaseous substances, nanosensors can also be used to detect airborne pathogens that may cause severe respiratory diseases. The nanosensors meant for these purposes have four major parts: analyte, biorecognition element, transducer, and signal readout/electrochemical tests. The analytes include bacteria, protozoans, mycoplasma, and viruses. Antibodies, proteins, nucleic acids, phage, aptamers, and MIP/CIP are part of the biorecognition element. The transducers are typically composed of metals, ceramic components, polymers, nanoparticles, and arrays. At last, the signal detections are based on amperometry, potentiometry, conductometry, capacitive, and conductance (Cesewski and Johnson 2020).

Lastly, some of the newly introduced nanobiosensors are derived from green chemistry and are made out of nanoparticles synthesized using metallic salts and plant extracts. Some of the prominent examples are CuO, MnO, Ni, $Fe_3SO_4$, and Pt, made using the leaf extracts of *Malva sylvestris*, *Citrus limon*, *Ocimum sanctum*, *Aloe vera*, and *Azadirachta indica* (Phumying et al. 2013; Awwad et al. 2015; Jayandran et al. 2015; Pandian et al. 2015; Thirumurugan et al. 2016).

## 1.6 Conclusion

The increasing demand for a clean environment has become a pressing need of the time. This chapter provides a brief overview of the components, types, and importance of biosensors in environmental quality monitoring. This chapter also describes the role of nanotechnology in advancing biosensors by providing a platform for the interaction of biological molecules and other biological elements in the synthesis and operation of nanobiosensors. The eco-friendly substances and the optical properties of nanoparticles make them portable, cost-effective, sensitive, and reusable. The application of cells, genetic material, and enzymes provides an additional edge to nanotechnology. However, despite the advantages, considerable challenges still exist, and thus, more studies need to be conducted. The chapter would provide valuable knowledge for further studies and the development of more precise nanobiosensors.

## References

Abraham KM, Roueinfar M, Ponce AT, Lussier ME, Benson DB, Hong KL (2018) In vitro selection and characterization of a single-stranded DNA aptamer against the herbicide atrazine. ACS Omega 3(10):13576–13583

Afsahi S, Lerner MB, Goldstein JM, Lee J, Tang X, Bagarozzi DA Jr, Pan D, Locascio L, Walker A, Barron F, Goldsmith BR (2018) Novel graphene-based biosensor for early detection of Zika virus infection. Biosens Bioelectron 100:85–88

Awwad AM, Albiss BA, Salem NM (2015) Antibacterial activity of synthesized copper oxide nanoparticles using *Malva sylvestris* leaf extract. SMU Med J 2(1):91–101

Bahadır EB, Sezgintürk MK (2017) Biosensor technologies for analyses of food contaminants. In: Nanobiosensors. Academic Press, pp 289–337

Balkrishna A, Rohela A, Kumar A, Mishra S, Arya V, Kala V, Thakur N, Thakur N, Kumari A, Khan N (2023) Elucidating the role of plant extracts mediated gold nanoparticles as smart antimicrobials: two-way attack. J Nanomater 2023:4085090

Balkrishna A, Kumari A, Kumar A, Arya V, Chauhan A, Upadhyay NK, Guleria I, Amarowicz R, Kumar D, Kuca K (2024) Biosensors for detection of pesticide residue, mycotoxins and heavy metals in fruits and vegetables: a concise review. Microchem J 205:111292

Cesewski E, Johnson BN (2020) Electrochemical biosensors for pathogen detection. Biosens Bioelectron 159:1–29

Chevalier P, Piccardo M, de Naurois GM, Gabay I, Katzir A, Capasso F (2018) In-water fiber-optic evanescent wave sensing with quantum cascade lasers. Sens Actuators B Chem 262:195–199

Chung SH, Kim H, Jeong SW (2018) Improved thermal conductivity of carbon-based thermal interface materials by high-magnetic-field alignment. Carbon 140:24–29

Dyussembayev K, Sambasivam P, Bar I, Brownlie JC, Shiddiky MJ, Ford R (2021) Biosensor technologies for early detection and quantification of plant pathogens. Front Chem 9:1–13

Etorki AM, Hillman AR, Ryder KS, Glidle A (2007) Quartz crystal microbalance determination of trace metal ions in solution. J Electroanal Chem 599(2):275–287

Gholivand MB, Akbari A, Norouzi L (2018) Development of a novel hollow fiber-pencil graphite modified electrochemical sensor for the ultra-trace analysis of glyphosate. Sens Actuators B Chem 272:415–424

Goodman RP, Schaap IA, Tardin CF, Erben CM, Berry RM, Schmidt CF, Turberfield AJ (2005) Rapid chiral assembly of rigid DNA building blocks for molecular nanofabrication. Science 310(5754):1661–1665

Hargrove WL, Heyman JM, Mayer A, Mirchi A, Granados-Olivas A, Ganjegunte G, Gutzler D, Pennington DD, Ward FA, Garnica Chavira L, Sheng Z, Kumar S, Villanueva-Rosales N, Walker WS (2023) The future of water in a desert river basin facing climate change and competing demands: a holistic approach to water sustainability in arid and semi-arid regions. J Hydrol Reg Stud 46:1–22

He J, Zhang L, Xu L, Kong F, Xu Z (2020) Development of nanozyme-labeled biomimetic immunoassay for determination of sulfadiazine residue in foods. Adv Polym Technol 2020(1):1–8

Huang X, Zhu Y, Kianfar E (2021) Nano biosensors: properties, applications and electrochemical techniques. J Market Res 12:1649–1672

Huang CW, Lin C, Nguyen MK, Hussain A, Bui XT, Ngo HH (2023) A review of biosensor for environmental monitoring: principle, application, and corresponding achievement of sustainable development goals. Bioengineered 14(1):58–80

Jaleh B, Nasrollahzadeh M, Eslamipanah M, Nasri A, Shabanlou E, Manwar NR, Zboril R, Fornasiero P, Gawande MB (2022) The role of carbon-based materials for fuel cells performance. Carbon 198:301–352

Jayandran M, Haneefa MM, Balasubramanian V (2015) Green synthesis and characterization of Manganese nanoparticles using natural plant extracts and its evaluation of antimicrobial activity. J Appl Pharm Sci 5(12):105–110

Kavita V (2017) DNA biosensors-a review. J Bioeng Biomed Sci 7(2):1–5

Ke Y, Meyer T, Shih WM, Bellot G (2016) Regulation at a distance of biomolecular interactions using a DNA origami nanoactuator. Nat Commun 7(1):1–8

Khiyami MA, Almoammar H, Awad YM, Alghuthaymi MA, Abd-Elsalam KA (2014) Plant pathogen nanodiagnostic techniques: forthcoming changes? Biotechnol Biotechnol Equip 28(5):775–785

Lau HY, Wang Y, Wee EJ, Botella JR, Trau M (2016) Field demonstration of a multiplexed point-of-care diagnostic platform for plant pathogens. Anal Chem 88(16):8074–8081

Lawrence NS, Deo RP, Wang J (2004) Electrochemical determination of hydrogen sulfide at carbon nanotube modified electrodes. Anal Chim Acta 517(1–2):131–137

Lee GJ, Lee HM, Rhee CK (2007) Bismuth nano-powder electrode for trace analysis of heavy metals using anodic stripping voltammetry. Electrochem Commun 9(10):2514–2518

Lesiak A, Drzozga K, Cabaj J, Bański M, Malecha K, Podhorodecki A (2019) Optical sensors based on II-VI quantum dots. Nanomaterials 9(2):1–24

Liu M, Zhao G, Tang Y, Yu Z, Lei Y, Li M, Zhang Y, Li D (2010) A simple, stable and picomole level lead sensor fabricated on DNA-based carbon hybridized TiO2 nanotube arrays. Environ Sci Technol 44(11):4241–4246

Liu M, Khan A, Wang Z, Liu Y, Yang G, Deng Y, He N (2019a) Aptasensors for pesticide detection. Biosens Bioelectron 130:174–184

Liu X, Chen W, Lian M, Chen X, Lu Y, Yang W (2019b) Enzyme immobilization on ZIF-67/MWCNT composite engenders high sensitivity electrochemical sensing. J Electroanal Chem 833:505–511

Long F, Zhu A, Shi H (2013) Recent advances in optical biosensors for environmental monitoring and early warning. Sensors 13(10):13928–13948

Mahmood Q, Shaheen S, Azeem M (2024) Nanobiosensors: application in healthcare, environmental monitoring and food safety. Asian J Agric Biol 2024(1):1–12

Mandal N, Adhikary S, Rakshit R (2020) Nanobiosensors: recent developments in soil health assessment. Soil analysis: recent trends and applications, In: Rakshit A, Ghosh S, Chakraborty S, Philip V, Datta A (eds) Soil analysis: recent trends and applications. Springer, Singapore, pp 285–304

Marco MP, Barcelo D (1996) Environmental applications of analytical biosensors. Meas Sci Technol 7(11):1547

McGhee CE, Loh KY, Lu Y (2017) DNAzyme sensors for detection of metal ions in the environment and imaging them in living cells. Curr Opin Biotechnol 45:191–201

Mohamed EF, Awad G (2020) Photodegradation of gaseous toluene and disinfection of airborne microorganisms from polluted air using immobilized TiO2 nanoparticle photocatalyst–based filter. Environ Sci Pollut Res 27:24507–24517

Mohamed EF, Do TO (2020) Synthesis of new hollow nanocomposite photocatalysts: sunlight applications for removal of gaseous organic pollutants. J Taiwan Inst Chem Eng 111:181–190

Muenchen DK, Martinazzo J, de Cezaro AM, Rigo AA, Brezolin AN, Manzoli A, de Lima Leite F, Steffens C, Steffens J (2016) Pesticide detection in soil using biosensors and nanobiosensors. Biointerface Res Appl Chem 6(6):1659–1675

Numan A, Gill AA, Rafique S, Guduri M, Zhan Y, Maddiboyina B, Li L, Singh S, Dang NN (2021) Rationally engineered nanosensors: a novel strategy for the detection of heavy metal ions in the environment. J Hazard Mater 409:124493

Omer N (2019) Water quality parameters. In Water quality-science, assessments and policy. Summers JK (ed) IntechOpen: London

Ozkan-Ariksoysal D, Kayran YU, Yilmaz FF, Ciucu AA, David IG, David V, Hosgor-Limoncu M, Ozsoz M (2017) DNA-wrapped multi-walled carbon nanotube modified electrochemical biosensor for the detection of *Escherichia coli* from real samples. Talanta 166:27–35

Pandian CJ, Palanivel R, Dhananasekaran S (2015) Green synthesis of nickel nanoparticles using *Ocimum sanctum* and their application in dye and pollutant adsorption. Chin J Chem Eng 23(8):1307–1315

Pérez-Fernández B, de la Escosura-Muñiz A (2022) Electrochemical biosensors based on nanomaterials for aflatoxins detection: a review (2015–2021). Anal Chim Acta 1212:1–15

Phumying S, Labuayai S, Thomas C, Amornkitbamrung V, Swatsitang E, Maensiri S (2013) *Aloe vera* plant-extracted solution hydrothermal synthesis and magnetic properties of magnetite (Fe3O4) nanoparticles. Appl Phys A 111:1187–1193

Rafique M, Sadaf I, Rafique MS, Tahir MB (2017) A review on green synthesis of silver nanoparticles and their applications. Artif Cells Nanomedicine Biotechnol 45(7):1272–1291

Ramanathan K, Rank M, Svitel J, Dzgoev A, Danielsson B (1999) The development and applications of thermal biosensors for bioprocess monitoring. Trends Biotechnol 17(12):499–505

Rangel AE, Chen Z, Ayele TM, Heemstra JM (2018) In vitro selection of an XNA aptamer capable of small-molecule recognition. Nucleic Acids Res 46(16):8057–8068

Salouti M, Derakhshan FK (2020) Biosensors and nanobiosensors in environmental applications. In: Biogenic nano-particles and their use in agro-ecosystems. Springer, pp 515–591

Sharma TK, Bruno JG, Dhiman A (2017) ABCs of DNA aptamer and related assay development. Biotechnol Adv 35(2):275–301

Shi X, Gu W, Li B, Chen N, Zhao K, Xian Y (2014) Enzymatic biosensors based on the use of metal oxide nanoparticles. Microchim Acta 181:1–22

Siontorou CG, Georgopoulos KN (2016) A biosensor platform for soil management: the case of nitrites. J Clean Prod 111:133–142

Soukarié D, Ecochard V, Salomé L (2020) DNA-based nanobiosensors for monitoring of water quality. Int J Hyg Environ Health 226:113485

Strosnider H (2003) Whole-cell bacterial biosensors and the detection of bioavailable arsenic. US Environmental Protection Agency

Sun D, Hussain HI, Yi Z, Siegele R, Cresswell T, Kong L, Cahill DM (2014) Uptake and cellular distribution, in four plant species, of fluorescently labeled mesoporous silica nanoparticles. Plant Cell Rep 33:1389–1402

Surya SG, Yuvaraja S, Varrla E, Baghini MS, Palaparthy VS, Salama KN (2020) An in-field integrated capacitive sensor for rapid detection and quantification of soil moisture. Sens Actuators B Chem 321:128542

Tetyana P, Morgan Shumbula P, Njengele-Tetyana Z (2021) Biosensors: design, development and applications. In: Nanopores. IntechOpen, pp 1–19

Thirumurugan A, Aswitha P, Kiruthika C, Nagarajan S, Christy AN (2016) Green synthesis of platinum nanoparticles using *Azadirachta indica*–an eco-friendly approach. Mater Lett 170:175–178

Turdean GL (2011) Design and development of biosensors for the detection of heavy metal toxicity. Int J Electrochem 2011(1):1–15

Vaseghi A, Safaie N, Bakhshinejad B, Mohsenifar A, Sadeghizadeh M (2013) Detection of pseudomonas syringae pathovars by thiol-linked DNA–Gold nanoparticle probes. Sens Actuators B Chem 181:644–651

Wang X, Lu X, Chen J (2014) Development of biosensor technologies for analysis of environmental contaminants. Trends Environ Anal Chem 2:25–32

Wen Y, Li FY, Dong X, Zhang J, Xiong Q, Chen P (2013) The electrical detection of lead ions using gold-nanoparticle-and DNAzyme-functionalized graphene device. Adv Healthcare Mater 2(2):271–274

Woldu A (2022) Biosensors and its applications in water quality monitoring. Int J Sci Eng Res 13(5):12–29

Zhang Y, Zhang J, Wu H, Guo S, Zhang J (2012) Glass carbon electrode modified with horseradish peroxidase immobilized on partially reduced graphene oxide for detecting phenolic compounds. J Electroanal Chem 681:49–55

Zhao WW, Xu JJ, Chen HY (2017) Photoelectrochemical enzymatic biosensors. Biosens Bioelectron 92:294–304

Zuo Q, Chen Y, Chen ZP, Yu RQ (2018) Quantification of cadmium in rice by surface-enhanced Raman spectroscopy based on a ratiometric indicator and conical holed enhancing substrates. Anal Sci 34(12):1405–1410

# Chapter 2
# Photodegradation of Environmental Contaminants Using Green Nanotechnology

**Nikesh Thakur, Pankaj Kumar, and Naveen Thakur**

**Abstract** Environmental pollution from industrial, agricultural, and domestic sources has raised significant global concerns, creating an urgent need for sustainable remediation strategies. Green nanotechnology offers a promising solution, focusing on eco-friendly synthesis methods, sustainability, and the development of advanced nanomaterials. This chapter explores the creation, mechanisms, and applications of green synthesized nanomaterials for the photodegradation of environmental contaminants. It covers various types of pollutants, including both organic and inorganic contaminants, and highlights sustainable synthesis strategies, such as utilizing renewable resources and non-toxic methods. The material properties of green nanomaterials, including their structural features and photocatalytic activities, are thoroughly examined. Additionally, the chapter discusses the efficiency of these materials in breaking down pollutants through photodegradation, emphasizing the advantages of green synthesis in terms of reduced environmental impact and enhanced performance. The potential industrial applications of green nanomaterials are also highlighted, demonstrating their relevance in various sectors, including wastewater treatment, air purification, and soil remediation. The chapter underscores the importance of renewable resources and non-toxic methodologies in advancing environmentally friendly technologies, with a focus on promoting global environmental health and sustainability.

N. Thakur
Department of Physics, University Institute of Engineering and Technology, Guru Nanak University, Hyderabad, Telangana, India

P. Kumar
Department of Physics, Akal University, Talwandi Sabo, Bathinda, Punjab, 151302, India

N. Thakur (✉)
Faculty of Allied and Health Care Sciences, Desh Bhagat University, Punjab, India
e-mail: naveenthakur2327@gmail.com

 
A. Balkrishna et al. (eds.), *Nanotechnology in Environmental Science and Healthcare*,
https://doi.org/10.1007/978-981-95-6714-0_2

## 2.1 Introduction

The growing presence of organic and inorganic pollutants in the environment, spanning air, water, and soil, has become a significant concern for ecosystems and human health. Contaminants, such as heavy metals, pesticides, industrial effluents, and volatile organic compounds, disrupt the ecological balance and pose substantial health risks, including respiratory issues, neurological damage, and an increased risk of cancer (Purkait et al. 2020). Although traditional remediation methods, such as thermal desorption, chemical oxidation, and landfill disposal, have had some success, they often require high energy inputs, are costly, and may generate secondary pollution. These limitations underscore the need for more sustainable and efficient alternatives for environmental cleanup. Green nanotechnology, an emerging interdisciplinary field that combines the principles of green chemistry with nanoscience, offers a promising solution to these challenges. It harnesses the unique properties of nanoscale materials to provide environmentally friendly and highly effective remediation strategies (Ashour et al. 2023). Green nanotechnology emphasizes minimizing waste, reducing the use of hazardous chemicals, and ensuring the sustainability of materials throughout their life cycle (Devi et al. 2024).

Nanomaterials, due to their high surface area-to-volume ratios, adjustable surface chemistry, and catalytic properties, are particularly well suited for interacting with contaminants on a molecular level, thereby enhancing the removal or breakdown of pollutants (Karthik et al. 2022). For example, zero-valent iron nanoparticles have proven effective in cleaning up groundwater contaminated with chlorinated compounds. In contrast, titanium dioxide nanoparticles are used in photocatalytic processes to break down organic pollutants under sunlight, converting harmful substances into harmless by-products, such as water and carbon dioxide. The use of green nanotechnology offers notable benefits, including reduced environmental and economic costs (Vasantharaj et al. 2021). Many nanotechnology-based methods can operate at ambient conditions, reducing energy consumption compared to traditional approaches. Additionally, green synthesis techniques, which utilize natural sources like plant extracts and microorganisms, eliminate the need for toxic reducing agents in nanoparticle production, further decreasing the environmental footprint (Kumar et al. 2024a). Beyond remediation, green nanotechnology also aids in pollutant detection and prevention. Nanosensors can detect contaminants at very low levels with high accuracy, enabling early intervention. Nanostructured membranes and adsorbents are also being developed to capture pollutants, such as industrial emissions or microplastics, before they enter the environment (Mirsadeghi et al. 2020). However, challenges remain, including scalability, cost-effectiveness, and the potential ecotoxicological risks of nanoparticles. Ongoing research is focused on understanding the life cycle impacts of nanomaterials and mitigating any unintended adverse effects. Ultimately, green nanotechnology offers a sustainable and transformative approach to environmental cleanup, paving the way for a cleaner future (Purkait et al. 2020).

### *2.1.1 Environmental Contaminants and Their Challenges*

Environmental contamination from both organic and inorganic compounds has become a global concern due to its widespread presence and harmful effects on ecosystems and human health. These pollutants include organic dyes, pharmaceuticals, pesticides, and heavy metals, each presenting unique challenges for their removal and management (Kumar et al. 2024b). Organic dyes, typically released from industries, such as textiles, paper, and printing, are among the most persistent pollutants in aquatic environments. Their intense color blocks sunlight, hindering photosynthesis in aquatic plants, and they often break down into toxic by-products that are harmful to both marine life and humans (El-saied et al. 2021). Pharmaceuticals, such as antibiotics, hormones, and painkillers, are another primary class of organic contaminants. Introduced into the environment through hospital wastewater, agricultural runoff, and improper disposal, these substances persist, causing toxicity to non-target organisms and contributing to the growing issue of antimicrobial resistance. Pesticides, widely used in agriculture to increase crop yields, are known for their long-lasting environmental presence and potential to accumulate in food chains (Kumar et al. 2024c). Chlorinated pesticides, such as DDT, are particularly concerning due to their ability to accumulate in the fatty tissues of organisms, leading to biomagnification and harmful effects on higher trophic levels (Hassan et al. 2015). Similarly, heavy metals like arsenic, cadmium, mercury, and lead, which are primarily released through industrial processes, mining, and improper waste management, are non-degradable and highly toxic. These metals can bind to biological molecules, disrupting enzymatic functions, damaging DNA, and causing chronic diseases in humans (Nagajyothi et al. 2020).

The impact of these contaminants goes beyond toxicity, as many are resistant to natural degradation due to their stable chemical structures, such as the aromatic rings in organic dyes and pharmaceuticals. This resistance allows them to persist in the environment, leading to long-term accumulation and detrimental effects on ecosystems (Kumar et al. 2024d). The bioaccumulation of these substances in organisms poses direct risks to biodiversity and food security, with examples such as mercury in fish making them unsafe for consumption and pesticide residues threatening both human health and agricultural ecosystems (Chen et al. 2007). Traditional remediation methods, including physical techniques, such as adsorption and filtration, chemical methods like oxidation and coagulation, and biological approaches like microbial degradation, all face limitations. These methods may generate secondary waste, require high energy inputs, or be slow and ineffective, particularly for emerging contaminants. As these pollutants become more complex and persistent, the need for innovative, sustainable solutions, such as green nanotechnology, becomes increasingly urgent to mitigate environmental pollution and its long-term impacts (Bhattacharjee et al. 2024).

### 2.1.2 Green Nanotechnology: Overview and Potential

Green nanotechnology is a transformative approach to addressing environmental issues by integrating the advanced capabilities of nanotechnology with the principles of sustainable green chemistry. It focuses on designing, developing, and applying environmentally friendly, economically feasible, and socially responsible nanomaterials and processes (Goutam et al. 2018). A key aspect of green nanotechnology is green synthesis, which emphasizes the use of renewable resources, non-toxic reagents, and energy-efficient methods. Unlike traditional nanoparticle production, which often relies on harmful chemicals and energy-intensive processes, green synthesis utilizes natural materials like plant extracts, microorganisms, and natural polymers. These biological agents act as reducing and stabilizing agents, enabling the eco-friendly production of nanomaterials (Kumar et al. 2024e). For example, plant-based extracts rich in antioxidants, alkaloids, or flavonoids can reduce metal ions to create nanoparticles without the need for toxic chemicals. Microorganisms, such as bacteria, fungi, and algae, also show great promise in sustainably producing nanoparticles (Rocha et al. 2024). By eliminating harmful reagents and minimizing waste, green synthesis aligns with sustainability and environmental conservation goals.

Green nanotechnology offers several advantages over conventional methods. Traditional nanoparticle production is often criticized for its use of hazardous substances and the generation of secondary pollutants. In contrast, green nanotechnology minimizes environmental impact throughout the life cycle of nanomaterials (Demarema et al. 2024). Green synthesis methods typically operate under ambient conditions, reducing energy consumption. These techniques also enhance the biocompatibility and safety of nanomaterials, making them suitable for sensitive applications in areas, such as water purification, agriculture, and biomedicine (Kumar et al. 2024f). Additionally, the use of natural, renewable materials in green synthesis helps conserve resources and reduce dependence on finite raw materials, addressing both environmental and economic concerns (Dihom et al. 2022). A promising application of green nanotechnology is photodegradation, where light, typically sunlight, is used to break down harmful pollutants into safer substances. Nanomaterials, such as titanium dioxide ($TiO_2$), zinc oxide (ZnO), and silver nanoparticles, possess exceptional photocatalytic properties, enabling the efficient degradation of contaminants, including organic dyes, pesticides, and pharmaceuticals. The high surface area and adjustable properties of nanomaterials enhance their interaction with pollutants, speeding up the degradation process (Muthukumar et al. 2022). Green nanotechnology further improves this by producing photocatalysts through eco-friendly methods, ensuring that the synthesis process itself does not harm the environment. For example, plant-synthesized titanium dioxide nanoparticles effectively degrade toxic dyes under sunlight, demonstrating high efficiency, stability, and reusability (Kumar et al. 2024g). Similarly, silver and gold nanoparticles made using green methods show promise in eliminating pharmaceutical pollutants and pesticide residues, offering sustainable solutions for environmental cleanup (Moradnia et al. 2020).

## 2.2 Green Synthesis of Photocatalytic Nanomaterials

The green synthesis of photocatalytic nanomaterials represents a significant step toward sustainable environmental remediation. Traditional methods for producing photocatalytic materials, such as $TiO_2$ and ZnO nanoparticles, often involve hazardous chemicals and high energy consumption and generate toxic by-products, which undermine their environmental benefits (Miri et al. 2023). Green synthesis addresses these challenges by adopting eco-friendly processes that minimize environmental impact while enhancing the functionality of the resulting nanomaterials. This approach utilizes natural, renewable, and non-toxic resources, such as plant extracts, microorganisms, and biodegradable polymers, as reducing and stabilizing agents during synthesis (Hosny et al. 2022). Plant-based extracts, for instance, are rich in phytochemicals like alkaloids, tannins, and flavonoids, which facilitate the reduction of metal precursors to form nanoparticles. These methods not only eliminate the need for toxic reagents but also operate under mild conditions of temperature and pressure, significantly reducing energy requirements. (Rana et al. 2024) The resulting green synthesized nanomaterials exhibit enhanced photocatalytic properties, including increased surface area, improved light absorption, and superior charge carrier dynamics as shown in Fig. 2.1. These attributes make them highly effective in degrading organic pollutants, such as dyes and pesticides, under sunlight (Rani et al. 2021). By aligning material functionality with environmental sustainability, green synthesis ensures that photocatalytic nanomaterials can be deployed for pollution remediation without contributing to further ecological harm (Thakur and Kumar 2024).

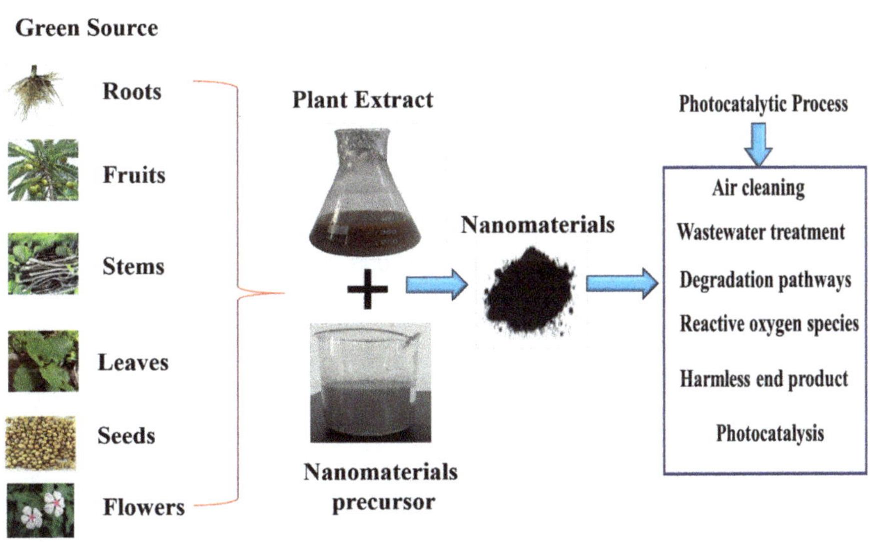


**Fig. 2.1** Green synthesis of photocatalytic nanomaterials

### 2.2.1 Biogenic Synthesis of Nanomaterials

Biogenic synthesis of nanomaterials harnesses biological systems and natural resources to produce nanoparticles in a sustainable and environmentally friendly manner. This method not only reduces the environmental impact associated with conventional chemical synthesis but also enhances the functional properties of the resulting nanomaterials. Among the various biogenic approaches, the use of plant extracts, microbial systems, and biowaste has garnered significant attention due to their efficiency and sustainability (Isa et al. 2021). Plant extracts serve as versatile reducing and stabilizing agents due to their rich content of bioactive compounds, including polyphenols, alkaloids, tannins, and enzymes. These phytochemicals facilitate the reduction of metal ions into nanoparticles, simultaneously preventing aggregation and ensuring stability (Thakur and Thakur 2023). For instance, extracts from plants, such as neem, tea, and eucalyptus, have been used to synthesize nanoparticles with enhanced catalytic, antimicrobial, and optical properties. The process is not only environmentally benign but also cost-effective, as it eliminates the need for hazardous chemicals and operates under mild conditions (Rostami-Vartooni et al. 2016).

Microbial systems, including bacteria, fungi, and algae, are another promising avenue for nanoparticle synthesis. Microorganisms naturally produce enzymes and metabolites capable of reducing metal ions and stabilizing nanoparticles. For example, bacteria, such as Escherichia coli and fungi like *Aspergillus niger,* have been employed to synthesize metallic and oxide nanoparticles (Thakur and Thakur 2024a). Algae, being photosynthetic and abundant, offer an additional advantage by contributing to large-scale and renewable production of nanoparticles (Ayodhya et al. 2016). Biowaste utilization further extends the concept of biogenic synthesis by transforming agricultural and industrial by-products into valuable nanomaterials. Materials like fruit peels, husks, and waste biomass contain organic compounds that act as reducing agents. This approach not only addresses the challenge of waste management but also provides a low-cost alternative for nanoparticle production (Thakur and Thakur 2024b). For instance, banana peels and sugarcane bagasse have been used to synthesize nanoparticles with photocatalytic and antimicrobial applications. Biogenic synthesis thus combines environmental sustainability with advanced functionality, offering a viable pathway for creating nanomaterials that align with the principles of green chemistry (Hariharan 2006).

### 2.2.2 Types of Photocatalytic Nanomaterials

Photocatalytic nanomaterials are a cornerstone of advanced environmental remediation technologies due to their ability to harness light energy for degrading pollutants. These materials come in various forms, each offering unique advantages for photodegradation applications. Among the most widely used types are metal oxide nanoparticles, bimetallic and doped nanoparticles, and carbon-based materials

(Phang et al. 2021). Metal oxide nanoparticles, such as $TiO_2$, ZnO, and iron oxide ($Fe_2O_3$), are among the most extensively studied photocatalysts (Thakur et al. 2023). $TiO_2$ is particularly notable for its high stability, non-toxicity, and strong photocatalytic activity under ultraviolet light (Lanjwani et al. 2024). ZnO offers similar benefits, with additional versatility in structural modifications to enhance performance (Quddus et al. 2023). $Fe_2O_3$, or hematite, stands out for its ability to absorb visible light due to its narrower bandgap, making it more efficient under solar irradiation. These metal oxides are highly effective in breaking down organic pollutants, such as dyes and pesticides, into less harmful by-products (Laouini et al. 2021).

Bimetallic and doped nanoparticles represent another class of photocatalytic materials designed to overcome limitations like poor light absorption and rapid charge recombination. By combining two metals or introducing dopants (e.g., nitrogen, sulfur, or transition metals), these nanoparticles exhibit enhanced charge separation and broadened light absorption into the visible spectrum (Thakur et al. 2024a). For example, bimetallic nanoparticles like Ag-$TiO_2$ or doped $TiO_2$ demonstrate superior photocatalytic efficiency compared to their single-metal counterparts, enabling the degradation of more complex pollutants under natural sunlight (Pinna et al. 2021). Carbon-based materials, including graphene, biochar, and carbon dots, have emerged as promising alternatives or complements to metal-based photocatalysts. Graphene, with its exceptional conductivity and high surface area, facilitates efficient electron transport, reducing charge recombination. Biochar, derived from biomass, offers affordability while exhibiting effective pollutant adsorption and photocatalytic properties (Abdullah et al. 2021). Carbon dots, nanoscale carbon particles with tunable optical properties, are particularly attractive for visible light–driven photocatalysis. Together, these materials provide diverse, efficient, and sustainable solutions for addressing environmental pollution (Emmanuel et al. 2023).

## 2.3 Mechanisms of Photodegradation

Photodegradation is a light-driven process that involves the breakdown of pollutants into less harmful or inert substances, and its efficiency hinges on understanding the underlying mechanisms. The process typically begins with the absorption of photons by a photocatalyst, such as a metal oxide or carbon-based nanomaterial (Ahmad et al. 2018). When exposed to light of sufficient energy, the photocatalyst's electrons are excited from the valence band to the conduction band, creating electron–hole pairs. This separation of charges is crucial for initiating subsequent chemical reactions. The photogenerated electrons and holes migrate to the catalyst surface, where they interact with water molecules and dissolved oxygen (Rani and Shanker 2022). Electrons reduce oxygen to form reactive oxygen species (ROS) like superoxide anions ($O_2{\bullet}^-$) and hydroxyl radicals (•OH), while holes oxidize water or hydroxide ions to generate additional •OH radicals. These ROS are highly reactive and can degrade organic pollutants by attacking their molecular structures, leading to the cleavage of bonds and the eventual mineralization of contaminants into harmless substances,

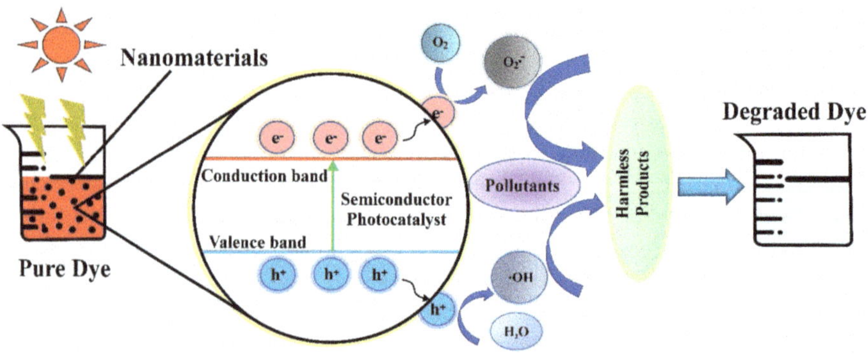


**Fig. 2.2** Photocatalytic mechanism of green synthesized nanomaterials for degradation of various pollutants. Adapted from (Patra et al. 2023) under the Creative Commons Attribution (CC BY) license (http://creativecommons.org/licenses/by/4.0/)

such as carbon dioxide and water (Eddy et al. 2024), as shown in Fig. 2.2. Factors, such as charge recombination, surface area, and light absorption efficiency, significantly influence the process. Advanced strategies, including doping and the use of co-catalysts, are employed to enhance charge separation and broaden light utilization, thereby optimizing photocatalytic material performance for effective pollutant degradation (Fardood et al. 2024).

### *2.3.1 Photocatalysis Basics*

Photocatalysis is a process in which light energy is used to accelerate a chemical reaction that typically would not occur or would occur at a much slower rate. The foundation of photocatalytic responses lies in the photoactivation of a photocatalyst, a material that absorbs light and undergoes electronic excitation (Luque et al. 2021). When a photocatalyst, such as $TiO_2$ or ZnO, is exposed to light of suitable energy, typically ultraviolet or visible light, its electrons in the valence band are excited to the conduction band. This creates electron–hole pairs: electrons in the conduction band and positively charged holes in the valence band (Thakur et al. 2024b). The generation of these electron–hole pairs is a critical step in photocatalysis. These charges then migrate to the surface of the photocatalyst, where they can interact with molecules in the surrounding environment, particularly water and oxygen, leading to the formation of highly reactive species (Hashemi et al. 2022). The electron in the conduction band typically reduces molecular oxygen to produce $O_2•^-$, while the hole in the valence band can oxidize water or hydroxide ions to generate •OH, both of which are highly reactive and serve as key agents in contaminant degradation. These reactive species hydroxyl radicals and superoxide ions are essential for the breakdown of organic contaminants, such as dyes, pesticides, and pharmaceuticals. They attack the molecular bonds of the pollutants, causing their decomposition into smaller, less harmful fragments (Vishnu et al. 2023). Over time, this process leads to the mineralization of the contaminants, converting them into non-toxic by-products like carbon dioxide, water, and inorganic salts. The overall result is a cleaner, less polluted environment, with the photocatalyst playing a crucial role in driving these reactions upon exposure to light.

### *2.3.2 Enhancing Photocatalytic Activity*

Enhancing the photocatalytic activity of nanomaterials is crucial for improving their efficiency in environmental remediation, particularly under visible light, as natural sunlight is predominantly composed of visible light. One of the key strategies for enhancing photocatalytic performance is bandgap engineering, which involves modifying the electronic structure of photocatalysts to enable them to absorb visible light more efficiently (Singh et al. 2019). Materials like $TiO_2$, which have a wide bandgap and primarily absorb ultraviolet light, can be modified by doping with elements, such as nitrogen, sulfur, or carbon. These modifications narrow the bandgap, enabling the photocatalyst to absorb lower-energy visible light and thus broaden the spectrum of light it can utilize for photocatalytic reactions. This modification enhances the efficiency of photocatalysts under sunlight and improves their applicability for outdoor pollution treatment (Ahmed et al. 2024). Another critical approach is surface functionalization, which involves modifying the surface properties of the photocatalytic material to improve its interaction with pollutants. Functional groups, such as carboxyl, amino, or hydroxyl groups, can be introduced to the surface of nanomaterials to increase their hydrophilicity, making it easier for water and contaminants to adsorb onto the surface. These surface modifications can also improve the stability and reactivity of the photocatalysts, preventing their aggregation and enhancing charge separation. By facilitating a stronger interaction with pollutants, surface functionalization ensures more effective degradation during the photocatalytic process (Iqbal et al. 2021). Lastly, synergistic effects in hybrid nanomaterials offer significant improvements in photocatalytic performance.

Combining different nanomaterials, such as metal oxides with carbon-based materials or semiconductors with noble metals, can create synergies that enhance electron transfer, reduce charge recombination, and broaden the range of light absorption (Miri et al. 2022). For instance, coupling $TiO_2$ with graphene or silver nanoparticles has shown enhanced photocatalytic degradation of pollutants due to the efficient transfer of electrons and increased surface area. These hybrid materials capitalize on the strengths of individual components, resulting in improved photocatalytic activity and greater efficiency in pollutant removal. Together, these strategies enable the development of advanced photocatalytic nanomaterials that are more efficient, versatile, and sustainable for environmental cleanup applications (Lu et al. 2019).

## 2.4 Applications of Green Nanotechnology in Photodegradation

Green nanotechnology leverages environmentally friendly synthesis methods to produce nanoparticles that can address environmental challenges, including photodegradation. Photodegradation involves breaking down harmful organic pollutants in water, air, and soil under light irradiation. Nanoparticles like $TiO_2$, ZnO, and doped metal oxides synthesized via green routes exhibit high photocatalytic activity. These nanoparticles utilize sunlight or UV light to ROS, which degrade pollutants, such as dyes, pesticides, and industrial effluents into less harmful by-products. Green synthesized nanoparticles reduce the use of hazardous chemicals during production, making the process more sustainable. Applications include wastewater treatment, air purification, and remediation of contaminated soils as shown in Fig. 2.3. They are also employed in self-cleaning surfaces and coatings. By combining efficiency with eco-friendly production, green nanotechnology in photodegradation offers a sustainable solution for mitigating environmental pollution, aligning with global efforts to promote cleaner technologies.

### *2.4.1 Water Treatment*

Water treatment is one of the most critical applications of photocatalytic nanomaterials, particularly in addressing the growing environmental concerns related to organic dyes from the textile industry and the presence of pharmaceuticals and endocrine-disrupting chemicals (EDCs) in water sources. These pollutants are persistent, toxic, and difficult to remove using conventional water treatment methods, making photocatalysis an attractive alternative for efficient and sustainable remediation (Babu et al. 2019). Degradation of organic dyes from textile industries is a

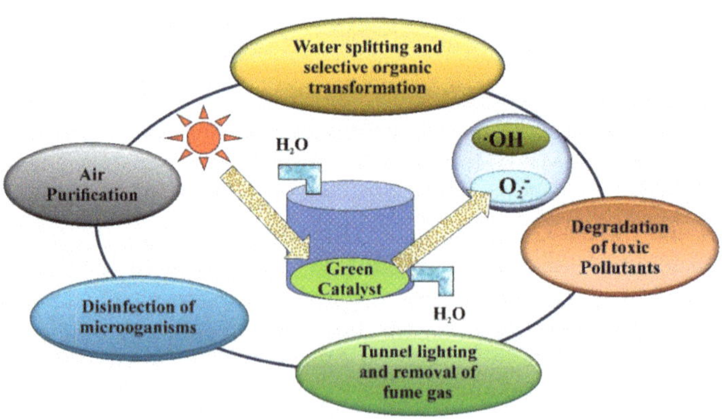


**Fig. 2.3** Applications of green synthesized nanomaterials in an environmental setting

significant challenge due to the wide range of dyes used in manufacturing, many of which are highly toxic, resistant to natural degradation, and pose long-term risks to aquatic ecosystems. Photocatalytic nanomaterials, such as $TiO_2$ and ZnO, are highly effective in breaking down these dyes. When exposed to light, these photocatalysts generate ROS like hydroxyl radicals and superoxide ions, which degrade the dye molecules into smaller, less toxic compounds. This process can lead to the mineralization of dyes into non-toxic by-products, such as carbon dioxide and water. Furthermore, photocatalysts can be tailored to target specific dyes by adjusting their surface properties, enhancing pollutant interaction, and improving the efficiency of degradation under visible light (Atchudan et al. 2018).

The removal of pharmaceuticals and endocrine-disrupting chemicals (EDCs) from water is another critical application of photocatalysis. Pharmaceuticals, including antibiotics, hormones, and painkillers, are commonly found in wastewater due to their widespread use and improper disposal. EDCs, found in substances, such as personal care products and pesticides, can interfere with the hormonal systems of aquatic organisms, leading to disruptions in reproductive health and biodiversity (Sarkar et al., 2020). Photocatalytic materials, particularly those doped or modified for enhanced light absorption, can efficiently degrade these pollutants in wastewater. By generating ROS, photocatalysts break down the complex molecular structures of pharmaceuticals and EDCs, rendering them less harmful or completely mineralizing them into benign compounds. In both cases, photocatalytic water treatment offers a sustainable and cost-effective solution, utilizing renewable energy sources, such as sunlight to degrade persistent and toxic contaminants, thereby ensuring cleaner water for both environmental and human health (Niveditha et al. 2023).

### 2.4.2 *Air Purification*

Air purification is a crucial area where photocatalysis plays a transformative role, particularly in addressing the pollution caused by volatile organic compounds (VOCs), nitrogen oxides (NOx), and sulfur oxides (SOx), which are major contributors to air pollution and associated health hazards (Palma et al. 2020). Photocatalytic nanomaterials offer an effective and sustainable approach for mitigating these pollutants, utilizing light energy to trigger oxidation and reduction reactions that break down harmful compounds. Photocatalytic oxidation of volatile organic compounds (VOCs), which encompass a wide range of chemicals including benzene, toluene, and xylene, is crucial in indoor and industrial air purification. These compounds are often released from industrial activities, vehicle emissions, and household products, and their accumulation in the atmosphere can lead to serious health problems and contribute to smog formation (Adeleke et al. 2018). Photocatalysts, such as $TiO_2$ and ZnO, facilitate the oxidation of VOCs when exposed to light, generating ROS like •OH. These ROS break down VOC molecules into smaller, less toxic substances, such as carbon dioxide and water. This process not only reduces the concentration of harmful VOCs but also prevents the formation of secondary pollutants, such as ozone, which is harmful to both humans and the environment (Nagaraja et al. 2024).

The reduction of NOx and SOx emissions is another critical application of photocatalysis in air purification. Nitrogen oxides (NOx) and sulfur oxides (SOx) are primarily produced by vehicle exhausts, industrial emissions, and power plants. These pollutants contribute to acid rain, smog, and respiratory diseases (Rambiyana et al. 2023). Photocatalytic materials can be engineered to facilitate the reduction of NOx and SOx by promoting reactions that convert these harmful compounds into less toxic forms, such as nitrogen ($N_2$) and sulfuric acid ($H_2SO_4$), under exposure to light. Green nanomaterials, such as doped $TiO_2$ or composite materials combining metal oxides with carbon-based materials, can significantly enhance photocatalytic activity, improving the efficiency of NOx and SOx removal (Pawar et al. 2015). By incorporating renewable resources and non-toxic reagents in the synthesis of these photocatalysts, green nanotechnology ensures that the process is environmentally friendly and sustainable. Overall, photocatalysis offers a powerful solution for air purification, harnessing the energy of light to reduce harmful emissions and enhance air quality in both outdoor and indoor environments (Zulfiqar et al. 2024).

### *2.4.3 Soil Remediation*

Soil contamination by pesticides and heavy metals poses a significant threat to ecosystem health and food safety. Traditional soil remediation techniques often face challenges due to their high costs, inefficiency, and potential to create secondary pollution. Photocatalytic processes, however, offer a promising alternative by harnessing light energy to break down harmful substances, thereby ensuring a sustainable and effective solution for soil remediation (Rafique et al. 2020). Photodegradation of pesticide residues is a critical aspect of soil remediation, especially given the widespread use of pesticides in agriculture. Many pesticides are persistent in the environment, with residues often remaining in soil long after their application, leading to soil and water contamination (Liu et al. 2020). Photocatalytic nanomaterials, such as titanium dioxide ($TiO_2$), have been extensively studied for their ability to degrade pesticide molecules through a process driven by light. When exposed to UV or visible light, photocatalysts produce ROS, such as •OH and $O_2$•$^-$, which attack the pesticide molecules, breaking down their complex structures into non-toxic by-products. This process not only reduces the toxicity of pesticide residues but also accelerates their natural degradation in the soil, improving soil health and preventing the accumulation of harmful chemicals in the food chain (Emmanuel et al. 2024).

Heavy metal detoxification is another key area where photocatalysis can be applied in soil remediation. Heavy metals, such as lead (Pb), cadmium (Cd), arsenic (As), and mercury (Hg), are often present in soil due to industrial activities, mining, and agricultural practices. These metals are highly toxic and can accumulate in soil, posing serious risks to plant and animal life (Saranya et al. 2018). Photocatalytic processes can aid in the detoxification of heavy metals by facilitating the reduction of metal ions into less toxic or insoluble forms. For instance, photocatalytic reactions can reduce toxic metal ions, such as Cr(VI), to less harmful Cr(III) or convert dissolved heavy metals into their oxide forms, which can then be removed from the soil more easily. Enhanced photocatalytic activity can be achieved by doping photocatalysts with specific elements or using hybrid nanomaterials, which further improve the efficiency of heavy metal removal (Asha et al. 2022). In conclusion, photocatalytic processes offer an innovative and eco-friendly approach for soil remediation, addressing the persistent challenges of pesticide degradation and heavy metal detoxification. By utilizing renewable light energy, these processes promote the sustainable restoration of contaminated soils, enhancing agricultural productivity and reducing environmental risks (Kaur et al. 2018).

## 2.5 Challenges

While green nanotechnology offers promising solutions for the photodegradation of environmental contaminants, several challenges must be addressed for its widespread adoption in industrial applications. One major challenge is scaling up green synthesis methods to produce nanomaterials in large quantities. Although green synthesis techniques, such as plant-based or microbial reduction, are eco-friendly and cost-effective at a small scale, they often face difficulties in maintaining consistent quality, reproducibility, and efficiency when scaled up for commercial use. Standardizing these processes to meet industrial demands is crucial for the success of green nanotechnology in large-scale environmental cleanup. Another challenge is the stability and recyclability of green nanomaterials. Over time, photocatalytic nanomaterials may degrade or lose their effectiveness due to surface poisoning, aggregation, or leaching of toxic components. Ensuring long-term stability and the ability to recycle or regenerate nanomaterials without significant loss of efficiency is essential for making these materials practical and economically viable for continuous environmental remediation processes. Additionally, regulatory and safety concerns regarding the use of green nanomaterials need to be addressed. The potential ecological and health risks associated with the release of nanoparticles into ecosystems, even those synthesized using natural sources, require rigorous testing and regulatory frameworks to ensure their safety. Understanding the long-term impact of these materials on human health and biodiversity is crucial for their safe implementation in environmental cleanup technologies.

## 2.6 Future Directions

The future of photodegradation using green nanotechnology holds exciting potential with advancements aimed at improving efficiency and sustainability. One key direction is the development of multifunctional nanomaterials. Researchers are working on designing nanomaterials that combine multiple functionalities, such as enhanced photocatalytic activity, self-cleaning properties, and the ability to degrade a wider range of contaminants. By incorporating diverse materials, such as metal oxides, carbon-based nanomaterials, and even biological components into hybrid nanomaterials, photocatalysts can be tailored to tackle multiple environmental pollutants simultaneously, offering greater versatility in real-world applications. Another promising area is the integration of renewable energy sources for photocatalysis. To enhance the practicality and sustainability of photocatalytic processes, there is a growing focus on harnessing renewable energy, such as solar power, to drive photocatalytic reactions. This shift toward renewable energy can reduce the carbon footprint of photodegradation technologies, making them more feasible for large-scale, low-energy-demand applications in wastewater treatment, air purification, and soil remediation. Ultimately, a life cycle assessment (LCA) approach is becoming increasingly crucial for quantifying the actual environmental benefits of green nanotechnology. By evaluating the entire life cycle of nanomaterials from production to disposal, LCA helps identify areas where ecological impacts can be minimized, ensuring that the technologies not only clean the environment but also do so in an eco-friendly and resource-efficient manner. These future advancements hold the potential to make green nanotechnology a cornerstone of sustainable environmental remediation.

## 2.7 Conclusion

Green nanotechnology offers a transformative and eco-friendly approach to environmental remediation, particularly through the photodegradation of contaminants. This field integrates the principles of green chemistry with nanoscience, emphasizing the use of renewable materials and non-toxic reagents in the synthesis of nanomaterials. By utilizing plant extracts, microorganisms, and waste materials as precursors, green synthesis methods not only reduce the environmental impact of nanoparticle production but also promote the sustainability of the process. These nanomaterials, including metal oxides, carbon-based materials, and hybrid composites, exhibit remarkable photocatalytic properties, enabling the efficient degradation of organic pollutants, heavy metals, and other contaminants upon exposure to light. One of the key advantages of green nanotechnology is its ability to harness renewable energy sources, such as sunlight, to drive photocatalytic reactions. This makes the process highly energy-efficient, cost-effective, and suitable for large-scale applications in wastewater treatment, air purification, and soil remediation. Moreover, the development of multifunctional nanomaterials that can target a broad spectrum of pollutants enhances the versatility and effectiveness of these technologies. However, significant challenges remain in realizing the full potential of green nanotechnology. Scalability is a major issue, as green synthesis methods frequently encounter challenges in maintaining consistency and quality when transitioning from laboratory-scale to industrial-scale production. Regulatory acceptance is another hurdle, as concerns over the safety and environmental impact of nanoparticles require rigorous testing and clear regulatory frameworks. Addressing these challenges will be crucial for the widespread adoption of green nanotechnology, thereby creating a cleaner and more sustainable environment.

## References

Abdullah FH, Bakar NA, Bakar MA (2021) Comparative study of chemically synthesized and low temperature bio-inspired *Musa acuminata* peel extract mediated zinc oxide nanoparticles for enhanced visible-photocatalytic degradation of organic contaminants in wastewater treatment. J Hazard Mater 406:124779

Adeleke JT, Theivasanthi T, Thiruppathi M, Swaminathan M, Akomolafe T, Alabi AB (2018) Photocatalytic degradation of methylene blue by $ZnO/NiFe_2O_4$ nanoparticles. Appl Surf Sci 455:195–200

Ahmad KS, Jaffri SB (2018) Phytosynthetic Ag doped ZnO nanoparticles: semiconducting green remediators: photocatalytic and antimicrobial potential of green nanoparticles. Open Chem 16(1):556–570

Ahmed AM, Nasr M, Aljaafreh MJ, Sayed S, Hamdy H, Shaban M, Marashdeh M, Al-Hmoud M, Zayed M (2024) Sodium titanate nanostructured modified by green synthesis of iron oxide for highly efficient photodegradation of dye contaminants. Nanotechnol Rev 13(1):1–17

Asha S, Bessy TC, Sherin JJ, Kumar CV, Bindhu MR, Sureshkumar S, Al-Khattaf FS, Hatamleh AA (2022) Efficient photocatalytic degradation of industrial contaminants by *Piper longum* mediated ZnO nanoparticles. Environ Res 208:112686

Ashour M, Mansour AT, Abdelwahab AM, Alprol AE (2023) Metal oxide nanoparticles' green synthesis by plants: prospects in phyto-and bioremediation and photocatalytic degradation of organic pollutants. Processes 11(12):1–24

Atchudan R, Edison TNJI, Perumal S, Vinodh R, Lee YR (2018) In-situ green synthesis of nitrogen-doped carbon dots for bioimaging and $TiO_2$ nanoparticles@ nitrogen-doped carbon composite for photocatalytic degradation of organic pollutants. J Alloy Compd 766:12–24

Ayodhya D, Venkatesham M, Santoshi Kumari A, Reddy GB, Ramakrishna D, Veerabhadram G (2016) Photocatalytic degradation of dye pollutants under solar, visible and UV lights using green synthesised CuS nanoparticles. J Exp Nanosci 11(6):418–432

Babu AT, Antony R (2019) Green synthesis of silver doped nano metal oxides of zinc & copper for antibacterial properties, adsorption, catalytic hydrogenation & photodegradation of aromatics. J Environ Chem Eng 7(1):102840

Bhattacharjee N, Som I, Saha R, Mondal S (2024) A critical review on novel eco-friendly green approach to synthesize zinc oxide nanoparticles for photocatalytic degradation of water pollutants. Int J Environ Anal Chem 104(3):489–516

Chen CC, Lu CS, Chung YC, Jan JL (2007) UV light induced photodegradation of malachite green on $TiO_2$ nanoparticles. J Hazard Mater 141(3):520–528

Demarema S, Nasr M, Ookawara S, Abdelhaleem A (2024) New insights into green synthesis of metal oxide based photocatalysts for photodegradation of organic pollutants: a bibliometric analysis and techno-economic evaluation. J Clean Prod 463:142679

Devi M, Sharma S, Kumar P, Thakur N, Kumar G, Sharma MV, Sharma AJ, Kumar K, Sharma A, Jeet K, Thakur N (2024) Antifungal, antibacterial and antioxidant activity of *Pinus roxburghii* mediated green synthesized zinc and gadolinium doped manganese oxide nanoparticles. Colloid Surf c Environ Aspects 2:100046

Dihom HR, Al-Shaibani MM, Mohamed RMSR, Al-Gheethi AA, Sharma A, Khamidun MHB (2022) Photocatalytic degradation of disperse azo dyes in textile wastewater using green zinc oxide nanoparticles synthesized in plant extract: a critical review. J Water Process Eng 47:102705

Eddy NO, Garg R, Garg R, Ukpe RA, Abugu H (2024) Adsorption and photodegradation of organic contaminants by silver nanoparticles: isotherms, kinetics and computational analysis. Environ Monit Assess 196(1):65

El-saied HAA, El-Fawal EM (2021) Green superabsorbent nanocomposite hydrogels for high-efficiency adsorption and photo-degradation/reduction of toxic pollutants from waste water. Polym Test 97:107134

Emmanuel SS, Adesibikan AA, Opatola EA, Olawoyin CO (2023) A pragmatic review on photocatalytic degradation of methyl orange dye pollutant using greenly biofunctionalized nanometallic materials: a focus on aquatic body. Appl Organomet Chem 37(7):e7108

Emmanuel SS, Esan AO, Afigo FSO, Adesibikan AA, Idris MO (2024) A review on eco-sustainable photocatalytic degradation of pharmaceutical pollutants using biosynthesized nanoparticles. J Chin Chem Soc 71(11):1–26

Fardood ST, Moradnia F, Aminabhavi TM (2024) Green synthesis of novel $Zn_{0.5}Ni_{0.5}FeCrO_4$ spinel magnetic nanoparticles: photodegradation of 4-nitrophenol and aniline under visible light irradiation. Environ Pollut 358:124534

Goutam SP, Saxena G, Singh V, Yadav AK, Bharagava RN, Thapa KB (2018) Green synthesis of $TiO_2$ nanoparticles using leaf extract of *Jatropha curcas* L. for photocatalytic degradation of tannery wastewater. Chem Eng J 336:386–396

Hariharan C (2006) Photocatalytic degradation of organic contaminants in water by ZnO nanoparticles: revisited. Appl Catal A 304:55–61

Hashemi Z, Mizwari ZM, Mohammadi-Aghdam S, Mortazavi-Derazkola S, Ebrahimzadeh MA (2022) Sustainable green synthesis of silver nanoparticles using *Sambucus ebulus* phenolic extract (AgNPs@ SEE): optimization and assessment of photocatalytic degradation of methyl orange and their in vitro antibacterial and anticancer activity. Arab J Chem 15(1):103525

Hassan SS, El Azab WI, Ali HR, Mansour MS (2015) Green synthesis and characterization of ZnO nanoparticles for photocatalytic degradation of anthracene. Adv Nat Sci Nanosci Nanotechnol 6(4):045012

Hosny M, Eltaweil AS, Mostafa M, El-Badry YA, Hussein EE, Omer AM, Fawzy M (2022) Facile synthesis of gold nanoparticles for anticancer, antioxidant applications and photocatalytic degradation of toxic organic pollutants. ACS Omega 7(3):3121–3133

Iqbal A, Haq AU, Cerrón-Calle GA, Naqvi SAR, Westerhoff P, Garcia-Segura S (2021) Green synthesis of flower-shaped copper oxide and nickel oxide nanoparticles via *Capparis decidua* leaf extract for synergic adsorption-photocatalytic degradation of pesticides. Catalysts 11(7):806

Isa EDM, Shameli K, Ch'ng HJ, Jusoh NWC, Hazan R (2021) Photocatalytic degradation of selected pharmaceuticals using green fabricated zinc oxide nanoparticles. Adv Powder Technol 32(7):2398–2409

Karthik KV, Raghu AV, Reddy KR, Ravishankar R, Sangeeta M, Shetti NP, Reddy CV (2022) Green synthesis of Cu-doped ZnO nanoparticles and its application for the photocatalytic degradation of hazardous organic pollutants. Chemosphere 287:132081

Kaur M, Mehta SK, Kansal SK (2018) Visible light driven photocatalytic degradation of ofloxacin and malachite green dye using cadmium sulphide nanoparticles. J Environ Chem Eng 6(3):3631–3639

Kumar P, Kaushik A, Kumar S, Thakur N (2024a) Chemical and green synthesized Co/Ni-doped hematite nanoparticles for enhancing the photocatalytic and antioxidant properties. Phys Scr 99(10):105960

Kumar P, Kumar S, Tapwal A, Thakur N (2024b) Chemical/green synthesized cobalt/copper-doped $\alpha$-$Fe_2O_3$ nanoparticles: potential for environmental remediation. J Mater Res 39(5):836–849

Kumar P, Kumar S, Tapwal A, Nimesh S, Thakur N (2024c) Water purification and biological efficacy of green synthesized Co/Zn-doped $\alpha$-$Fe_2O_3$ nanoparticles. Sustain Chem Environ 8:100160

Kumar P, Pathak D, Thakur N (2024d) Trimetallic doped hematite ($\alpha$-$Fe_2O_3$) nanoparticles using biomolecules of *Azadirachta indica* leaf extract for photocatalytic dye removal: insights into catalyst stability and reusability. Emergent Mater 7:2505–2521

Kumar P, Tapwal A, Kumar S, Thakur N (2024e) Improved photocatalytic and free radical scavenging studies of synthesized polymer PVP/*Azadirachta indica* leave extract-mediated Ni-Zn doped hematite nanoparticles. Adv Nat Sci Nanosci Nanotechnol 15(2):025014

Kumar P, Thakur N, Kumar K, Kumar S, Dutt A, Thakur VK, Gutiérrez-Rodelo C, Thakur P, Navarrete A, Thakur N (2024f) Catalyzing innovation: exploring iron oxide nanoparticles-origins, advancements and future application horizons. Coord Chem Rev 507:215750

Kumar R, Kaushal S, Verma N, Kumar P, Thakur N, Kumar A, Kumar S, Umar A, Almas T, Pal K, Baskoutas S (2024g) Nano bioaugmentation for textile dye remediation: A sustainable approach for health and environment management. J Mol Liquids 415(A):126254

Lanjwani MF, Tuzen M, Khuhawar MY, Saleh TA (2024) Trends in photocatalytic degradation of organic dye pollutants using nanoparticles: a review. Inorg Chem Commun 159:111613

Laouini SE, Bouafia A, Soldatov AV, Algarni H, Tedjani ML, Ali GA, Barhoum A (2021) Green synthesized of $Ag/Ag_2O$ nanoparticles using aqueous leaves extracts of *Phoenix dactylifera* L. and their azo dye photodegradation. Membranes 11(7):468

Liu X, Li X, Liu X, He S, Jin J, Meng H (2020) Green preparation of Ag-ZnO-rGO nanoparticles for efficient adsorption and photodegradation activity. Colloids Surf A 584:124011

Lu J, Batjikh I, Hurh J, Han Y, Ali H, Mathiyalagan R, Ling C, Ahn JC, Yang DC (2019) Photocatalytic degradation of methylene blue using biosynthesized zinc oxide nanoparticles from bark extract of *Kalopanax septemlobus*. Optik 182:980–985

Luque PA, Chinchillas-Chinchillas MJ, Nava O, Lugo-Medina E, Martínez-Rosas ME, Carrillo-Castillo A, Vilchis-Nestor AR, Madrigal-Muñoz LE, Garrafa-Gálvez HE (2021) Green synthesis of tin dioxide nanoparticles using *Camellia sinensis* and its application in photocatalytic degradation of textile dyes. Optik 229:166259

Miri A, Sarani M, Najafidoust A, Mehrabani M, Zadeh FA, Varma RS (2022) Photocatalytic performance and cytotoxic activity of green-synthesized cobalt ferrite nanoparticles. Mater Res Bull 149:111706

Miri A, Sedighi AS, Najafidoust A, Khatami M, Sarani M (2023) Study of photodegradation performance and ability of lead removal of green synthesised maghemite nanoparticles. Int J Environ Anal Chem 103(19):8369–8383

Mirsadeghi S, Zandavar H, Yousefi M, Rajabi HR, Pourmortazavi SM (2020) Green-photodegradation of model pharmaceutical contaminations over biogenic $Fe_3O_4$/Au nanocomposite and antimicrobial activity. J Environ Manage 270:110831

Moradnia F, Fardood ST, Ramazani A, Gupta VK (2020) Green synthesis of recyclable $MgFeCrO_4$ spinel nanoparticles for rapid photodegradation of direct black 122 dye. J Photochem Photobiol A 392:112433

Muthukumar H, Palanirajan SK, Shanmugam MK, Arivalagan P, Gummadi SN (2022) Photocatalytic degradation of caffeine and *E. coli* inactivation using silver oxide nanoparticles obtained by a facile green co-reduction method. Clean Technol Environ Policy 24(4):1087–1098

Nagajyothi PC, Prabhakar Vattikuti SV, Devarayapalli KC, Yoo K, Shim J, Sreekanth TVM (2020) Green synthesis: photocatalytic degradation of textile dyes using metal and metal oxide nanoparticles-latest trends and advancements. Crit Rev Environ Sci Technol 50(24):2617–2723

Nagaraja K, Arunpandian M, Hwan OT (2024) Enhanced photocatalytic degradation of organic pollutants by green-synthesized gold nanoparticles using polysaccharide for environmental remediation. Int J Biol Macromol 269:131866

Niveditha BS, Sushmitha L, Usha BK, Nagaraju G (2023) Facile green synthesis of Cu-doped $MoO_3$ nanoparticles and its application for the photocatalytic degradation of hazardous organic pollutants. Nano-Struct Nano-Objects 36:101066

Palma TL, Vieira B, Nunes J, Lourenço JP, Monteiro OC, Costa MC (2020) Photodegradation of chloramphenicol and paracetamol using $PbS/TiO_2$ nanocomposites produced by green synthesis. J Iran Chem Soc 17:2013–2031

Patra R, Dash P, Panda PK, Yang PC (2023) A breakthrough in photocatalytic wastewater treatment: The incredible potential of g-C3N4/Titanate perovskite-based nanocomposites. Nanomaterials 13(15):2173

Pawar RC, Pyo Y, Ahn SH, Lee CS (2015) Photoelectrochemical properties and photodegradation of organic pollutants using hematite hybrids modified by gold nanoparticles and graphitic carbon nitride. Appl Catal B 176:654–666

Phang YK, Aminuzzaman M, Akhtaruzzaman M, Muhammad G, Ogawa S, Watanabe A, Tey LH (2021) Green synthesis and characterization of CuO nanoparticles derived from papaya peel extract for the photocatalytic degradation of palm oil mill effluent (POME). Sustainability 13(2):1–15

Pinna M, Binda G, Altomare M, Marelli M, Dossi C, Monticelli D, Spanu D, Recchia S (2021) Biochar nanoparticles over $TiO_2$ nanotube arrays: a green co-catalyst to boost the photocatalytic degradation of organic pollutants. Catalysts 11(9):1048

Purkait P, Bhattacharyya A, Roy S, Maitra S, Das G, Ghosh Chaudhuri M (2020) Green synthesis of $TiO_2$ nanoparticle: its characterization and potential application in Zoxamide photodegradation. J Water Environ Nanotechnol 5(3):191–203

Quddus F, Shah A, Iftikhar FJ, Shah NS, Haleem A (2023) Environmentally benign nanoparticles for the photocatalytic degradation of pharmaceutical drugs. Catalysts 13(3):511

Rafique M, Shafiq F, Gillani SSA, Shakil M, Tahir MB, Sadaf I (2020) Eco-friendly green and biosynthesis of copper oxide nanoparticles using *Citrofortunella microcarpa* leaves extract for efficient photocatalytic degradation of Rhodamin B dye form textile wastewater. Optik 208(4):164053

Rambiyana T, Mahlaule-Glory LM, Ngoepe NM, Mathipa MM, Mpelane S, Mketo N, Hintsho-Mbita NC (2023) Photocatalytic degradation of dyes, pharmaceutical and the removal bacterial pollutants using Rod shaped green derived CuO nanoparticles. Int J Environ Anal Chem 103(19):8063–8079

Rana A, Kumar P, Thakur N, Kumar S, Kumar K, Thakur N (2024) Investigation of photocatalytic, antibacterial and antioxidant properties of environmentally green synthesized zinc oxide and yttrium doped zinc oxide nanoparticles. Nano-Struct Nano-Objects 38:101188

Rani M, Shanker U (2022) Efficient removal of plastic additives by sunlight active titanium dioxide decorated Cd–Mg ferrite nanocomposite: green synthesis, kinetics and photoactivity. Chemosphere 290:133307

Rani M, Keshu and Shanker U (2021) Sunlight-induced photocatalytic degradation of organic pollutants by biosynthesized hetrometallic oxides nanoparticles. Environ Sci Pollut Res 28(43):61760–61780

Rocha V, Franco S, Bertão AR, Neves IC, Tavares T (2024) Recycling of natural and waste materials as supports for green silver nanoparticles as efficient catalysts in photodegradation of organic pollutants. Environ Technol Innov 34:103576

Rostami-Vartooni A, Nasrollahzadeh M, Salavati-Niasari M, Atarod M (2016) Photocatalytic degradation of azo dyes by titanium dioxide supported silver nanoparticles prepared by a green method using *Carpobrotus acinaciformis* extract. J Alloy Compd 689:15–20

Saranya KS, Vellora Thekkae Padil V, Senan C, Pilankatta R, Saranya K, George B, Wacławek S, Černík M (2018) Green synthesis of high temperature stable anatase titanium dioxide nanoparticles using gum kondagogu: characterization and solar driven photocatalytic degradation of organic dye. Nanomaterials 8(12):1–19

Sarkar S, Ponce NT, Banerjee A, Bandopadhyay R, Rajendran S, Lichtfouse E (2020) Green polymeric nanomaterials for the photocatalytic degradation of dyes: a review. Environ Chem Lett 18(5):1569–1580

Singh J, Kumar V, Kim KH, Rawat M (2019) Biogenic synthesis of copper oxide nanoparticles using plant extract and its prodigious potential for photocatalytic degradation of dyes. Environ Res 177:108569

Thakur N, Kumar P (2024) Effect of shape and size on synthesized triple (Co/Ni/Zn)-doped $\alpha$-$Fe_2O_3$ nanoparticles on their photocatalytic and scavenging properties. Int J Nanosci 23(6):2450010

Thakur N, Thakur N (2023) Removal of organic dyes and free radical assay by encapsulating polyvinylpyrrolidone and *Tinospora cordifolia* in dual (Co–Cu) doped $TiO_2$ nanoparticles. Environ Pollut 335:122229

Thakur N, Thakur N (2024a) Degradation of textiles dyes and scavenging activity of spherical shape obtained anatase phase of Co–Ni-doped $TiO_2$ nanocatalyst. J Mater Sci Mater Electron 35(2):134

Thakur N, Thakur N (2024b) Photocatalytic adsorption and scavenging potential of chemical and green encapsulated anatase phase of coupled doped Zn-Co $TiO_2$ nanoparticles. J Dispers Sci Technol 1–16

Thakur N, Kumar A, Thakur N (2023) *Tinospora cordifolia* and polyvinylpyrrolidone encapsulated dual doped Ni-Cu $TiO_2$ emerging nanocatalyst for the removal of organic dyes from wastewater and its free radical assay activity. Hybrid Adv 4:100086

Thakur N, Thakur N, Kumar A, Thakur VK, Kalia S, Arya V, Kumar A, Kumar S, Kyzas GZ (2024a) A critical review on the recent trends of photocatalytic, antibacterial, antioxidant and nanohybrid applications of anatase and rutile $TiO_2$ nanoparticles. Sci Total Environ 914:169815

Thakur S, Kumar P, Thakur N, Kumar K, Jeet K, Kumar S, Thakur N (2024b) Photocatalytic, antibacterial and antioxidant potential of spheroidal shape chromium and yttrium doped cobalt oxide nanoparticles: a green approach. J Indian Chem Soc 101(8):101199

Vasantharaj S, Sathiyavimal S, Senthilkumar P, Kalpana VN, Rajalakshmi G, Alsehli M, Elfasakhany A, Pugazhendhi A (2021) Enhanced photocatalytic degradation of water pollutants using bio-green synthesis of zinc oxide nanoparticles (ZnO NPs). J Environ Chem Eng 9(4):105772

Vishnu G, Singh S, Naik TSK, Viswanath R, Ramamurthy PC, Bhadrecha P, Naik HB, Singh J, Khan NA, Zahmatkesh S (2023) Photodegradation of methylene blue dye using light driven photocatalyst-green cobalt doped cadmium ferrite nanoparticles as antibacterial agents. J Clean Prod 404:136977

Zulfiqar N, Nadeem R, Musaimi OA (2024) Photocatalytic degradation of antibiotics via exploitation of a magnetic nanocomposite: a green nanotechnology approach toward drug-contaminated wastewater reclamation. ACS Omega 9(7):7986–8004

# Chapter 3
# Biogenic Nanocomposites for the Removal of Dyes and Micropollutants from Wastewater

**Kuldeep Kumar, Ravi Kumar, and Naveen Thakur**

**Abstract** Water pollution caused by dyes and micropollutants from industrial and agricultural activities poses significant environmental and health risks. Conventional water treatment methods often prove inadequate in addressing these contaminants due to their persistent and complex nature. In recent years, biogenic nanocomposites have emerged as a promising solution for wastewater remediation. These materials, synthesized from naturally occurring precursors, offer eco-friendly, cost-effective, and sustainable alternatives to synthetic adsorbents and catalysts. This review highlights the potential of biogenic nanocomposites for the efficient removal of dyes and micropollutants from wastewater. The unique properties of these nanocomposites, including large surface area, high reactivity, and tunable surface functionalities, make them ideal for adsorption, photocatalysis, and degradation of hazardous contaminants. Various biogenic sources, such as agricultural waste, microorganisms, and plant extracts, are explored in the synthesis of these materials. We also discuss the mechanisms of dye and pollutant removal, along with the challenges and future perspectives for large-scale applications. The integration of biogenic nanocomposites into wastewater treatment technologies holds great potential for achieving cleaner water and promoting environmental sustainability.

K. Kumar (✉) · R. Kumar
Department of Chemistry, Career Point University, Hamirpur, HP, India

Centre for Nano-Science and Technology, Career Point University, Hamirpur, HP, India

R. Kumar
Department of Chemistry, School of Sciences, Baddi University of Emerging Sciences and Technology, Baddi, Solan, HP, India

N. Thakur
Faculty of Allied and Health Care Sciences, Desh Bhagat University, Fatehgarh Sahib, Punjab, India



A. Balkrishna et al. (eds.), *Nanotechnology in Environmental Science and Healthcare*,
https://doi.org/10.1007/978-981-95-6714-0_3

## 3.1 Introduction

Water pollution is one of the major environmental concerns of the twenty-first century, driven by industrialization, urbanization, and agricultural activities. Contaminants, such as dyes and micropollutants (including pharmaceuticals, personal care products, and pesticides), have emerged as significant threats to both aquatic ecosystems and human health. These substances are often resistant to conventional wastewater treatment methods due to their complex chemical structures and recalcitrant nature. The development of advanced materials, particularly nanomaterials, offers promising solutions for the efficient removal of these pollutants. Among them, "biogenic nanocomposites," derived from natural sources and waste materials, present an eco-friendly and cost-effective approach. Water contamination by organic dyes and micropollutants has become a critical environmental concern due to the harmful effects these pollutants have on aquatic ecosystems and human health. Dyes, commonly used in industries, such as textiles, printing, paper, and plastics, are often non-biodegradable and resistant to conventional wastewater treatment methods (Rehman et al. 2023). Micropollutants, including pharmaceuticals, pesticides, and personal care products, are present in trace amounts but have significant ecological and toxicological effects. These contaminants pose challenges not only due to their persistence in water but also their ability to bioaccumulate, leading to long-term environmental hazards (Gautam et al. 2019).

Dyes are compounds that impart color to materials, such as textiles and plastics, and can be classified based on their source, application, and chemical structure. Based on their source, dyes can be natural, derived from plants and animals, or synthetic, produced through chemical processes. Application-wise, they include direct dyes, which are applied directly to fibers; acid dyes, used with acidic conditions on wool and silk; basic dyes, known for bright colors on acrylic fibers; reactive dyes, which form strong bonds with fabrics; disperse dyes for synthetic fibers; vat dyes applied by reduction and oxidation; and sulfur dyes, known for dull shades on cotton. Chemically, dyes are classified, such as azo dyes, which contain the -N = N- group and are widely used; anthraquinone dyes for blue and green shades; phthalocyanine dyes, which are used for blue and green; and triphenylmethane dyes, which are vibrant but less stable. Drugs, on the other hand, are used in the diagnosis, treatment, or prevention of diseases. They can be classified based on therapeutic use, chemical structure, or mechanism of action. Therapeutically, they include analgesics for pain relief, antibiotics for bacterial infections, antivirals, antifungals, antipyretics to reduce fever, antihistamines for allergies, and sedatives. Structurally, drugs include groups, such as alkaloids, found in plants and containing nitrogen; steroids, characterized by a four-ring structure; glycosides, which contain sugar molecules; and peptide-based drugs, like insulin. Mechanistically, drugs can function as enzyme inhibitors, blocking specific enzymes; receptor agonists or antagonists that interact with cell receptors; ion channel modulators for cardiac or neural conditions; and DNA/RNA modulators in cancer treatments and antiviral drugs. This classification facilitates understanding of their applications and benefits across medical and industrial fields.

In response to the limitations of traditional wastewater treatment technologies, the development of innovative materials capable of effectively removing both dyes and micropollutants has gained significant attention. Among these, nanocomposites have emerged as promising materials for environmental remediation due to their large surface area, tunable surface chemistry, and ability to adsorb or degrade pollutants. Specifically, biogenic nanocomposites, which are synthesized from natural and renewable resources, offer a green and sustainable solution for wastewater treatment. Derived from materials, such as plant biomass, microorganisms, and agricultural waste, biogenic nanocomposites are eco-friendly, are cost-effective, and exhibit remarkable efficiency in pollutant removal (Islam et al. 2021).

Biogenic nanocomposites combine the unique properties of nanoparticles, such as high reactivity and catalytic potential, with the environmental sustainability of biogenic materials. These composites can be tailored to target specific pollutants, and their multifunctional characteristics allow them to act as adsorbents, photocatalysts, or even antimicrobial agents. For instance, biogenic materials, such as cellulose, chitosan, and biochar, have been effectively combined with metal and metal oxide nanoparticles, resulting in hybrid nanocomposites that demonstrate enhanced removal capacities for a wide range of contaminants (Nandasana and Ghosh 2024). Further, several metal and metal oxide nanoparticles exhibited dye degradation potential (Thakur et al. 2022; Thakur et al. 2024).

The textile and chemical industries are major sources of dyes and drugs. Some of the pharmaceuticals and dyes are depicted in Fig. 3.1. The textile industry extensively uses synthetic dyes, including azo dyes, anthraquinone dyes, and sulfur dyes, to color fabrics. These dyes are valued for their vivid colors, durability, and resistance to fading. However, the wastewater from textile industries often contains dye residues that can be harmful to the environment, requiring proper treatment before disposal. In the chemical and pharmaceutical industries, synthetic processes are widely employed to produce drugs. Bulk manufacturing of pharmaceutical ingredients relies on organic synthesis to create complex compounds, such as analgesics, antibiotics, and antihypertensives. Industrial advancements in drug synthesis have enabled the development of active pharmaceutical ingredients (APIs) with high precision, optimization of dosage forms, and effective quality control. However, these processes can generate chemical waste, necessitating the adoption of sustainable practices to mitigate environmental impact.

Furthermore, the application of biogenic nanocomposites aligns with the principles of green chemistry, focusing on reducing waste, minimizing the use of hazardous substances, and utilizing renewable materials. Their low toxicity and biodegradability make them a superior alternative to synthetic materials, which may contribute to secondary pollution. In addition, the use of agricultural by-products and waste in the synthesis of biogenic nanocomposites adds value to waste materials while addressing environmental pollution (Sanni et al. 2024). Biological synthesis and characterization methods for nanoparticles are highlighted in Fig. 3.2.

This chapter focuses on the development and application of biogenic nanocomposites for removing dyes and micropollutants from wastewater. By examining recent advancements in the synthesis, characterization, and performance of these materials,

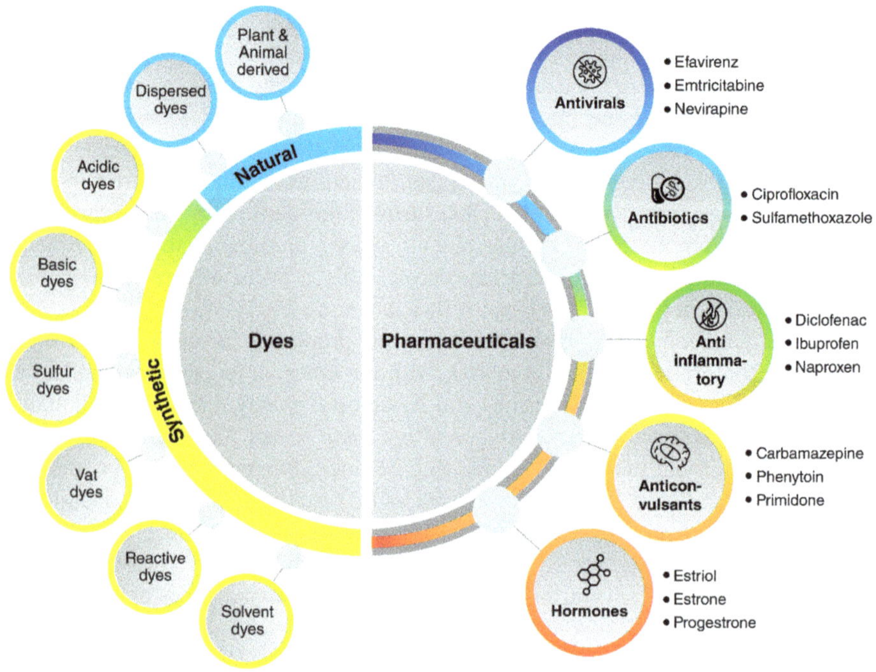


**Fig. 3.1** Classes of dyes and pharmaceuticals

we aim to highlight their potential as efficient, sustainable, and scalable solutions for wastewater remediation. The role of various biogenic sources and the synergistic effects of combining them with nanomaterials are discussed, alongside the challenges and future perspectives in this rapidly evolving field.

## 3.2 Dyes and Micropollutants

### *3.2.1 Dyes in Wastewater*

Dyes in wastewater are a significant environmental concern due to their widespread use in industries like textiles, leather, paper, cosmetics, and food processing. These industries discharge large amounts of dye-laden effluents, contributing to severe water pollution. The dyes found in wastewater are typically synthetic and highly stable, making them difficult to remove through conventional treatment methods. Common types of dyes include azo, reactive, acid, basic, and disperse dyes, each posing unique challenges for removal. These dyes not only impart color to water but also introduce toxic chemicals that can harm aquatic life, disrupt ecosystems, and pose risks to human health (Arabkhani et al. 2022). Certain dyes, particularly azo

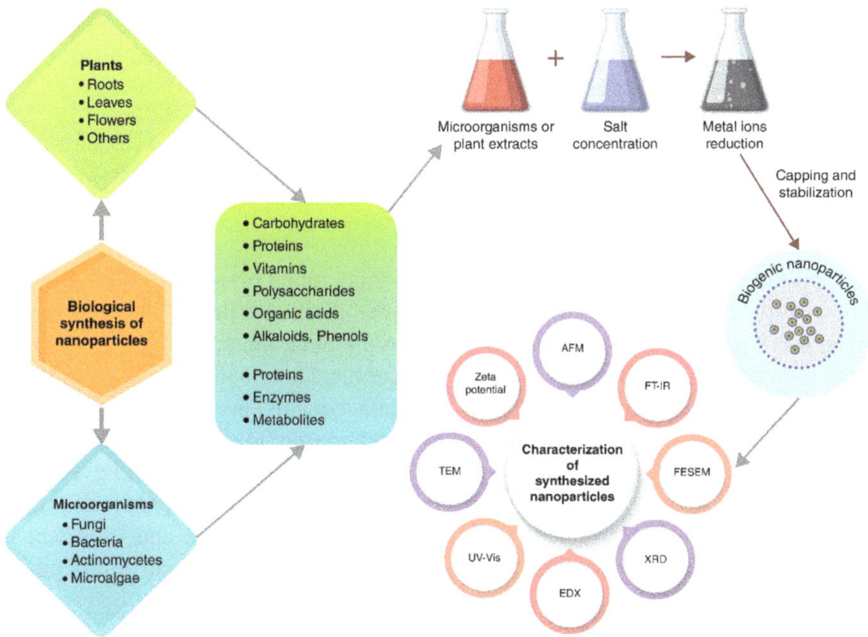


**Fig. 3.2** Various biological sources for synthesis of nanoparticles and their characterization methods

dyes, degrade into carcinogenic compounds, posing long-term environmental and health risks. Additionally, the presence of dyes in water reduces light penetration, which affects photosynthesis in aquatic plants and leads to oxygen imbalances.

Treating dye-contaminated wastewater is a challenging task, and various methods have been developed to address this issue. Physical processes, such as adsorption and membrane filtration, are commonly used; however, they can be expensive and generate waste. Chemical treatments, including oxidation and electrochemical processes, offer efficient dye removal but often require high energy input and may produce harmful by-products (Kathalingam et al. 2023). Biological methods, like biodegradation using microorganisms and constructed wetlands, are environmentally friendly but tend to be slower and less effective for highly toxic dyes. Emerging technologies, particularly those involving nanomaterials, show great promise due to their high efficiency in adsorbing and degrading dyes. Biogenic nanocomposites, in particular, have attracted attention for their eco-friendly nature and potential to remove a wide range of dyes and other micropollutants from wastewater (Ali et al. 2019).

Despite advancements in treatment technologies, large-scale application remains challenging due to high costs, sludge generation, and the complexity of dye mixtures in industrial effluents. As a result, future research is focused on developing more sustainable and cost-effective solutions, such as using renewable biogenic materials and integrating advanced oxidation processes with biological treatments. Overall, addressing the issue of dyes in wastewater requires a combination of innovative technologies and sustainable practices to minimize their environmental impact and ensure cleaner water resources.

### *3.2.2 Micropollutants*

Micropollutants are a diverse group of contaminants found in water at trace concentrations, typically ranging from nanograms to micrograms per liter. Despite their low levels, these pollutants pose significant risks to the environment and human health due to their persistence, toxicity, and potential for bioaccumulation. Micropollutants originate from various sources, including pharmaceuticals, personal care products, pesticides, industrial chemicals, heavy metals, and household cleaning agents (Sanni et al. 2024). Pharmaceuticals, such as antibiotics, hormones, and pain relievers, are commonly detected in wastewater due to human consumption and improper disposal. Similarly, personal care products, such as shampoos, fragrances, and sunscreens, enter water systems through domestic use and laundry. Agricultural runoff introduces pesticides and herbicides, while industrial activities contribute synthetic chemicals and heavy metals (Bharathi et al. 2024). These pollutants can persist in the environment for extended periods, resulting in widespread contamination of water bodies.

The environmental impact of micropollutants is significant, even at trace levels. They are toxic to aquatic organisms, disrupting their biological functions, including reproduction, growth, and behavior. Endocrine-disrupting chemicals (EDCs), a class of micropollutants, can interfere with hormone systems, leading to feminization in fish and other reproductive abnormalities in wildlife. Many micropollutants also have the potential to bioaccumulate, meaning they can build up in the tissues of organisms and magnify up the food chain, ultimately affecting humans who consume contaminated fish and seafood. Furthermore, the presence of antibiotics in wastewater contributes to the development and spread of antibiotic-resistant bacteria, posing a major public health risk (Law et al. 2023).

Treating micropollutants in wastewater is challenging because traditional treatment plants are not designed to effectively remove them. Advanced treatment methods, such as adsorption, membrane filtration, and advanced oxidation processes (AOPs), are often necessary to address micropollutant contamination (Yadav et al. 2024). Physical methods, such as adsorption using activated carbon, can trap organic molecules, while membrane technologies, like reverse osmosis, are effective but costly and prone to fouling. Chemical processes, including AOPs and electrochemical treatments, degrade micropollutants into less harmful substances but may produce toxic by-products and require high energy input. Biological methods, such as biodegradation by microorganisms and constructed wetlands, offer eco-friendly alternatives; however, they are slower and may not be effective for all micropollutants. Emerging technologies, including nanotechnology and photocatalysis, hold promise for enhancing micropollutant removal efficiency (Dihom et al. 2022). Despite these advancements, the treatment of micropollutants on a large scale remains a challenge due to the diversity of pollutants, high costs, and secondary pollution issues. Future research is focused on developing more sustainable, cost-effective methods, such as biogenic nanocomposites and hybrid treatment systems that combine multiple processes. Addressing micropollutants is crucial for ensuring water quality and protecting ecosystems and human health from long-term harm.

## 3.3 Biogenic Nanocomposites

Biogenic nanocomposites are hybrid materials that combine biologically derived components with nanoparticles, offering a sustainable and eco-friendly approach to nanotechnology. These composites utilize natural sources, including plant extracts, microorganisms, and agricultural waste, to produce nanoparticles in a "green" manner, thereby avoiding the use of toxic chemicals and harsh conditions typically required in traditional synthesis methods (Wang et al. 2023). This not only makes the production process safer but also cost-effective, especially when utilizing waste materials.

The biological components in these nanocomposites help control the size, shape, and properties of the nanoparticles, which enhances their functionality. One of the key advantages of biogenic nanocomposites is their biocompatibility and low toxicity, making them highly suitable for biomedical applications, such as drug delivery and wound healing (Rajkumar et al. 2024). Additionally, they possess a high surface area and reactivity, which make them effective in environmental remediation processes, such as water purification and the degradation of pollutants. The synergy between biogenic and nanomaterial components often results in enhanced performance, including improved mechanical strength and increased adsorption capacity. Overall, biogenic nanocomposites present a versatile, sustainable, and efficient solution for applications in energy storage, environmental remediation, and health care (Martinez et al. 2024).

### *3.3.1 Sources of Biogenic Nanocomposites*

Sources of biogenic nanocomposites primarily stem from naturally occurring biological materials, which provide eco-friendly, renewable, and sustainable alternatives for nanoparticle synthesis. Common sources include plant-based materials, such as extracts from leaves, fruits, and roots, which contain natural reducing agents like polyphenols, flavonoids, and alkaloids that facilitate the synthesis of nanoparticles. Agricultural waste, such as coconut shells, rice husks, and sugarcane bagasse, is also widely utilized for producing biogenic nanocomposites, providing a cost-effective method for recycling biomass (Ahmad et al. 2024). Additionally, microorganisms, such as bacteria, fungi, and algae, play a significant role in biosynthesis, as they can biologically reduce metal salts to form nanoparticles. For example, certain bacteria and fungi secrete enzymes that help reduce metal ions and stabilize the nanoparticles. Algae, particularly microalgae, are also emerging as a valuable source due to their rich biochemical composition and ability to synthesize metal nanoparticles under mild conditions. Even animal-derived materials, such as chitosan from crustacean shells and proteins from silk or wool fibers, are used to create biogenic nanocomposites (Alarfai et al. 2024). These natural sources not only contribute to green chemistry but also open up possibilities for sustainable, scalable production of nanocomposites for various applications.

## 3.4 Synthesis of Biogenic Nanocomposites

The synthesis of biogenic nanocomposites involves environmentally friendly and sustainable methods that utilize biological materials to reduce, stabilize, and functionalize nanoparticles. This "green synthesis" approach often begins with the extraction of biological components from sources, such as plants, microorganisms, or agricultural waste, which contain natural reducing agents like proteins, polysaccharides, enzymes, flavonoids, and phenolic compounds (Chellapandi et al. 2024). These biomolecules act as capping and stabilizing agents during the nanoparticle formation process. In the case of plant-based synthesis, plant extracts are prepared by boiling or grinding plant parts (such as leaves, fruits, or roots) in solvents like water or ethanol, which are then combined with a metal precursor solution. The plant extract facilitates the reduction of metal ions (such as silver, gold, or copper) into nanoparticles, and the biomolecules in the extract help stabilize the nanoparticles, preventing agglomeration (Gautam et al. 2022).

For microbial synthesis, bacteria, fungi, or algae are cultured under specific conditions to promote the secretion of enzymes or metabolites that reduce metal ions into nanoparticles. Microbial systems often have the advantage of controlling nanoparticle size and morphology due to the enzymatic pathways involved in the reduction process. Fungi, for instance, can produce nanoparticles extracellularly through enzymes like nitrate reductase, which help in metal ion reduction. Algae, on the other hand, can absorb and bioreduce metal ions on their cell surfaces or intracellularly, leading to the synthesis of nanoparticles (Malakootian et al. 2020).

Agricultural waste, such as coconut shells, rice husks, or sugarcane bagasse, can be carbonized to produce biochar or activated carbon, which can be further functionalized with nanoparticles to form biogenic nanocomposites. This process typically involves pyrolysis of the biomass to produce carbon-rich materials, followed by impregnation with metal salts and reduction to form metal nanoparticles embedded within the carbon matrix. This method not only adds value to agricultural waste but also creates highly efficient adsorbents for environmental applications (Agasti et al. 2022).

Once synthesized, biogenic nanocomposites are often characterized using techniques, such as UV-visible spectroscopy, X-ray diffraction (XRD), scanning electron microscopy (SEM), and transmission electron microscopy (TEM), to confirm the presence, size, and morphology of the nanoparticles. The advantage of biogenic synthesis lies in its ability to produce nanocomposites under mild, non-toxic conditions without the need for hazardous chemicals or extreme temperatures. This sustainable approach is gaining traction for applications in environmental remediation, catalysis, energy storage, and biomedical fields due to its eco-friendly nature and potential for large-scale production.

## 3.5 Dye and Micropollutant Removal

The removal of dyes and micropollutants from wastewater involves several mechanisms that can be broadly categorized into physical, chemical, and biological processes. **Physical processes** include adsorption, which is a widely used method in which contaminants adhere to the surface of adsorbent materials, such as activated carbon, biogenic nanocomposites, or clay minerals. The effectiveness of adsorption is influenced by factors, such as the surface area, pore size, and surface charge of the adsorbent, as well as the properties of the dye or micropollutant, including molecular size, polarity, and solubility (Lakkaboyana et al. 2021). **Chemical processes**, on the other hand, involve reactions that alter the chemical structure of contaminants. Techniques, such as advanced oxidation processes (AOPs), utilize strong oxidants, including ozone, hydrogen peroxide, or UV light, to generate reactive species that can degrade complex organic compounds into simpler, less toxic forms. For instance, hydroxyl radicals generated during AOPs are highly reactive and can effectively break down a wide range of organic dyes and micropollutants through oxidation reactions (Gabelica et al. 2021).

**Biological processes** play a significant role in removing dyes and micropollutants, often employing microorganisms, such as bacteria and fungi, to metabolize or transform contaminants. These microorganisms can degrade organic compounds through various pathways, including aerobic and anaerobic respiration, where they utilize pollutants as carbon sources, or through enzymatic reactions, such as lignin peroxidase and manganese peroxidase, which are produced by certain fungi. This biodegradation process is particularly effective for dyes and micropollutants that are resistant to chemical methods, providing a more sustainable and eco-friendly approach to wastewater treatment (Malakootian et al. 2020).

In recent years, **hybrid methods** that combine physical, chemical, and biological mechanisms have gained popularity for their enhanced efficiency in pollutant removal. For example, biogenic nanocomposites may leverage both adsorption and catalytic degradation, wherein the nanoparticles facilitate the degradation of dyes and micropollutants while also providing a surface for adsorption. Overall, the choice of removal mechanism depends on the specific characteristics of the pollutants, the composition of the wastewater, and the desired treatment efficiency, highlighting the need for tailored approaches in wastewater management. The integration of these mechanisms can lead to improved treatment outcomes, contributing to the effective removal of dyes and micropollutants from wastewater and ensuring environmental protection (Yin and Deng 2015).

## 3.6 Applications of Biogenic Nanocomposites in Wastewater Treatment

Biogenic nanocomposites have emerged as promising materials for wastewater treatment due to their unique properties, sustainability, and effectiveness in removing various contaminants, including dyes and micropollutants. Their applications in wastewater treatment can be broadly categorized into adsorption, catalysis, and bioremediation. The mechanism of dye degradation is illustrated in Fig. 3.3.

### *3.6.1 Adsorption*

One of the most significant applications of biogenic nanocomposites in wastewater treatment is their use as adsorbents. The high surface area and porosity of these materials allow them to effectively capture and retain organic pollutants, heavy metals, and dyes. For instance, biogenic nanocomposites made from agricultural waste, such as coconut shell-derived activated carbon or chitosan-based composites, have demonstrated excellent adsorption capacities for various organic dyes, including methylene blue and rhodamine B (Ghosh and Webster 2021). The presence of functional groups, such as hydroxyl, carboxyl, and amine groups on the surface of these composites,

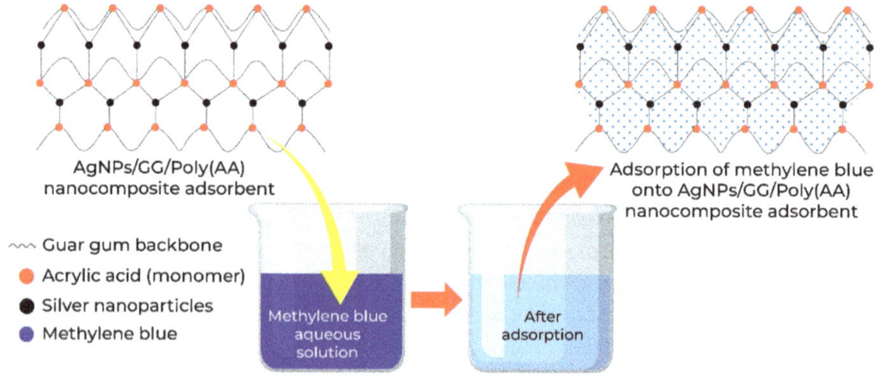


**Fig. 3.3** Mechanism of adsorption of MB dye onto AgNPs/GG/Poly(AA) nanocomposite. Reproduced from Rehman et al. (2023) under the Creative Commons Attribution (CC BY) license (http://creativecommons.org/licenses/by/4.0/)

enhances their affinity for adsorbing contaminants through electrostatic interactions, hydrogen bonding, and van der Waals forces. Additionally, the use of biogenic materials in the synthesis of nanocomposites contributes to a more sustainable approach to wastewater treatment, as it helps to utilize waste resources and reduces reliance on synthetic adsorbents.

### *3.6.2 Catalysis*

Biogenic nanocomposites also play a critical role in catalytic degradation processes. They can be employed in advanced oxidation processes (AOPs), where reactive species generated by the nanocomposites facilitate the breakdown of complex organic compounds into less harmful substances. For example, metal nanoparticles, such as silver or copper, can be incorporated into biogenic matrices, thereby enhancing their catalytic activity. These biogenic nanocomposites can effectively degrade dyes and micropollutants through oxidative reactions, offering a significant advantage over traditional treatment methods. Their ability to catalyze the degradation of persistent pollutants, such as pharmaceuticals and personal care products, makes them highly valuable in addressing emerging contaminants in wastewater (Viet et al., 2024).

### *3.6.3 Bioremediation*

The bioremediation potential of biogenic nanocomposites is another essential application in wastewater treatment. The incorporation of microorganisms within biogenic matrices can enhance the biodegradation process of organic pollutants. These composites can provide a suitable environment for microbial growth while also protecting the microorganisms from harsh wastewater conditions. For instance, nanocomposites infused with bacteria or fungi can metabolize dyes and micropollutants, converting them into less toxic by-products. This synergistic approach combines physical and biological treatment mechanisms, leading to more efficient removal of contaminants (Zulfiqar et al. 2024).

### *3.6.4 Removal of Heavy Metals*

Biogenic nanocomposites are effective in removing heavy metals from wastewater due to their high surface area and the presence of functional groups that can bind metal ions. Composites made from materials, such as chitosan or biochar, can selectively adsorb heavy metals, including lead, cadmium, and mercury, through ion exchange and complexation processes. This is crucial for industries that generate wastewater containing toxic metals, as it helps meet regulatory standards for effluent discharge and minimizes environmental impact (Tanzifi et al. 2020).

The use of biogenic nanocomposites in wastewater treatment is also advantageous from an economic perspective. They can be produced using low-cost raw materials, such as agricultural waste, and synthesized through simple and environmentally friendly processes. This cost-effectiveness, combined with their scalability, makes biogenic nanocomposites an attractive option for large-scale wastewater treatment applications, particularly in developing regions where resources are limited. As research in this field continues to advance, the potential applications of biogenic nanocomposites in wastewater treatment are expected to expand further. Ongoing studies are focused on enhancing the performance of these materials through surface modifications, functionalization, and the integration of novel nanomaterials. Furthermore, the combination of biogenic nanocomposites with emerging technologies, such as membrane filtration and photocatalysis, holds promise for developing more efficient and sustainable wastewater treatment solutions. In conclusion, biogenic nanocomposites offer a multifaceted approach to wastewater treatment, addressing the challenges of dye and micropollutant removal while promoting sustainability and environmental responsibility. Their versatility, effectiveness, and eco-friendly nature make them valuable assets in the pursuit of cleaner water resources and enhanced wastewater management practices.

Different polymer chains are interconnected through van der Waals forces and covalent bonds, forming a three-dimensional polymer matrix. Silver nanoparticles are embedded within this matrix, enhancing the surface area and surface energy of the resulting adsorbent. These nanoparticles also introduce additional functional sites that improve the binding affinity for dye molecules (Rehman et al. 2023). As a result, methylene blue (MB) dye molecules are effectively adsorbed onto the surface of the nanocomposite.

## 3.7 Challenges and Future Perspectives

The use of biogenic nanocomposites for removing dyes and micropollutants from wastewater presents several challenges that need to be addressed to enhance their practical application. One major challenge is scalability and economic viability; while these materials show promise in laboratory settings, scaling up their production for industrial use poses significant hurdles. Establishing cost-effective methods for mass production that maintain the performance characteristics of biogenic nanocomposites is crucial. Additionally, the performance consistency of these materials can vary depending on the source material and synthesis conditions, resulting in variability in their chemical composition and effectiveness. Therefore, standardizing protocols for synthesis and characterization is essential for ensuring reliable performance. Environmental stability is another concern, as factors, such as pH, temperature, and competing ions, can affect the adsorption capacity and catalytic activity of biogenic nanocomposites in complex wastewater environments. Furthermore, although research has focused on dye removal, there is comparatively less information available regarding their effectiveness against a broader range of micropollutants, necessitating further investigation. Regulatory and safety considerations must also be taken into account to ensure that these novel materials are safe for the environment and human health.

Looking ahead, several future perspectives can enhance the application of biogenic nanocomposites in wastewater treatment. Research should focus on optimizing the functionalization of these materials to improve their performance, such as modifying their surface properties to enhance adsorption capacity and selectivity for specific contaminants. Additionally, integrating biogenic nanocomposites with advanced technologies, such as photocatalysis, membrane filtration, and electrochemical methods, could lead to more efficient and sustainable treatment solutions. Long-term performance studies examining the reusability and impact of multiple treatment cycles will provide valuable insights for practical applications. Conducting comprehensive life cycle assessments (LCA) can evaluate the environmental impact of biogenic nanocomposites compared to traditional treatment methods, facilitating informed decision-making regarding their implementation. Ultimately, increasing public awareness and acceptance of the benefits of biogenic nanocomposites is crucial for their widespread adoption, and engaging with stakeholders can help foster support for these innovative materials in environmental applications. Overall, addressing these challenges and exploring future perspectives will be crucial for the successful implementation of biogenic nanocomposites in sustainable wastewater treatment practices.

## 3.8 Conclusion

In conclusion, biogenic nanocomposites represent a promising and sustainable solution for the effective removal of dyes and micropollutants from wastewater. Their unique properties, derived from naturally sourced materials, enable efficient adsorption, catalytic degradation, and bioremediation of a wide range of contaminants. Despite the significant potential these materials hold, challenges, such as scalability, performance consistency, environmental stability, and regulatory considerations, must be addressed to facilitate their practical application in industrial wastewater treatment. Future research should focus on optimizing the functionalization of biogenic nanocomposites, integrating them with advanced treatment technologies, and conducting long-term performance assessments to enhance their effectiveness. Moreover, comprehensive life cycle assessments will be essential in evaluating their environmental impact compared to conventional methods. By fostering public awareness and stakeholder engagement, the adoption of biogenic nanocomposites can contribute to more sustainable wastewater management practices. Ultimately, the continued exploration and development of biogenic nanocomposites will not only address pressing environmental challenges but also pave the way for innovative solutions in the field of water treatment and resource recovery.

## References

Agasti N, Gautam V, Pandey N, Genwa M, Meena PL, Tandon S, Samantaray R (2022) Carbon nanotube based magnetic composites for decontamination of organic chemical pollutants in water: A review. Appl Surf Sci Adv 10:1–25

Ahmad N, Rahman QI, Iqbal A, Siddiqui MR, Wabaidur SM, Ahmad N, Coulon F (2024) Design of novel Sb2S3/Polythiophene heterojunction for efficient adsorption and methanol sensing. Sep Purif Technol 350:127826

Alarfaj N, Al Musayeib N, Amina M, El-Tohamy M (2024) Synthesis and characterization of polysiphonia/cerium oxide/nickel oxide nanocomposites for the removal of toxins from contaminated water and antibacterial potential. Environ Sci Pollut Res 31(11):17064–17096

Ali I, Peng C, Khan ZM, Naz I, Sultan M, Ali M, Abbasi IA, Islam T, Ye T (2019) Overview of microbes based fabricated biogenic nanoparticles for water and wastewater treatment. J Environ Manag 230:128–150

Arabkhani P, Asfaram A, Aghaei-Jazeh M, Ateia M (2022) Plant-mediated green synthesis of nanocomposite-based multifunctional adsorbent with antibacterial activity and high removal efficiency of micropollutants from contaminated waters. J Water Process Eng 49:103025

Arumugham T, Krishnamoorthy R, Hasan SW, Banat F (2022) Recent advances in bioceramic-based nanocomposites for membrane and adsorption separation processes: A review. J Water Process Eng 49:103152

Bharathi D, Albeshr MF, Mythili R, Vasantharaj S, Lee J (2024) Chitosan-loaded biogenic silver nanocomposite for photocatalytic remediation of dye pollutants and antibacterial activity. Environ Geochem Health 46(7):250

Chellapandi T, Madhumitha G, Avinash J (2024) Ultrasonication-assisted synthesis of CuO-decorated montmorillonite K30 nanocomposites for photocatalytic removal of emerging contaminants: A response surface methodology approach. Environ Res 259:119574

Dihom HR, Al-Shaibani MM, Mohamed RMSR, Al-Gheethi AA, Sharma A, Khamidun MHB (2022) Photocatalytic degradation of disperse azo dyes in textile wastewater using green zinc oxide nanoparticles synthesized in plant extract: A critical review. J Water Process Eng 47:102705

Gabelica I, Ćurković L, Mandić V, Panžić I, Ljubas D, Zadro K (2021) Rapid microwave-assisted synthesis of $Fe_3O_4/SiO_2/TiO_2$ core-2-layer-shell nanocomposite for photocatalytic degradation of ciprofloxacin. Catalysts 11(10):1136

Gautam PK, Shivalkar S, Singh A, Shivapriya Pingali M, Chaudhary S, Banerjee S, Varadwaj PK, Samanta SK (2022) Bio-inspired Nanocomposites for Remediation of Pharmaceutical Pollutants. Nanotechnol Environ Remediat:323–339

Gautam PK, Singh A, Misra K, Sahoo AK, Samanta SK (2019) Synthesis and applications of biogenic nanomaterials in drinking and wastewater treatment. J Environ Manag 231:734–748

Ghosh S, Webster TJ (2021) Nanotechnology for water processing. In: The Future of Effluent Treatment Plants. Elsevier, pp 335–360

Islam T, Ali I, Naz I, Peng C, Khan ZM, Amjed MA (2021) Bionanocomposites for wastewater treatment. In: In: Aquananotechnology. Elsevier, pp 249–272

Kathalingam A, Santhoshkumar P, Ramesh S, Sivanesan I, Kim HS (2023) Biogenic polymer nanoparticles to remove hydrophobic organic contaminants from water. Water Environ Res 95(10):e10935

Lakkaboyana SK, Soontarapa K, Marella RK, Kannan K (2021) Preparation of novel chitosan polymeric nanocomposite as an efficient material for the removal of Acid Blue 25 from aqueous environment. Int J Biol Macromol 168:760–768

Law CKY, Kundu K, Bonin L, Peñacoba-Antona L, Bolea-Fernandez E, Vanhaecke F, Rabaey K, Esteve-Núñez A, De Gusseme B, Boon N (2023) Electrochemically assisted production of biogenic palladium nanoparticles for the catalytic removal of micropollutants in wastewater treatment plants effluent. J Environ Sci 128:203–212

Malakootian M, Khatami M, Ahmadian M, Asadzadeh SN (2020) Biogenic silver nanoparticles/hydrogen peroxide/ozone: efficient degradation of reactive blue 19. BioNanoScience 10(1):34–41
Martinez EG, Villarreal NE, Delgado DM, Delgado FP, Castaño VM (2024) Bionanomining: bioengineered CuO nanoparticles from mining and organic waste for photo-catalytic dye degradation. Environ Geochem Health 46(9):327
Nandasana M, Ghosh S (2024) Biogenic nanoparticles for dye removal. In: Advances in Chemical Pollution, Environmental Management and Protection, vol 10. Elsevier, pp 1–33
Rajkumar M, Presley SD, Thiyagarajulu N, Thangaraj R, Kannan S (2024) Recent Advances in Nanotechnology: A New Paradigm toward Wastewater Treatment. Bioremediation Emerg Contam Water. 2:89–110
Rehman MU, Taj MB, Carabineiro SA (2023) Biogenic adsorbents for removal of drugs and dyes: a comprehensive review on properties, modification and applications. Chemosphere 338:1–19
Sanni SE, Oni BA, Okoro EE, Pandya SR (2024) Recent advances in the use of biogenic nanomaterials and photocatalysts for wastewater treatment: challenges and future prospects. Front Nanotechnol 6:1469309
Tanzifi M, Yaraki MT, Beiramzadeh Z, Saremi LH, Najafifard M, Moradi H, Mansouri M, Karami M, Bazgir H (2020) Carboxymethyl cellulose improved adsorption capacity of polypyrrole/CMC composite nanoparticles for removal of reactive dyes: Experimental optimization and DFT calculation. Chemosphere 255:127052
Thakur N, Arya V, Kumar A, Thakur N (2024) Catalytic efficiency of encapsulated coupled doped Zn-Ni $TiO_2$ nanocatalyst for the removal of organic pollutants and free radical assay. Int J Environ Anal Chem 104(20):9754–9771
Thakur N, Thakur N, Kumar K, Arya V, Kumar A, Kalia S (2022) Photocatalytic efficiency of green synthesized ZnO nanoparticles for the degradation of methyl orange dye: A review. Biomater Polym Horiz 1(4):1–11
Viet NM, Ha MN, Trang PHT, Trieu NH, Son NT, Van Noi N, Ha TTV, Phuong NM (2024) Graphene-Based Materials for Degradation of Organic Pollutants in Wastewater. In: In: Graphene-Based Photocatalysts: From Fundamentals to Applications. Springer Nature Switzerland, Cham, pp 297–315
Wang G, Hambly AC, Zhao D, Wang G, Tang K, Andersen HR (2023) Peroxymonosulfate activation by suspended biogenic manganese oxides for polishing micropollutants in wastewater effluent. Sep Purif Technol 306:122501
Yadav S, Chander S, Gupta A, Kataria N, Khoo KS (2024) Biogenic engineered zinc oxide nanoparticle for sulfur black dye removal from contaminated wastewater: comparative optimization, simulation modeling, and isotherms. Bioengineered 15(1):2325721
Yin J, Deng B (2015) Polymer-matrix nanocomposite membranes for water treatment. J Membr Sci 479:256–275
Zulfiqar R, Munir R, Bashir MZ, Albasher G, Samreen T, Hafiz I, Murtaza MA, Noreen S (2024) Synthesis of polymeric sunscreen photocatalyst $ZnO_2$, $CuO_2$, $PbO_2$, and $CdO_2$ using ethylene glycol for reactive blue Dye removal from textile waste water. Catal Surv Asia 28(2):159–185

# Chapter 4
# Nanomaterials in Water Purification: From Pathogens to Microplastics and Heavy Metals

**Pankaj Kumar, Jasmeet Kaur, Rahul, Puneet Kaur, and Sonia Malik**

**Abstract** Recently, nanotechnology has been applied in various fields, including physics, chemistry, materials science, biology, and medicine. Nanotechnology is also applied in the removal of multiple impurities/pollutants, including pathogenic bacteria and viruses, from water and other water bodies. Among the various methods used for wastewater treatment, adsorption is one of the more effective methods due to its low cost and simplicity of operation. Moreover, many methods of NP synthesis are expensive and not environmentally sustainable. The utilization of biomass as a source for the production of NPs has the potential to mitigate issues pertaining to cost, sustainability, and pollution. The utilization of bio-based nanomaterials (bio-NMs) sourced from biomass has garnered attention in the field of water purification due to their cost-effectiveness, biocompatibility, and biodegradability. Several research studies have been conducted to efficiently produce NPs (both inorganic and organic) from biomass for applications in wastewater treatment. Recently, nanoadsorbents have been explored for removing various pollutants from wastewater. In this chapter, the utilization of NMs as an effective adsorbent for removing heavy metals, organic compounds, pesticides, pharmaceuticals, and microplastic pollutants is reviewed and discussed. It is to be mentioned that the adsorption capacities of the adsorbents vary depending on the characteristics of the adsorbents, the extent of chemical modification, and the concentration of adsorbates. There are also a few issues and drawbacks in the utilization of nanoadsorbents that have been addressed. To determine the practical application of NMs as low-cost adsorbents on a commercial scale, further

---

P. Kumar · J. Kaur
Department of Physical and Mathematical Sciences, Baba Farid College of Engineering and Technology, Bathinda, Punjab, India

P. Kumar
Department of Physics, Akal University, Talwandi Sabo, Bathinda 151302, Punjab, India

Rahul · P. Kaur
Department of Physics, National Cheng Kung University, Tainan, Taiwan (R.O.C.)

S. Malik (✉)
Department of Biotechnology, Baba Farid College, Bathinda, Punjab, India
e-mail: soniamalik@babafaridgroup.edu.in



A. Balkrishna et al. (eds.), *Nanotechnology in Environmental Science and Healthcare*,
https://doi.org/10.1007/978-981-95-6714-0_4

research in this direction is needed. This chapter demonstrates that NMs can be effectively utilized for wastewater treatment and may be novel and promising adsorbents/materials for future water treatment applications.

## 4.1 Introduction

Emerging contaminants found in aquatic environments include pesticides, medications, personal hygiene products, industrial additives, household products, microplastics, heavy metals, and garbage (Talreja et al. 2024). Desalination and wastewater treatment appear promising at present, but several obstacles need to be addressed before these procedures can be made commercially available. In an industrial setting, efficiency decreases (Kumar et al. 2024a, b, c, d, e, f, g). Furthermore, these methods are expensive, and even the less expensive ones produce secondary waste. These factors make water desalination and wastewater purification extremely difficult, to the point where some emerging nations have irreparably harmed their access to clean drinking water (Sun et al. 2025). Therefore, a comprehensive approach is needed to develop cost-effective and environmentally friendly technologies that could prevent future water contamination and overcome the limitations of wastewater treatment systems. In the light of this notion, the use of nanomaterials (NMs) stands out among advancements in water and wastewater treatment due to their increased efficacy in removing pollutants (Kumar et al. 2025a, b).

Because of their vast surface area and their catalytic, optical, electrical, magnetic, hydrophilic, hydrophobic, and antibacterial qualities, nanoparticles (NPs) provide significant advantages in wastewater treatment. To improve the effectiveness of the water treatment procedures, NMs are integrated into the membrane (Mitra and Nayak 2025). The properties of NPs can be tailored to target specific pollutants for removal using a range of production techniques and experimental setups. NPs are often synthesized using either a top-down or bottom-up method. As a result, components of bacteria, fungi, actinomycetes, yeasts, algae, and plants can all create NPs and serve as stabilizing and reducing agents during the synthesis of NPs (Irfan et al. 2025). Despite having distinct morphologies, sizes, and distributions, any organism may produce the same amount of NP. Generally speaking, the properties of the NPs produced during biosynthesis are influenced by temperature, pH, reaction media, biological source type, solvent, and buffer used for NP dispersion, as well as surface charge (Kumar 2024). The process of creating NMs from biological sources is safe and environmentally benign. Additionally, it offers a different approach to efficiently handling biowaste. The environmental component of these benefits is the main reason why the scientific and industrial sectors are interested in them.

To eliminate contaminants from water, traditional treatment techniques have included filtering, flocculation, coagulation, and the activated sludge process (Esteki et al. 2024). Both human health and environmental health depend on water. Water bodies have been severely contaminated by agriculture, urbanization, and industrialization, endangering the environment and human health. They are not appropriate for novel contaminants, such as industrial solvents, medications, and chemicals that affect hormones (Talreja et al. 2025a, b). Advanced therapy technologies have garnered attention due to these difficulties, and nanotechnology is considered the most promising response. NMs are better suited for eliminating pollutants from wastewater due to their physical, chemical, and biological properties, such as surface area and reactivity (Kumar and Kumar 2024). Because NPs may adsorb, degrade, and neutralize contaminants such as heavy metals, microplastics, organic chemicals, and pathogens, they can readily supplement conventional treatment technologies. These materials employ various pollutant removal techniques, including photocatalysis, adsorption, and catalysis, all of which are crucial in the complex process of wastewater treatment (Akinnike et al. 2024).

## 4.2 Types of Water Contaminants

### *4.2.1 Pathogens*

Waterborne pathogen footprints and pathogen levels in water resources are frequently evaluated using indicator species. Water quality monitoring is the primary factor in the prevention of illness and infections caused by microorganisms. In the twentieth century, due consideration has been given to water quality monitoring for pathogen detection. The various detection methods for biological contaminants are discussed here in detail (Fig. 4.1). One popular method for estimating the possible pathogen burdens in ambient water bodies is to track the numbers of indicator organisms. *Faecal coliforms* and *E. coli* levels in rivers, lakes, estuaries, and coastal waterways have been used for decades by public health officials and scientists to assess the quality of water (Ravenscroft et al. 2025). The ability of current indicator species to detect the possible presence of harmful bacteria is under discussion, additionally determining the pathogen's origin. Microbial source tracing (MST), a relatively recent method, may be used to identify the source of *faecal coliform.* In the past, antibiotic resistance analysis employed the MST approach to assess the impact of cattle on watershed-scale water quality (Fulke et al. 2025). Based on a phenotypic approach, the authors proposed that host-origin libraries help identify the origins of the pathogens. However, a lot of MST techniques are predicated on the idea that certain bacterial strains are exclusive to a particular animal species or group. Regarding the common faecal bacterium *E. coli,* this assumption may be questionable. As a result, employing *E. coli* for source tracking requires caution. Furthermore, before using the MST approach at the watershed scale, consideration should be given to the costs of developing libraries, putting in place the substantial sampling programmes required for MST method verification, and calculating technique uncertainties (Hachad et al. 2024).

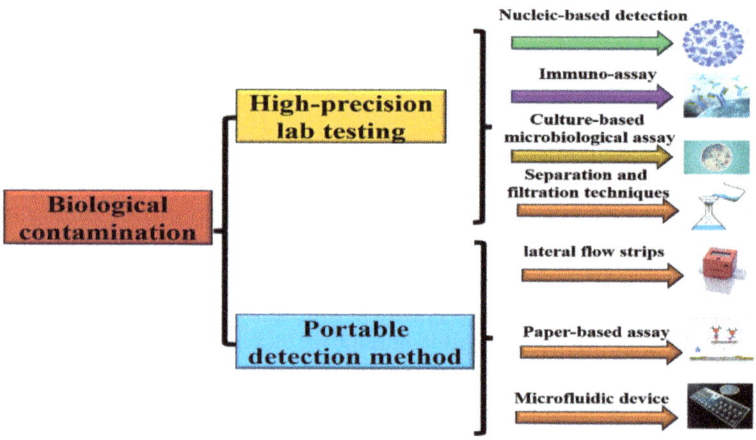


**Fig. 4.1** Classification of detection approaches for biological contaminants

## *4.2.2 Heavy Metals*

Heavy metal contamination of aquatic environments is extremely harmful because these elements cannot be broken down and accumulate inside living organisms (Singh et al. 2024). The primary means by which heavy metals enter aquatic ecosystems is through industrial waste, and the concentration of these metals in water varies depending on the type of wastewater treatment employed. Figure 4.2 schematically illustrates the sources of heavy metals. When metal particles are introduced into the water system, negative effects can be seen. The concentration of several metals in tissue is slow, and their concentration should be kept within a narrow range, as higher concentrations can harm the biological development of the plant (Kumar et al. 2025a, b). Fish pollutants can be transferred to humans through the food chain (Ray and Vashishth 2024). Water contamination can come from a variety of sources, including industrial wastewater, water tank leaks, dumping next to ships, radioactive waste, and atmospheric deposition. Heavy metals and industrial waste are dumped into rivers and lakes, where they collect and have detrimental effects on both people and wildlife. Toxins are the cause of immune system suppression, infertility, and acute poisoning (Kumar et al. 2024a, b, c, d, e, f, g). The nutrition of plants or animals at that time is then directly harmed, which has an impact on human health. Marine creatures like fish, molluscs, marine birds, and other sea creatures are being killed by the contaminants that are contaminating the water (Doshi et al. 2024).

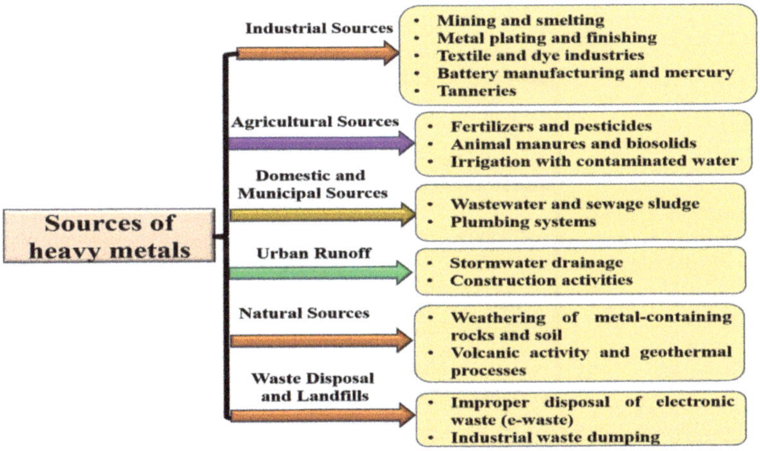


**Fig. 4.2** Key sources contributing to heavy metal contamination in water

### *4.2.3 Microplastics*

Surface runoff and sewage/wastewater treatment plant effluent discharge are the main ways that microplastics (primary and secondary) contaminate drinking water sources. Numerous sectors employ microplastics for a range of purposes, including cosmetics and pharmaceuticals. These major microplastics are wiped off after use and end up in household wastewater. The effluent discharged by sewage/wastewater treatment plants contains a significant amount of microplastics because these facilities are not set up to fully remove microplastics (Patil et al. 2024). Microplastics enter the fresh/drinking water supply chain when this wastewater is mixed with freshwater sources. For instance, the effluent from nearby wastewater treatment plants has been implicated in the rise in the concentration of microplastics in the water. It's also crucial to remember that a lot of the parts of water treatment facilities and water distribution systems are typically composed of plastics like high-density polyethylene, polyvinyl chloride, polypropylene, etc., which further contributes to the production of microplastics in the water they transport (Ullah et al. 2024).

### *4.2.4 Dyes and Organic Pollutants*

The occurrence of coloured effluent in textile wastewater is a serious issue. Organic pollutants infiltrate wastewater from the textile sector, leather processing, and residential effluents, as well as from agricultural runoff, food processing, and pharmaceutical discharge (Fig. 4.3). Even while we can state that not all dyes are as harmful, their evident pollution has a negative aesthetic impact. The aquatic life is impacted by the coloured water's ability to block sunlight from entering the bodies of water.

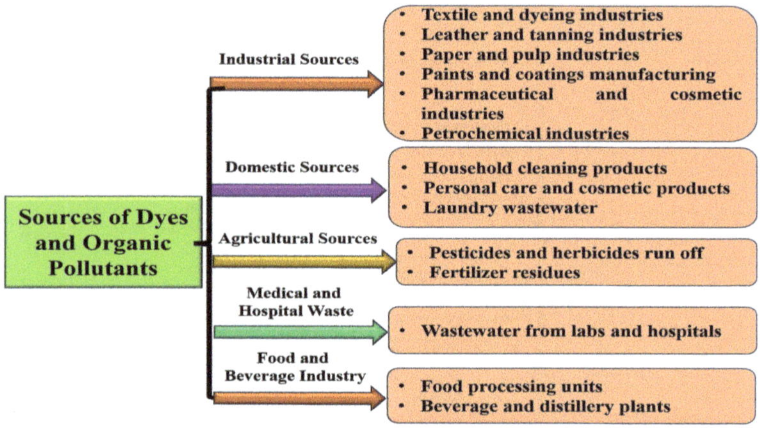


**Fig. 4.3** Diverse sources of dye-based and organic pollutants in wastewater

In many instances, a noticeable water colouring results from a dye concentration below 1 ppm. The dyeing and finishing operations are the main cause of wastewater contamination in the textile industry. Highly suspended solids (SS), chemical and biochemical oxygen demand (BOD and COD), heat, colour, acidity, basicity, and other inorganic contaminants are the main pollutants. Water-insoluble or unreactive dyes, such as disperse and vat dyes, typically have good exhaustion qualities (Taha et al. 2025). They can be eliminated through physical processes like flocculation and have fibre-binding qualities. However, traditional methods of colour removal are not very effective when dealing with water-soluble or reactive colours as effluents. Coloured chemicals known as reactive dyes are used to colour cellulose fibres, which are often identified by N = N azo linkages.

Azo dyes are adsorbed onto the cellulose and then react with the fibre as part of the process. The dye molecule and the fibre form a covalent link throughout the procedure, making the fibre far more resilient to odd circumstances (Kumar et al. 2024a, b, c, d, e, f, g). The cellulose substrate's ionized hydroxyl groups react with the dye's reactive components. The hydroxyl groups in the dye bath, however, compete with those on the cellulose fibre in an alkaline solution, forming hydrolysed dyes that no longer react with the fibre and producing a strongly coloured effluent. Finding a suitable and efficient wastewater treatment technique to eliminate colour from textile effluents becomes essential. Currently, various methods have been used, including chemical and physical processes, such as chemical precipitation and separation of pollutants, electrocoagulation, and elimination by adsorption on activated carbon (Rashid et al. 2024). But all these methods have a common drawback that they are not destructive; instead, they transfer the contamination from one phase to another, causing a new type of pollution. It becomes necessary to find a new method for further treatment of these effluents.

### 4.2.5 Nanomaterials for Water Purification

Research and technologies related to water treatment have expanded due to the distinct chemical, physical, and optical characteristics of NMs and their prudent application through nanotechnology (Fig. 4.4). Functional NMs' improved capacity to absorb light made them a great platform for photothermal water treatment in solar desalination, membrane distillation based on nano-photonics, and disinfection applications. Furthermore, conventional water treatment techniques use hazardous chemicals as disinfectants, which present serious risks due to the production of carcinogenic by-products and large volumes of coagulants based on Fe and Al that are usually disposed of in landfills after use and hardly ever regenerate. Using NMs in water treatment not only lowers possible toxicity footprints but also improves sustainability and total operating capacity through multicycle regeneration with unchanged performance. The development of technologies for wastewater treatment and pollution removal relies heavily on NMs because of their huge surface area, reactivity, and other characteristics (Bibi et al. 2024). Metal and metal oxide NPs, carbon-based NMs (such as graphene and carbon nanotubes), nanocomposites, hybrid materials, and other NMs are the main ones that are frequently utilized in the wastewater treatment process. Table 4.1 presents an analysis of several NMs according to their target pollutant, mode of action, benefits, and drawbacks.

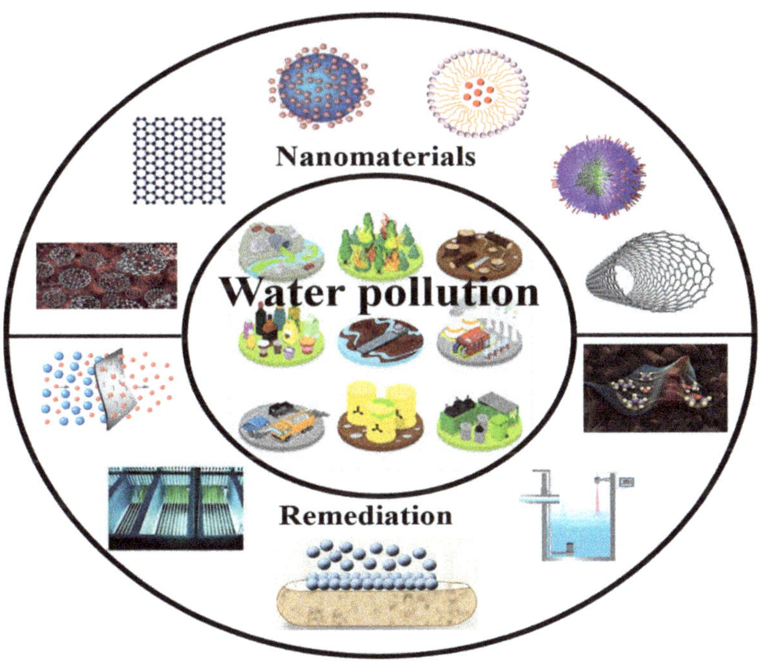


**Fig. 4.4** Nanomaterials and their potential applicability in different water remediation processes

**Table 4.1** Nanomaterials for water purification applications

| Nanomaterials | Targeted pathogens/ pollutants | Mechanism | Advantages | Limitations | References |
|---|---|---|---|---|---|
| AgNPs | Bacteria, viruses (pathogens) | Disrupts cell membranes; generates ROS | Strong antimicrobial; works at low concentrations | Cytotoxicity; cost; potential leaching | Sati et al. (2024) |
| ZnO NPs | Pathogens, dyes | Photocatalysis, ROS generation | Antibacterial, photocatalytic, easy synthesis | Light-dependent activity | Nawaz et al. (2024) |
| $TiO_2$ NPs | Dyes, organics, pathogens | UV-activated photocatalysis | Stable; reusable; cost-effective | Requires UV light; low efficiency in dark | Omran et al. (2025) |
| Graphene oxide (GO) | Heavy metals ($Pb^{2+}$, $Cr^{6+}$), dyes | Adsorption via functional groups | High surface area; versatile | Aggregation; regeneration issue | Sultana et al. (2025) |
| CNTs | Heavy metals, organics | Adsorption, $\pi$–$\pi$ interaction | High surface area; tunable | Expensive; toxicity concerns | Tundwal et al. (2024) |
| FeO NPs | Heavy metals, nitrates, dyes | Redox reactions, magnetic separation | Magnetic recovery; good for $Cr^{6+}$/$As^{5+}$ | Prone to oxidation; short shelf life | Kumar et al. (2024a, b, c, d, e, f, g) |
| MOFs | Heavy metals, VOCs | Pore adsorption, chelation | High porosity; selective removal | Moisture-sensitive; costly synthesis | Tu et al. (2024) |
| Chitosan-based NCs | Pathogens, dyes, metals | Adsorption, ion exchange | Biocompatible; renewable | Low mechanical strength | Rostami and Khodaei (2024) |
| g-$C_3N_4$ (Graphitic carbon nitride) | Dyes, pesticides | Visible-light photocatalysis | Metal-free; eco-friendly | Needs doping for efficiency | Khan et al. (2025) |
| MXenes ($Ti_3C_2Tx$) | Heavy metals, organics, dyes | Adsorption, intercalation | High conductivity; layered structure | Stability in aqueous phase | Hassan et al. (2024) |
| Electrospun nanofibers | Microplastics, bacteria | Filtration + in situ photocatalysis | Flexible membranes; customizable | Fouling risk; scale-up challenge | Das et al. (2024) |

### 4.2.6 Metal and Metal Oxide Nanoparticles

Metal and metal oxide (MONPs) have a large surface area, are reactive, and can adsorb or catalyse contaminants, they're beneficial for treating wastewater. The removal of organic pollutants, heavy metals, microplastics, and germs that cause disease are the ideal applications for these NPs. The biomineralization process used to create NPs requires a variety of species, such as bacteria, fungus, yeasts, algae, and higher plants (Eddy et al. 2025). Among the many exceptional qualities of nanosized metal oxides are their great heavy metal removal capability and selectivity. They have enormous potential as prospective heavy metal adsorbents. Manganese oxides, nanosized iron oxides, titanium oxides, cerium oxides, ZnOs, magnesium oxides, aluminium oxides, and zirconium oxides are examples of metal oxide-based NMs (Yesilay et al. 2023; Bokolia et al. 2024). These substances can eliminate disease-causing germs, heavy metals, organic pollutants, and microplastics. Researchers have discovered that Ag NPs are effective against gram-positive and gram-negative bacteria, viruses, and other organisms in drinking water treatment because of their high surface area to volume ratio, unique physical, chemical, and biological properties, and ease of deposition on solid materials (Deshmukh et al. 2019). MONP performance is dependent on shape, size, and aggregation for both antibacterial activity and wastewater dye removal.

### 4.2.7 Carbon-Based Nanomaterials (Nanotubes, Graphene)

Carbon-based NMs like graphene and carbon nanotubes are useful for a variety of applications because of their high surface area, conductivity, and adsorption properties. These materials have a lot of potential for wastewater treatment because of their special qualities, which include high specific surface area, mesoporous structure, and excellent chemical stability (Rahchamandi et al. 2024). Even at high temperatures and concentrations, they can survive the extreme conditions found in wastewater, such as acidic, basic, and salty environments. Despite a wealth of studies on the effectiveness of carbon-based NMs in wastewater systems, commercialization is hampered by toxicity, engineering, and environmental issues (Mecha et al. 2025). Because of their distinct mechanical, chemical, and structural characteristics, these materials are referred to as designed sorbents. The chapter emphasizes how well these substances work as sorbents to extract organic matter, dyes, heavy metals, and microplastics from contaminated water released by different industrial processes.

### 4.2.8 Nanocomposites and Hybrid Materials

Nanocomposite materials improve wastewater treatment plant (WWTP) performance by combining NPs with other materials like metal oxides, polymers, or natural minerals (Panigrahy et al. 2024). These materials are a concern due to their severe environmental and health effects. To address wastewater treatment issues, advancements in adsorption using nanoadsorbents and composites are being introduced. Nanocomposites (NCs) effectively remove bacteria, viruses, and pollutants from wastewater through precise binding actions like chelation, absorption, and ion exchange. Various NCs, including metal NCs, metal oxide NCs, carbon NCs, polymer NCs, and membrane NCs, are active in water purification (Natarajan et al. 2023). NCs are crucial for remediation due to their large surface area, size, reactivity, and active sites. To enhance pollutant adsorption, NCs are treated by increasing electron density, surface area, and adding functional groups. They can eliminate hazardous pollutants, pathogens, and heavy metals from wastewater more effectively (Upadhyay et al. 2014). This therapy approach using nanotechnology has a variety of regenerative efficiencies and doesn't produce any harmful by-products. Adsorption is thought to be one of the best ways to remediate heavy metals, and functionalized adsorbents can make the process even more efficient. This chapter aims to gather current information on the study and development of various adsorbents for the removal of dyes and heavy metals from water, with a focus on the adsorption capacities, pH effects, isotherm and kinetic models, removal efficiencies, and contact times of each adsorbent.

### *4.2.9 Other Emerging Nanomaterials*

In addition to expanding their research on the materials mentioned above, a number of other promising NMs are now being developed for wastewater treatment. These materials typically offer characteristics that can be intriguing for different kinds of pollutant removal.

- Due to their huge surface area and pore volume, nanofibers have been used extensively in water filtration. Polyvinyl alcohol (PVA) and other polymer nanofibers may be very useful for removing heavy metals and organic compounds from industrial water wastes (Thakur and Kumar 2024).
- For higher levels of filtration, silica or graphene oxide-based nanostructured membranes are used. Many of these membranes are also capable of filtering dissolved materials, organisms, and fine particles.
- Metal–Organic Frameworks (MOFs) are a novel class of porous materials that have demonstrated exceptional adsorptive capacity for a range of contaminants, including oils, dyes, and heavy metals (Marghade et al. 2024). Nevertheless, MOFs are utilized for selective adsorption in wastewater treatment due to their highly ordered and tunable pore systems.

Different forms of NMs include the above, where each form is more favourable for the removal of pollutants in wastewater. Problems related to water pollution can be solved by using metal and metal oxide NPs, carbon-based materials, composites, and some new materials. But despite these advantages, there are still some issues associated with material stability, material costs, and possible impacts on the environment that must be solved to introduce these technologies into practice.

### 4.2.10 *Natural Nanoparticles in Water Purification*

Water filtration has long been accomplished with natural polymers. In the past, charcoal has been utilized with success. Activated carbon is commonly used as an adsorbent to eliminate heavy metal pollution from water. Activated carbon's use is usually limited by its adsorption capacity and powder form, since it lacks specialized selectivity for metal ions. Polysaccharides are another well-liked type of natural polymer (Farghal et al. 2025).

- **Polysaccharide nanofibrous membranes**

One of the most well-known classes of biomaterials on the planet is polysaccharides. Polysaccharides, such as cellulose and chitin, have positive effects on the environment. Because of their abundance in nature, ease of functionalization, biodegradability, sustainability, and structural diversity, polysaccharide nanofibers have drawn a lot of attention. There is a distinct structural hierarchy in cellulose fibres. They are composed of nanofiber assemblies that range in diameter from 2 to 20 nm and lengths of several micrometers. Other names for these single cellulose fibres, which have a diameter of roughly one nanometre, include microfibril aggregates, whiskers, nanowhiskers, microcrystals, and microfibrillated cellulose (Dong et al. 2024).

- **Nanocellulose**

The ecologically benign material known as nanocellulose (NC) is found in nanocrystals and nanoscale fibrils. It is composed of repeated units of D-glucopyranose connected by 1–4 glycoside bonds. Usually, cell-free enzyme systems, aquatic animals, and microbes create it. Additionally, it is usually extracted by mechanical, chemical, or enzymatic means from natural cellulose sources like cotton linters and plant cell walls. Recently, nanocellulose has gained recognition as one of the most inventive materials, with uses across numerous industries (Castellanos et al. 2025). Because of their favourable physicochemical properties, such as inherent abundance, strength, stiffness, renewability, biodegradability, biocompatibility, and sustainability, researchers are taking NCs into consideration. Furthermore, a range of methods have been used to extract very pure nanofibers from cellulose sources. Chemical, biological, and mechanical therapies are some of these methods.

- **Clay nanoparticles used in water purification**

Clay is a fine substance made from natural rocks or soil that has a lot of minerals, organic matter, and trace amounts of metal oxides. Alkaline and alkaline earth elements replace iron and magnesium in clay, which is chemically composed of small crystallites of alumino-silicates in a range of ratios. Based on their chemical composition and particle shape, clays are classified into different classes, such as smectite, chlorite, kaolinite, iolite, and halloysite (Hussain et al. 2024). Numerous clays, such as bentonite, kaolinite, illite, sepiolite, montmorillonite, and pyrophyllite, are efficient at eliminating impurities from water. Because of their vast availability, low cost, and minimal environmental impact, nano-clays have been studied and produced for a range of uses (Naseer et al. 2025). The primary method by which nano-clays remove impurities from water is adsorption. Polymer/nano-clay composites have a high adsorption capability for water treatment because of their easy processing, effective cation exchange, large surface area, and relatively low cost and toxicity.

- **Chitosan nanomaterials**

The remarkable biocompatibility, biodegradability, and adsorption capacity of chitosan, a linear polymer derived from chitin found in insect exoskeletons and crustacean shells, are its defining characteristics. Its structure's amino and hydroxyl groups enable it to bind to pollutants like dyes, heavy metal ions, and other organic molecules with efficiency (Jiang et al. 2024a, b). In order to enhance water treatment procedures, this biopolymer is employed as an adsorbent, flocculant, or in the creation of composite materials. Further supporting their promise in water purification, recent studies have shown that chitosan membranes coated with graphene oxide NPs have an exceptional ability to adsorb organic pollutants. In a similar vein, chitosan has been used in conjunction with inorganic minerals like montmorillonite to create compounds that have better mechanical and thermal properties, making them more viable in harsh environments (Lahbib et al. 2025).

In addition, bio-based NMs including lignin, pectin, and chitin nanofibers offer sustainable substitutes for conventional water treatment methods. These materials are suited for a number of uses, including the selective adsorption of particular pollutants or their integration into sophisticated filtration systems, because of their high specific surface area, biodegradability, and functional groups (Al-Gethami et al. 2024). It has been shown that combining biopolymers like chitosan with clays like bentonite or kaolinite can enhance the adsorption of dyes and heavy metals. In addition, bio-based NMs, including lignin, pectin, and chitin nanofibers, offer sustainable substitutes for conventional water treatment methods. These materials are suited for a number of uses, including the selective adsorption of particular pollutants or their integration into sophisticated filtration systems, because of their high specific surface area, biodegradability, and functional groups (Islam et al. 2025). It has been shown that combining biopolymers like chitosan with clays like bentonite or kaolinite can enhance the adsorption of dyes and heavy metals. Using a complementary strategy, hydrogels with better mechanical and adsorption qualities have been created by combining synthetic polymers like polyacrylamide or polyvinyl alcohol with biopolymers like chitosan or alginate (Et-tanteny et al. 2024). Synthetic polymers' stability and processability combined with biopolymers' biodegradability and biocompatibility provide a synergistic impact that enhances these materials' performance. Increased adsorption capacity, regulated swelling behaviour, and improved feasibility for reuse in real-world applications are the outcomes of this strategy.

## 4.3 Mechanisms of Contaminant Removal by Nanomaterials

By utilizing their surface area, reactivity, and other characteristics, NMs have provided wastewater treatment techniques with cost-effectiveness. NMs can help a contaminant degrade through a variety of mechanisms, including adsorption, photocatalytic activity, reduction, oxidation, and, finally, membrane filtration (Asghar et al. 2024). The distinct properties of NPs are used in all of these techniques to extract contaminants from wastewater.

### *4.3.1 Adsorption and Surface Interaction*

Because of their high surface area and reactive sites, retention is a typical strategy used in wastewater treatment processes, including NPs. Pollutants such as heavy metals, dyes, and organic compounds are adsorbed using magnetic NPs, activated carbon, and metal oxides (Ahmaruzzaman et al. 2024). According to the aforementioned characteristics, MS has a high adsorption capacity because of its wide surface exposure; further functionalization with groups like amino or carboxyl enhances its ability to remove pollutants. Some of magnetic NPs include $Fe_3O_4$ which have been used in the adsorption of toxic heavy metals which include lead and cadmium from wastewater can easily be recovered through magnetic separation (Kumar et al. 2024a, b, c, d, e, f, g). For this reason, they are very suitable for application in large-scale systems. It also complements the removal of organic contaminants with carbon nanomaterials achievements high adsorption capacity of compounds such as phenols.

### *4.3.2 Photocatalysis and Photocatalytic*

Deterioration, one of the more sophisticated oxidation techniques, is photocatalysis, which uses NMs like $TiO_2$, ZnO, and CdS to break down organic contaminants in wastewater while receiving light support. These NMs absorb light when exposed to light, forming electron–hole pairs that react with oxygen and water to produce Reactive Oxygen Species (ROS) like OH, which break down organic contaminants into comparatively innocuous chemicals (Bi et al. 2024). $TiO_2$ excellent photocatalytic activity and UV stability, in particular, have attracted a lot of interest and research. However, because of its low photocatalytic activity in visible light, $TiO_2$ has several technical issues when used. Modifications such as metal and/or non-metal doping ($TiO_2$-N) and linked semiconductor systems ($TiO_2$-$NCu_2O$) have been used to get around this restriction (Dey et al. 2025). These modifications improve the ability to trap light and the effectiveness of electron–hole separation, which raises photocatalytic activity in the presence of visible light.

### 4.3.3 Catalytic Reduction and Oxidation Processes

The two main methods for eliminating harmful substances from wastewater are catalytic reduction and oxidation. Catalysis is a significant application of NMs; metal NPs, such as those made of gold, platinum, or palladium, have high activity in reduction and oxidation reactions (Zhuang et al. 2024). NMs aid in the transmission of electrons from a reductant to the pollutant during catalytic reduction, converting the latter into a less hazardous form. For instance, using the electron transfer process, palladium NPs are used to reduce organic pollutants like azo dyes. Additionally, organic pollutants are broken down into smaller, usually harmless molecules by nanocatalysts, taking part in oxidation reactions. According to earlier research, both Au and Ag NPs have an improved capacity for oxidation to break down chemical substances, including hydrocarbons and phenols (Devadoss et al. 2024). Because of their high catalytic activity, energy efficiency, and reusability, these NMs are not thought to be appropriate for the effective use of sustainable wastewater treatment.

### 4.3.4 Membrane Filtration and Separation

Due to its great selectivity in the process of separating contaminants, membrane filtration in conjunction with NMs has become significant for the application process. The membrane's permeability, selectivity, and fouling resistance are enhanced by graphene oxide, carbon nanotubes (CNTs), and metal–organic frameworks (MOFs) (Habert et al. 2024). For instance, because of the hydrophilic nature of graphene oxide and the membranes' pore size of several nanometres, graphene oxide membranes function exceptionally well in the desalination of water and in the separation of other organic contaminants. In addition to boosting water transport and preventing organic pollutants and bacteria from fouling the membrane, CNTs added to membranes also improve filtration capabilities by expanding the area of surface contact (Wang et al. 2024). Additionally, research has shown that adding membranes based on MOFs has a high chance of removing salts, heavy metals, and other contaminants from water. NMs can successfully treat wastewater in a number of methods, including membrane filtering, photocatalytic degradation, adsorption, and catalytic reduction/oxidation. NMs are very effective at removing many types of pollutants from water because of their huge surface area, reactivity, low particle size, and versatility (Sable et al. 2024). Thus, research progresses to produce a superior nanomaterial that will enhance wastewater treatment systems' future performance.

## 4.4 Applications and Case Studies

### *4.4.1 Nanomaterials in Pathogen Removal*

More uses for NMs with antibacterial qualities are evident. The active properties of NPs against bacterial, viral, and fungal strains, as well as overall wastewater treatment, are promising as Fig. 4.5. These NPs neutralize pathogens by releasing metal ions that damage microbial cell walls and disrupt metabolic processes (Foo et al. 2024). Because harmful bacteria are almost always present in wastewater from the industrial and healthcare sectors, this capability is extremely important. Although there are certain issues, research indicates that NPs have a lot of promise for treating wastewater. Among these are the harmful effects of NPs on humans and water, difficulties in implementing nanomaterial-based technologies on a broad scale, and the high cost of manufacture (Kumar et al. 2024a, b, c, d, e, f, g). Table 4.2 lists the explanations of each of the popular methods for removing heavy metals from water. To create sustainable, green NMs that will lessen the impact of nanotechnology, writers are also investigating NMs interact with the environment. Future research will focus on integrating nanotechnology into traditional treatment systems, reusing NMs, and creating composite NMs to address many pollutants simultaneously (Devarajan 2025). The effectiveness of nanotechnologies for a wide range of wastewater treatment applications will depend heavily on advances in recycling expertise and manufacturing cost reductions.

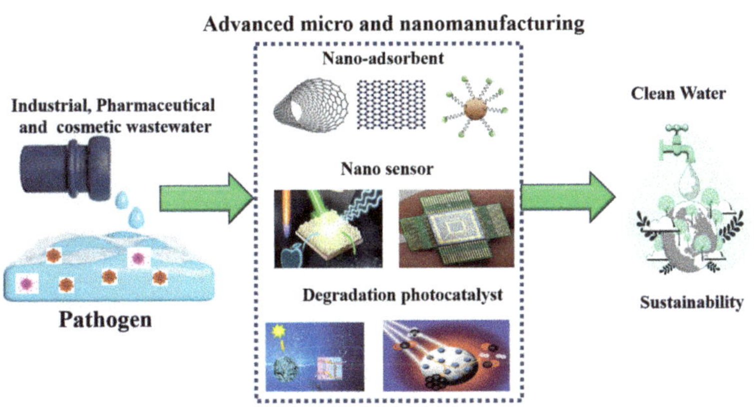


**Fig. 4.5** Advancements in micro and nanotechnology-based manufacturing for pathogen removal from wastewater

**Table 4.2** Common techniques for pathogen removal in wastewater

| Process | Description | Advantages | Limitations | References |
|---|---|---|---|---|
| Chlorination | Disinfection is achieved by adding chlorine or hypochlorite to kill microorganisms | Effective, low-cost, widely used | Produces harmful disinfection by-products; not effective on some protozoa | Jiang et al. (2024a, b) |
| UV irradiation | Uses ultraviolet light to inactivate pathogens by damaging DNA | Chemical-free, rapid action, effective against viruses | No residual effect; turbidity reduces efficiency | Monika et al. (2025) |
| Ozonation | Strong oxidizing agent that inactivates bacteria, viruses, and protozoa | Broad-spectrum disinfection improves taste & odour | Expensive, unstable ozone; needs on-site generation | Cai et al. (2025) |
| Membrane filtration (UF/NF/RO) | Physical removal of pathogens via pore-size exclusion | High removal efficiency, no chemicals | Fouling, high-energy demand, expensive membranes | Burke et al. (2025) |
| Slow sand filtration | Biological and physical filtration using a sand bed and biofilm | Low-tech, cost-effective, sustainable | Slow flow rate; large footprint needed | Bai et al. (2024) |
| Boiling | Heat treatment to kill pathogens, commonly used at the household level | Simple, effective on all microorganisms | Fuel consuming; not feasible for large-scale use | Liang et al. (2024) |
| Advanced oxidation processes (AOPs) | Generation of hydroxyl radicals (•OH) for oxidative degradation and disinfection | Destroys a wide range of pollutants, removes micropollutants | High cost; complex operation, may require post-treatment | Dai et al. (2024) |
| Nanomaterial-based disinfection | Use of antimicrobial NPs to inactivate pathogens | High surface area, multifunctional, reusable | Environmental toxicity; regulatory concerns | Luo et al. (2024) |

### *4.4.2 Heavy Metal Adsorption*

With the expansion of industry and human activity, such as the plating and electroplating industry, batteries, pesticides, mining, rayon, tanning, metal rinse, fluidized bed bioreactors, textile, metal smelting, petrochemicals, paper manufacturing, and electrolysis applications, the number of heavy metals in wastewater has been rising. Wastewater tainted with heavy metals enters the environment and endangers both the ecosystem and human health (Essalmi et al. 2024). Excessive levels of heavy metals in water could have a major detrimental effect on the health of living things because they are not biodegradable and can cause cancer. Lead (Pb), zinc (Zn), mercury (Hg), nickel (Ni), cadmium (Cd), copper (Cu), chromium (Cr), arsenic (As), and zinc (Zn) are the most common heavy metals (Kumar et al. 2024a, b, c, d, e, f, g; Mansouri et al. 2024). Even though traces of these heavy metals can be found, they are still dangerous. A comparison of methods for removing heavy metals from aqueous solutions is given in Table 4.3, which also highlights a number of benefits, drawbacks, and uses. Numerous important criteria are as follows: operation costs, initial metal ion concentration, environmental impact, pH levels, chemicals added, removal efficiency, and economic viability. Adsorption treatments influence the choice of the best method for removing heavy ions from wastewater, as shown in Fig. 4.6. Membrane treatments, chemical treatments, electric treatments, ion exchange, and photocatalysis are the categories into which these techniques fall.

- **Mechanisms for the Removal of Heavy Metals**

Most existing methods for removing these pollutants rely on sorption, sorption-reduction, or photocatalytic degradation. To better understand their potential environmental impact and explore alternative removal pathways, it's necessary to investigate the physicochemical properties of these contaminants and combine modern characterization techniques with molecular-level theoretical calculations. One of the simplest ways to remove heavy metal ions from solutions is through sorption, which involves chemical interactions between functional groups (such as those containing oxygen, hydroxyl, or carbonyl groups) and metal ions (Guo et al. 2024). This process creates strong surface complexes through hydrophobic contacts, π–π- donor–donor–acceptor interactions, hydrogen bond interactions, and electrostatic attraction. Similarly, sorption-reduction is a method for immobilizing and reducing high-valent metal ions to low-valent metal ions (İlyasoglu et al. 2025). High-valent metal ions are first reduced, forming more dense particles or clusters that readily precipitate. As a result, low-valent metal ions are significantly more biocompatible in the natural environment. This process is commonly attributed to the fact that high-valent metal ions are much more mobile than low-valent metal ions. Lastly, photocatalytic degradation is often used to remove various organic contaminants, such as persistent organic pollutants, but it has also been extensively applied to remove low concentrations of heavy metal and metalloid ions from solutions. This technique relies on photocatalytic reactions, which heavily depend on the catalyst's morphology, mass transport mechanisms, visible-light absorption, and active site surface distribution (Lotfi et al. 2024). The mechanism for removing heavy metal ions may vary depending on the type of metal ions and the light source (Thakur et al. 2023).

**Table 4.3** Common techniques for the remediation of heavy metals from wastewater

| Process/ Materials | Description | Advantages | Limitations | References |
|---|---|---|---|---|
| Activated carbon adsorption | Adsorption of heavy metals onto high-surface-area carbonaceous materials | Readily available, high surface area, reusable | Low selectivity, poor performance for some metals, and expensive regeneration | Parlayıcı et al. (2024) |
| Ion exchange | Exchange of metal ions with harmless ions on synthetic resins or zeolites | High selectivity, regenerable, effective at low concentrations | Sensitive to competing ions; expensive resins | Somya and Varshney (2025) |
| Biosorption | Use of biological materials (algae, fungi, agricultural waste) to adsorb heavy metals | Eco-friendly, cost-effective, good for low concentrations | Limited capacity, disposal of biomass, and slower kinetics | Karnwal (2024) |
| Nanomaterial Adsorption | NPs are used to adsorb or react with metal ions | High efficiency, tunable properties, and large surface area | Stability issues, possible environmental toxicity | Mehta et al. (2024) |
| Clay minerals | Natural adsorbents with negatively charged surfaces for metal ion binding | Abundant, low-cost, good metal affinity | Low regeneration potential, slow kinetics | Yang et al. (2024) |
| Metal–organic frameworks (MOFs) | Porous crystalline materials with metal centres and organic linkers for specific metal capture | Extremely high surface area, selective adsorption | Expensive synthesis, poor water stability for some MOFs | Lin et al. (2024) |
| Magnetic NPs | Used for adsorption and magnetic separation of heavy metals | Easy recovery, good $Cr^{6+}$/$As^{5+}$ removal, reusable | Aggregation, oxidation, and long-term instability | Zhang et al. (2024) |
| Functionalized CNTs | CNTs functionalized with oxygen or nitrogen groups for enhanced metal binding | High adsorption capacity, fast kinetics | Toxicity concerns, high cost, limited scalability | Li et al. (2024) |

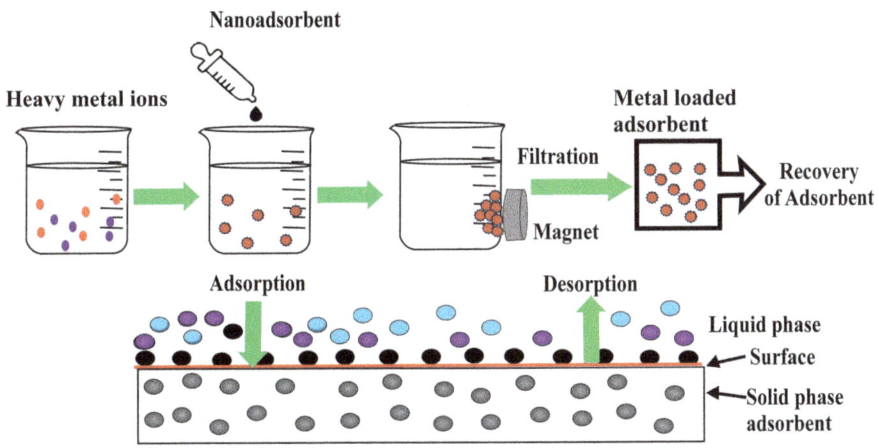


**Fig. 4.6** Heavy metals recovery using nanoadsorbents

### *4.4.3 Microplastic Filtration*

The increasing amount of microplastics in wastewater is a serious environmental problem that calls for creative solutions that go beyond traditional treatment techniques, as Fig. 4.7. The removal of microplastics is limited by current wastewater treatment frameworks, mainly because of low adsorption capacity, poor selectivity, and insufficient removal effectiveness, as Table 4.4. Furthermore, these methods frequently fail to accelerate the rate at which microplastics biodegrade, resulting in long-lasting environmental contamination (Pizzichetti et al. 2021). To address the limitations of traditional treatment approaches, this chapter proposes an integrated approach that synergistically combines sophisticated filtering materials, nanotechnology applications, and bioremediation techniques. Because of their inherent high adsorption capacity, biodegradability, and affinity for microplastic particles, bio-based filter media more especially chitosan and alginate beads are used in this innovative approach. CNTs and magnetic NPs, such as iron oxide varieties like magnetite or maghemite, are used to harness nanotechnology (Kumar et al. 2023). To increase selectivity towards particular microplastic kinds, CNTs are functionalized. By using their magnetic properties, magnetic NPs enable the quick removal of adsorbed microplastics from water. Microbial remediation and enzyme-based degradation are two ways that bioremediation is integrated. Filtration materials immobilize enzymes such as laccase and manganese peroxidase, which catalyse the degradation of microplastics into less toxic forms (Srivastava et al. 2025). At the same time, the incorporation of microorganisms that can break down plastic strengthens the biodegradation process. With improved adsorption capacity and 30–50% biodegradation rates, the suggested approach increases the effectiveness of microplastic removal to above 95%. With a specificity rate of over 80%, it also exhibits enhanced selectivity for a range of microplastics (Duan et al. 2025). With its scalable, effective, and

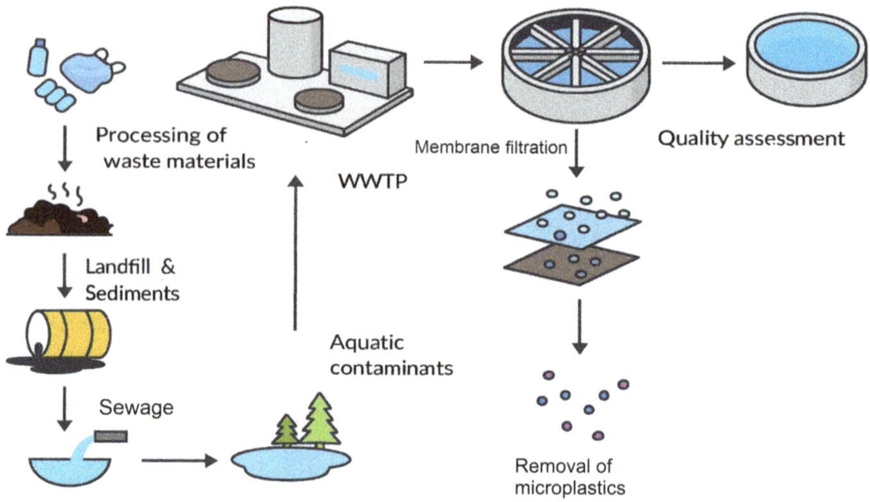


**Fig. 4.7** Microplastics removal from wastewater

environmentally responsible solution to the microplastics problem, this integrative approach represents a significant breakthrough in wastewater treatment. Its effects go beyond simple environmental cleanup; it may promote more robust ecosystems and protect public health, which would greatly aid in global environmental sustainability initiatives for current situations.

## 4.5 Future Scope and Limitations

While NMs bring many benefits to water treatment, they also have limitations, such as membrane aggregation, susceptibility to pathogens, and complex synthesis. However, they're especially useful in developing countries to improve water quality. Nanoparticle-based water treatment technologies have the potential to address microbial contamination and improve water quality, but there are challenges that need to be overcome for sustainable implementation. Future research should focus on overcoming these obstacles. Nanotechnology has huge potential for water monitoring, treatment, and recycling, with the potential for removing heavy metals from wastewater (Khdair et al. 2025). Nanomaterial-based membranes have potential for commercial use, but public awareness, engagement, and transparent communication are crucial for building trust and responsible implementation. Plastic pollution is getting worse due to its high stability and widespread use, making it difficult to decay and constantly compromising the health of the environment and living things. This includes, among other things, finding effective methods for locating, measuring, and evaluating nanoplastic presence in the environment, understanding degradation rates and mechanisms, and studying nanoplastic contamination and transmission in

**Table 4.4** Advanced techniques for the microplastic removal from wastewater

| Process | Description | Advantages | Limitations | References |
|---|---|---|---|---|
| Membrane filtration | Physical filtration of microplastics using porous membranes based on size exclusion | High removal efficiency | Fouling, expensive, and membrane replacement are required | Ojha et al. (2024) |
| Coagulation-flocculation | Uses metal salts to aggregate microplastics into flocs for sedimentation | Cost-effective, scalable, and can remove small particles | May require pre-treatment, residual sludge generation | Mustafa et al. (2025) |
| Dissolved air flotation | Microbubbles attach to microplastics and float them to the surface for removal | Suitable for low-density plastics, energy-efficient | Less effective on high-density or nano-sized plastics | Jia et al. (2025) |
| Electrocoagulation | Applies electric current to generate coagulants in situ for microplastic aggregation | No chemical addition; low sludge | High energy demand, electrode maintenance | Sedaghat et al. (2025) |
| Sand and granular filtration | Gravity-based or pressurized filtration using fine sand or granular media | Simple, cost-effective, long-standing use | Limited efficiency for particles <10 μm | Lapointe et al. (2025) |
| Magnetic nanocomposite adsorbents | NMs bind to microplastics for magnetic separation | High selectivity; easy recovery | Needs surface modification; scalability issue | Kumar and Thakur (2025) |

(continued)

**Table 4.4** (continued)

| Process | Description | Advantages | Limitations | References |
|---|---|---|---|---|
| Bio-based flocculants | Natural polymers trap and aggregate microplastics via hydrogen bonding or electrostatic forces | Biodegradable, eco-friendly | Less effective for highly hydrophobic plastics | Ahmed et al. (2024) |
| Advanced oxidation | Use of hydroxyl radicals (•OH) to break down plastic surfaces or additives | Targets microplastic-associated pollutants | Not effective for full plastic degradation, expensive | Adeoye et al. (2024) |

the environment (Muzamil et al. 2025). Additionally, research is needed on using efficient technologies to eliminate microplastics and nanoplastics. Several issues related to the uses and toxicity of NMs on the environment and human health have also emerged. Furthermore, the stability, antifouling, dispersibility, and compatibility of NMs can be widely studied. Future research should focus on using these materials for large-scale commercial applications. With these limitations in mind, new research should explore the use of NMs, including their reuse and biodegradation capabilities.

## 4.6 Conclusion

There's a growing global momentum behind using nanotechnology to treat wastewater. The unique properties of NMs, combined with existing technologies, have the potential to transform the way we deal with these issues. Heavy metal pollution has surged in recent decades, making it crucial to develop new strategies, policies, technologies, and materials to tackle this problem. Nanotechnology offers a promising solution, as engineered NMs can effectively remove toxic metal ions from polluted water. Various NPs materials, such as carbon-based, zeolite, polymer, magnetic, metal oxide, and silica-based materials, have shown high removal efficiency for different heavy metals. However, challenges include the instability of these materials, their small size, and long-term performance. Pathogens also pose a significant threat to water availability, and traditional disinfection methods have their limitations. NMs like metal oxide NPs, fullerenes, and photocatalysts hold promise for water treatment and contaminant removal. By combining these materials, it's possible to eradicate pathogens and impurities, paving the way for hybrid water treatment techniques. The versatility of these NMs allows for sustainable and robust technology. Microplastic pollution is another major concern, with significant risks to aquatic ecosystems through bioaccumulation and ecological damage. While current water and wastewater treatment technologies can partially remove microplastics, more work is needed to boost efficiency. NMs, with their multifunctional and flexible properties, have been increasingly used for water and wastewater treatment, serving as high-performance adsorbents and photocatalysts for microplastic removal. This chapter explores NMs can enhance the efficiency of removing pathogens, heavy metals, and microplastics, and contribute to sustainable cleanup efforts.

## References

Adeoye JB, Tan YH, Lau SY, Tan YY, Chiong T, Mubarak NM, Khalid M (2024) Advanced oxidation and biological integrated processes for pharmaceutical wastewater treatment: a review. J Environ Manage 353:120170

Ahmaruzzaman M, Roy S, Singha A, Rtimi S, Aminabhavi TM (2025) Emerging nanotechnologies in adsorption of dyes: a comprehensive review of carbon and metal oxide-based nanomaterials. Adsorption 31(2):34

Ahmed GE, Hassan GK, Gomaa EH, Aly SA, Salem SY, Badr EE, Fify AA (2024) Using a novel bio-based cationic flocculant for food industry wastewater treatment. Sci Rep 14(1):19423

Akinnike A, Agboola O, Edith A, Ezinne I (2024) The role of catalyst in the adsorption of dye: homogeneous catalyst, heterogeneous catalyst, and advanced catalytic activated carbon, critical review. Desalination Water Treat 100780

Al-Gethami W, Qamar MA, Shariq M, Alaghaz ANM, Farhan A, Areshi AA, Alnasir MH (2024) Emerging environmentally friendly bio-based nanocomposites for the efficient removal of dyes and micropollutants from wastewater by adsorption: a comprehensive review. RSC Adv 14(4):2804–2834

Asadi AMS, Cichocki Ł, Atamaleki A, Hashemi M, Lutze H, Imran M, Boczkaj G (2024) Catalysts for advanced oxidation processes: deep eutectic solvents-assisted synthesis–a review. Water Resour Ind 31:100251

Asghar N, Hussain A, Nguyen DA, Ali S, Hussain I, Junejo A, Ali A (2024) Advancement in nanomaterials for environmental pollutants remediation: a systematic review on bibliometrics analysis, material types, synthesis pathways, and related mechanisms. J Nanobiotechnology 22(1):26

Bai X, Samari-Kermani M, Schijven J, Raoof A, Dinkla IJ, Muyzer G (2024) Enhancing slow sand filtration for safe drinking water production: interdisciplinary insights into Schmutzdecke characteristics and filtration performance in mini-scale filters. Water Res 262:122059

Bi Z, Wang W, Zhao L, Wang X, Xing D, Zhou Y, Chen C (2024) The generation and transformation mechanisms of reactive oxygen species in the environment and their implications for pollution control processes: a review. Environ Res 119592

Bibi N, Qazi RA, Kanwal A, Jamila N, Khattak R, Hassan W, Wasil Z (2024) Nanomaterials in water purification/desalination. In: Handbook of nanomaterials, vol 2. Elsevier, pp 549–578

Bokolia M, Baliyan D, Kumar A, Das R, Kumar R, Singh B (2024) Biogenic synthesis of nanoparticles using microbes and plants: Mechanisms and multifaceted applications. Int J Environ Anal Chem 1–29

Burke M, Wells E, Larison C, Rao G, Bentley MJ, Linden YS, Linden KG (2025) Systematic review of microorganism removal performance by physiochemical water treatment technologies. Environ Sci & Technol

Cai Y, Zhao Y, Wang C, Yadav AK, Wei T, Kang P (2024) Ozone disinfection of waterborne pathogens: a review of mechanisms, applications, and challenges. Environ Sci Pollut Res 31(51):60709–60730

Castellanos HG, Aryanfar Y, Mohtaram S, Keçebaş A, Karaca-Dolgun G, Ahmad S, Islam S (2025) The efficacy of nano-cellulose-based composites in heavy metal removal from wastewater: a comprehensive review. J Chem Technol Biotechnol 100(2):291–312

Dai M, Niu Q, Wu S, Lin Y, Biswas JK, Yang C (2024) Hydroxyl radicals in ozone-based advanced oxidation of organic contaminants: a review. Environ Chem Lett 22(6):3059–3106

Das KP, Chauhan P, Staudinger U, Satapathy BK (2024) Exploring sustainable adsorbents to mitigate micro-/nano-plastic contamination: perspectives on electrospun fibrous constructs, biochar, and aerogels. Environ Sci: Adv

Deshmukh SP, Patil SM, Mullani SB, Delekar SD (2019) Silver nanoparticles as an effective disinfectant: a review. Mater Sci Eng, C 97:954–965

Devadoss AJ, Sharma N, Gadiyaram S (2024) Metal nanoparticles for visual detection of organic pollutants. In: Nanomaterials in environmental analysis. Elsevier, pp 9–24

Devarajan Y (2025) Nanomaterials-based wastewater treatment: addressing challenges and advancing sustainable solutions. BioNanoScience 15(1):1–14

Dey A, Vashishtha L, Gogate PR (2025) Ultrasound-assisted sol–gel synthesis of N-$TiO_2$, Fe-$TiO_2$, and $TiO_2$ with application in treatment of a commercial effluent. J Chem Technol & Biotechnol

Dong Y, Abbasi A, Mohammadnejad S, Nasrollahzadeh M, Sheibani R, Otadi M (2024) Recent progresses in bentonite/lignin or polysaccharide composites for sustainable water treatment. Int J Biol Macromol 134747

Doshi M, Rabari V, Patel A, Yadav VK, Sahoo DK, Trivedi J (2024) A systematic review on microplastic contamination in marine Crustacea and Mollusca of Asia: current scenario, concentration, characterization, polymeric risk assessment, and future prospectives. Water Environ Res 96(5):e11029

Duan X, Chen X, Shi L, Cao Y, Liang Y, Huang C, Cao Y (2025) Functionality-dependent removal efficiency and mechanisms of polystyrene microplastics by a robust magnetic biochar. J Environ Chem Eng 115509

Eddy NO, Ukpe RA, Garg R, Garg R, Odiongenyi A, Ameh P, Nyambi Akpet I (2025) Enhancing water purification efficiency through adsorption and photocatalysis: models, applications, and challenges. Int J Environ Anal Chem 105(7):1577–1594

Essalmi S, Lotfi S, BaQais A, Saadi M, Arab M, Ahsaine HA (2024) Design and application of metal organic frameworks for heavy metals adsorption in water: a review. RSC Adv 14(13):9365–9390

Esteki R, Ehrampoush MH, Nasab H, Ebrahimi AA (2024) Investigating the improvement of the quality of industrial effluents for reuse with added processes: coagulation, flocculation, multi-layer filter and UV. Sci Rep 14(1):3971

Et-tanteny R, El Amrani B, Benhamou M (2024) Investigation and modeling of physicochemical properties of bentonite-chitosan composites versus the concentration of chitosan added by intercalation. Chem Phys Impact 8:100611

Farghal HH, Nebsen M, Mostafa A, El-Sayed MM (2025) Biopolymer-coated nanoparticles for water purification. Biopolym Water Purif 439–464

Foo WH, Chia WY, Ende S, Chia SR, Chew KW (2024) Nanomaterials in aquaculture disinfection, water quality monitoring, and wastewater remediation. J Environ Chem Eng 113947

Fulke AB, Mamidala S, Nikalje P (2025) Advances in chemical and microbial source tracking (MST) of fecal pollution in coastal waters: a review. Environ Monit Assess 197(2):1–12

Guo Y, Feng H, Zhang L, Wu Y, Lan C, Tang J, Tang L (2024) Insights into the mechanism of selective removal of heavy metal ions by the pulsed/direct current electrochemical method. Environ Sci Technol 58(12):5589–5597

Habert AC, Borges CP, de Araujo Kronemberger F, Conceição FH, Nobrega R (2024) Membrane separation processes. In: Purification of biotechnological products. CRC Press, pp 82–126

Hachad M, Burnet JB, Sylvestre É, Duy SV, Villemur R, Sauvé S, Dorner S (2024) β-D-glucuronidase activity triggered monitoring of fecal contamination using microbial and chemical source tracking markers at drinking water intakes. Water Res 254:121374

Hassan NS, Jalil AA, Bahari MB, Izzuddin NM, Fauzi NAFM, Jusoh NWC, Tehubijuluw H (2024) A critical review of MXene-based composites in the adsorptive and photocatalysis of hexavalent chromium removal from industrial wastewater. Environ Res 119584

Hussain E, Ahtesham A, Shahadat M, Ibrahim MNM, Ismail S (2024) Recent advances of Clay/polymer-based nanomaterials for the treatment of environmental contaminants in wastewater: a review. J Environ Chem Eng 112401

İlyasoğlu G, Abdullah T, Okay O, Koyuncu İ (2025) Design of Electrospun Hydrophobically modified polyacrylic acid hydrogel nanofibers and their application for removal of ciprofloxacin. J Polym Environ 1–17

Irfan M, Sana A, Maryam A, Naveed M (2025) Exploring how microbial extracellular metabolites drive nanoparticle synthesis: a bioinformatics approach. BioNanoScience 15(2):310

Islam MA, Hasan MN, Evan MSH, Uddin MJ, Tulin WS, Islam MS, Chowdhury FI (2025) Chitin nanofibers: recent advances in preparation and applications in biomedical and beyond. RSC Adv 15(19):14655–14690

Jia M, Farid MU, Ho YW, Ma X, Wong PW, Nah T, An AK (2024) Advanced nanobubble flotation for enhanced removal of sub-10 μm microplastics from wastewater. Nat Commun 15(1):9079

Jiang Y, Qiao Y, Jin R, Jia M, Liu J, He Z, Liu Z (2024b) Application of chlorine dioxide and its disinfection mechanism. Arch Microbiol 206(10):400

Jiang R, Xiao M, Zhu HY, Zhao DX, Zang X, Fu YQ, Liu H (2024) Sustainable chitosan-based materials as heterogeneous catalyst for application in wastewater treatment and water purification: an up-to-date review. Int J Biol Macromol 133043

Karnwal A (2024) Unveiling the promise of biosorption for heavy metal removal from water sources. Desalination Water Treat 100523

Khan MD, Fareed I, Maqsood T, Nawaz F, Sandali Y, Alzahrani A, Butt FK (2025) Innovative Z-scheme $MnV_2O_6$/g-$C_3N_4$ nanocomposite: photocatalytic, electrocatalytic and biosensing properties. Mater Chem Phys 330:130159

Khdair AI, Aburumman GA, Gholipour S, Afrand M (2025) Nanoparticles in water purification: multifunctional roles, challenges, and sustainable applications. Environ Sci: Nano

Kumar P (2024) Synthesis, characterization, and photocatalytic properties of lanthanide-doped hematite nanoparticles. Int J Trendy Res Eng Technol 8(2):23–29

Kumar P, Thakur N, Tapwal A, Kumar S (2023) Enhancing the adsorption capacity of green/chemical synthesized hematite nanoparticles by copper doping: removal of toxic Congo red dye and antioxidant activity. Appl Nanosci 13(9):6591–6604

Kumar P, Kaushik A, Kumar S, Thakur N (2024a) Chemical and green synthesized Co/Ni-doped hematite nanoparticles for enhancing the photocatalytic and antioxidant properties. Phys Scr 99(10):105960

Kumar P, Kumar S, Tapwal A, Thakur N (2024b) Chemical/green synthesized cobalt/copper-doped α-$Fe_2O_3$ nanoparticles: potential for environmental remediation. J Mater Res 39(5):836–849

Kumar P, Kumar S, Tapwal A, Nimesh S, Thakur N (2024c) Water purification and biological efficacy of green synthesized Co/Zn-Doped α-$Fe_2O_3$ nanoparticles. Sustain Chem Environ 8:100160

Kumar P, Singh K, Singh L, Singh J, Gupta S (2024e) Biomedical potential of nanotechnology: a review of emerging applications and future prospects. Far Western Review 2(2):175–196

Kumar P, Tapwal A, Kumar S, Thakur N (2024f) Improved photocatalytic and free radical scavenging studies of synthesized polymer PVP/*Azadirachta indica* leave extract-mediated Ni-Zn doped hematite nanoparticles. Adv Nat Sci: Nanosci Nanotechnol 15(2):025014

Kumar P, Thakur N, Kumar K, Kumar S, Dutt A, Thakur VK, Thakur N (2024g) Catalyzing innovation: exploring iron oxide nanoparticles-Origins, advancements, and future application horizons. Coord Chem Rev 507:215750

Kumar P, Arya V, Kumar A, Thakur N (2025a) Polymer/phytochemical mediated eco-friendly synthesis of Cu/Zn doped hematite nanoparticles revealing biological properties and photocatalytic activity. Int J Mater Res 116(1):30–49

Kumar R, Kumar P (2024) Exploring the diverse biomedical applications of hybrid nanomaterials. In: Innovations and applications of hybrid nanomaterials. IGI Global, pp 1–15

Kumar P, Thakur N (2025) Examination of seed germination process and assess phytotoxicity in *Vigna radiata* (Mung bean) and *Allium cepa* (Onion) using triple-doped green hematite nanoparticles and their photocatalytic potential. Environ Sci Pollut Res 1–17

Kumar P, Pathak D, Thakur N (2024) Trimetallic doped hematite (α-Fe2O3) nanoparticles using biomolecules of Azadirachta indica leaf extract for photocatalytic dye removal: insights into catalyst stability and reusability. Emergent Mater 1–17

Kumar R, Kaushal S, Kumar P, Kumar V, Thakur N, Kumar K (2025) Industrial waste-derived carbon nanomaterials. In: Waste derived carbon nanomaterials, vol 1. American Chemical Society, pp 179–201

Lahbib M, Mejri C, Bejaoui M, Oueslati A, Oueslati W (2025) Hybrid nanocomposite design: exploring the interplay between microstructure, dielectric properties, and photocatalytic activities in Chitosan/Na-Montmorillonite/Fe3O4 nanoparticles. Emergent Mater 1–24

Lapointe M, Jahandideh H, Alimi OS, Farner JM, Tufenkji N (2025) Sustainable granular materials improve removal of natural organic matter, turbidity and microplastics during adsorption, ballasted flocculation and granular filtration. J Environ Chem Eng 13(2):116012

Li Q, Xu Y, Pedersen A, Wang M, Zhang M, Feng J, Jones CR (2024) Investigating the role of Fe-pyrrolic N4 configuration in the oxygen reduction reaction via covalently bound porphyrin functionalized carbon nanotubes. Adv Func Mater 34(16):2311086

Liang Z, Zhu Y, Leonard W, Fang Z (2024) Recent advances in edible insect processing technologies. Food Res Int 114137

Lin H, Yang Y, Hsu YC, Zhang J, Welton C, Afolabi I, Zhou HC (2024) Metal–organic frameworks for water harvesting and concurrent carbon capture: a review for hygroscopic materials. Adv Mater 36(12):2209073

Lotfi S, Ouardi ME, Ahsaine HA, Assani A (2024) Recent progress on the synthesis, morphology and photocatalytic dye degradation of BiVO4 photocatalysts: a review. Catal Rev 66(1):214–258

Luo L, Huang W, Zhang J, Yu Y, Sun T (2024) Metal-based nanoparticles as antimicrobial agents: a review. ACS Appl Nano Mater 7(3):2529–2545

Mansouri B, Rezaei A, Sharafi K, Azadi N, Pirsaheb M, Rezaei M, Nakhaee S (2024) Mixture effects of trace element levels on cardiovascular diseases and type 2 diabetes risk in adults using G-computation analysis. Sci Rep 14(1):5743

Marghade D, Shelare S, Prakash C, Soudagar MEM, Khan TY, Kalam MA (2024) Innovations in metal-organic frameworks (MOFs): pioneering adsorption approaches for persistent organic pollutant (POP) removal. Environ Res 119404

Mecha CA, Monchari C, Chollom MN, Bakare BF (2025) Applications of solid waste-derived carbon nanomaterials in water treatment. Waste Deriv Carbon Nanomater 2:85–101

Mehta P, Chelike DK, Rathore RK (2024) Adsorption-based approaches for exploring nanoparticle effectiveness in wastewater treatment. Chem Sel 9(25):e202400959

Mitra AK, Nayak S (2025) Sustainable nanomaterials in wastewater remediation. In: Nanomaterials in wastewater research. Springer, Singapore, pp 159–188

Monika, Madugula SK, Kondabagil K, Kunwar A (2025) Far-UVC (222 nm) irradiation effectively inactivates ssRNA, dsRNA, ssDNA, and dsDNA viruses as compared to germicidal UVC (254 nm). Photochem Photobiol 101(1):147–156

Mustafa M, Epelle EI, Macfarlane A, Cusack M, Burns A, Yaseen M (2025) Innovative approaches to greywater micropollutant removal: AI-driven solutions and future outlook. RSC Adv 15(16):12125–12151

Muzamil M, Banday RUZ, Rasool S, Gul D, Rasool S, Wani MH, Gurjar B (2025) Micro-and nanoplastics: impact on crop plants and agricultural practices. In: Micro-nano plastics exposure, environmental degradation and public health crisis. Springer, Singapore, pp 175–186

Naseer A, Younas F, Munir R, Sardar MF, Noreen S, Cui Z (2025) Clay-based nanocomposite materials used in treatment of wastewater: recent advancements, cost investigation and future perspectives. Sep Purif Rev 54(2):149–166

Natarajan B, Kannan P, Rather JA, Sheikh RA (2023) Recent developments in metal nanoparticles functionalized nanocomposite adsorbents for heavy metals removal from wastewaters. J Taiwan Inst Chem Eng 147:104942

Nawaz A, Farhan A, Maqbool F, Ahmad H, Qayyum W, Ghazy E, Fathi-karkan S (2024) Zinc oxide nanoparticles: pathways to micropollutant adsorption, dye removal, and antibacterial actions-A study of mechanisms, challenges, and future prospects. J Mol Struct 138545

Ojha PC, Satpathy SS, Ojha R, Dash J, Pradhan D (2024) Insight into the removal of nanoplastics and microplastics by physical, chemical, and biological techniques. Environ Monit Assess 196(11):1055

Omran BA, Abdel-Salam MO, Farghal HH, El-Sayed MM, Baek KH (2025) Synthesis of $TiO_2$–CuO@ graphene oxide hybrid bionanocomposite with enhanced antibacterial and organic dye degradation activities. Mater Adv 6(8):2654–2676

Panigrahy SK, Nandha A, Chaturvedi M, Mishra PK (2024) Novel nanocomposites with advanced materials and their role in waste water treatment. Next Sustain 4:100042

Parlayıcı Ş, Bahadir M, Pehlivan E (2024) Nanoporous carbonaceous materials (biochar and activated carbon): recent progress and potential applications for arsenic removal. J Dispers Sci Technol 1–22

Patil PM, Mahamuni-Badiger P, Ingavale RR, Patel PR, Dhanavade MJ (2024) Usage of microplastic beads in pharmaceuticals and cosmetics industry: a review. Microplastic Pollut 51–72

Pizzichetti ARP, Pablos C, Álvarez-Fernández C, Reynolds K, Stanley S, Marugán J (2021) Evaluation of membranes performance for microplastic removal in a simple and low-cost filtration system. Case Stud Chem Environ Eng 3:100075

Rahchamandi SYR, Mirhadi E, Gheybi F, Kazemi-Beydokhti A, Jaafari MR, Mostafavi E, Alavizadeh SH (2024) Engineering carbon-based nanomaterials for the delivery of platinum compounds: an innovative cancer disarming frontier. Environ Res 119933

Rashid MM, Abir N, Kamal SAB, Al-Amin M, Ahamed MA, Islam MT, Iqbal MI (2024) Functionalized cellulose for textile organic pollutant treatment: a comprehensive review. Water Conserv Sci Eng 9(1):11

Ravenscroft J, Eftim S, Soller J, Jones K, Ichida A, Marion J, Lee J (2025) Applying epidemiology and quantitative microbial risk assessment to ambient water quality evaluation: case study of fecal contaminated water in a US inland lake. Hum Ecol Risk Assess Int J 31(3–4):459–488

Ray S, Vashishth R (2024) From water to plate: Reviewing the bioaccumulation of heavy metals in fish and unraveling human health risks in the food chain. Emerg Contam 100358

Rostami MS, Khodaei MM (2024) Recent advances in chitosan-based nanocomposites for adsorption and removal of heavy metal ions. Int J Biol Macromol 132386

Sable H, Kumar V, Singh V, Rustagi S, Chahal S, Chaudhary V (2024) Strategically engineering advanced nanomaterials for heavy-metal remediation from wastewater. Coord Chem Rev 518:216079

Sati A, Ranade TN, Mali SN, Ahmad Yasin HK, Pratap A (2025) Silver nanoparticles (AgNPs): comprehensive insights into bio/synthesis, key influencing factors, multifaceted applications, and toxicity–a 2024 update. ACS Omega 10(8):7549–7582

Sedaghat O, Bahramifar N, Nowrouzi M, Younesi H (2025) Life cycle assessment of industrial wastewater treatment: evaluating the environmental impact of electrocoagulation technologies. J Water Process Eng 71:107257

Singh V, Ahmed G, Vedika S, Kumar P, Chaturvedi SK, Rai SN, Kumar A (2024) Toxic heavy metal ions contamination in water and their sustainable reduction by eco-friendly methods: isotherms, thermodynamics and kinetics study. Sci Rep 14(1):7595

Somya A, Varshney AP (2025) Ion exchange and zeolites: sustainable approaches to water pollution mitigation. In: Advancements in zeolites and micro-meso porous hierarchical materials. IGI Global Scientific Publishing, pp 151–180

Srivastava A, Devnani GL, Gupta P (2025) Magnetic separation and degradation approaches for effective microplastic removal from aquatic and terrestrial environments. Mater Adv 6(10):3043–3057

Sultana S, Ali MH, Dutta SK, Al Mamun MS, Chakrabarty S, Hasan MK (2025) MXenes: advances in synthesis, characterisation, and applications in water purification. Environ Technol Rev 14(1):273–309

Sun H, Wang S, Wang S, Song Y, Ling Z, Zhang L (2025) Rapid processing of dispersed hydrate-based purification for high-yield seawater desalination and wastewater treatment. Desalination 604:118741

Taha A, Singh N, Farooq N, Hashmi AA (2025) Facile synthesis of Schiff base palladium (II) complex for efficient catalytic reduction of aromatic nitro compounds and organic dyes for wastewater remediation. J Mol Liq 426:127343

Talreja N, Hegde C, Kumar EM, Chavali M (2025) Emerging environmental contaminants: sources, consequences and future challenges. Green Technol Ind Contam 119–149

Thakur N, Kumar P, Tapwal A, Jeet K (2023) Degradation of malachite green dye by capping polyvinylpyrrolidone and Azadirachta indica in hematite phase of Ni doped Fe2O3 nanoparticles via co-precipitation method. Nanofabrication 8

Thakur N, Kumar P (2024) Effect of shape and size on synthesized triple (Co/Ni/Zn)-doped α-$Fe_2O_3$ nanoparticles on their photocatalytic and scavenging properties. Int J Nanosci 23(06):2450010

Tu TN, Tran NT, Nguyen QH, Le VN, Kim J (2024) Metal-organic frameworks for aromatic-based VOC decomposition. Korean J Chem Eng 41(9):2461–2476

Tundwal A, Kumar H, Binoj BJ, Sharma R, Kumari R, Yadav A, Kumar P (2024) Conducting polymers and carbon nanotubes in the field of environmental remediation: sustainable developments. Coord Chem Rev 500:215533

Ullah Z, Peng L, Lodhi AF, Kakar MU, Mehboob MZ, Iqbal I (2024) The threat of microplastics and microbial degradation potential: a current perspective. Sci Total Environ 177045

Upadhyay RK, Soin N, Roy SS (2014) Role of graphene/metal oxide composites as photocatalysts, adsorbents and disinfectants in water treatment: a review. RSC Adv 4(8):3823–3851

Wang L, Li Z, Fan J, Han Z (2024) The intelligent prediction of membrane fouling during membrane filtration by mathematical models and artificial intelligence models. Chemosphere 349:141031

Yang X, Zhou Y, Hu J, Zheng Q, Zhao Y, Lv G, Liao L (2024) Clay minerals and clay-based materials for heavy metals pollution control. Sci Total Environ 176193

Yesilay G, Dos Santos OAL, Hazeem LJ, Backx BP, Kamel AH, Bououdina M (2023) Impact of pathogenic bacterial communities present in wastewater on aquatic organisms: application of nanomaterials for the removal of these pathogens. Aquat Toxicol 261:106620

Zhang Y, Zhang Y, Wu A (2024) Design and construction of magnetic nanomaterials and their remediation mechanisms for heavy metal-contaminated soil. Sci Total Environ 175369

Zhuang Z, Guan B, Chen J, Zheng C, Zhou J, Su T, Huang Z (2024) Review of nitrous oxide direct catalytic decomposition and selective catalytic reduction catalysts. Chem Eng J 150374

# Chapter 5
# Emerging Nano-Based Therapeutics to Mitigate Environmental Antimicrobial Resistance

**Priyanka Chaudhary, Divya Joshi, Navdeep Sharma, and Vedpriya Arya**

**Abstract** Antimicrobial resistance (AMR) has emerged as one of the most urgent global health threats, primarily driven by the overuse and misuse of antibiotics and environmental contamination through water, soil, and food. These factors create an environment where resistant pathogens proliferate, facilitated by transferring resistance genes through both vertical and horizontal transmission. This results in the rapid emergence and spread of multidrug-resistant (MDR) bacteria, which are increasingly difficult to treat with conventional antibiotics. In response to this escalating crisis, innovative solutions are urgently needed, and one such promising approach lies in the use of nanoparticles (NPs). NPs offer a promising solution to combat AMR by enhancing drug delivery and disrupting bacterial cell walls, biofilms, and intracellular components. While concerns about toxicity and biocompatibility exist, plant-based NPs may offer a safer alternative. Recent advances in nanoscience and machine learning may further improve the design and effectiveness of NPs, making them more efficient for targeted treatments and contributing to innovative solutions for tackling AMR.

## 5.1 Introduction

Antibiotic research has brought about a complete change in the medical industry. Nevertheless, the frequent and negligent use of antibiotics has led to the rapid emergence of harmful strains that have become resistant to them. The development of

P. Chaudhary (✉) · D. Joshi · N. Sharma · V. Arya
Patanjali Herbal Research Department, Patanjali Research Foundation, Haridwar, India
e-mail: priyanka.c@patanjali.res.in

D. Joshi
e-mail: divya.joshi@patanjali.res.in

N. Sharma
e-mail: navdeep.sharma@patanjali.res.in

V. Arya
e-mail: vedpriya.arya@patanjali.res.in



A. Balkrishna et al. (eds.), *Nanotechnology in Environmental Science and Healthcare*,
https://doi.org/10.1007/978-981-95-6714-0_5

AMR arises naturally as a result of genetic alterations in pathogens. However, the indiscriminate use of antibiotics has contributed majorly to its emergence and spread. Biofilms produced by microorganisms further aggravate the issue by compounding the effect many folds. AMR is a critical global health issue, responsible for a substantial number of infectious disease cases and fatalities worldwide. Unfortunately, more than one million people died each year from AMR-related infections between 1990 and 2021. Alarmingly, projections suggest that if current trends continue, AMR could lead to as many as 39 million deaths globally over the next 25 years (Murray et al. 2022; Naghavi 2024).

Concerns are mounting among researchers about the environment's role as a major reservoir for antimicrobial resistance genes (ARGs) and its influence in spreading resistance. It acts as a breeding ground for antibiotic-resistant bacteria (ARB). Wild animals and environmental factors such as wastewater, manure, and agricultural runoff are potential sources and transmission routes for resistance. Contamination of water, soil, and food products, particularly from animal waste, can facilitate the spread of resistant pathogens to humans. Livestock presents a significant source of resistance genes due to the use of antibiotics in farming, while wildlife can carry resistant bacteria from contaminated environments. These bacteria can subsequently proliferate through soil, water, and food, potentially exposing people (Samreen et al. 2021). It is now widely recognized that antibiotic resistance genes from the environment can be transmitted through the consumption of contaminated food, posing significant public health risks (Chaudhary and Singh 2015; Hernando-Amado et al. 2019). Numerous studies have reported a high prevalence of antimicrobial-resistant foodborne pathogens in various food samples, including packaged foods (Chaudhary and Singh 2014; Gdoura-Ben Amor et al. 2018). Each year, unsafe food leads to 600 million cases of foodborne illnesses and 420,000 deaths globally, according to a WHO report (WHO 2024). Developing new antibiotics demands financial resources and labor, and is a prolonged process (Lee et al. 2019). This has impelled global initiatives to discover new and effective antimicrobial agents, and innovative drug delivery and targeting methods.

Recent advances in nanotechnology developing novel formulations based on drug delivery systems and antimicrobial nanoparticles open up new possibilities for combating microbial resistance in response to these issues (Thakur et al. 2019; Kumar et al. 2024; Sharma et al. 2025). By leveraging the unique properties of NPs, these innovations enable targeted drug delivery, which improves the efficiency of medications by directing them precisely to infection sites and enhances therapeutic outcomes with reduced adverse effects. Additionally, NPs are increasingly used to coat medical equipment and surfaces, providing antimicrobial properties that help prevent bacterial growth and biofilm formation, a significant concern in healthcare settings (Shahriari et al. 2019; Ahmadi et al. 2020). Further, nano-enabled artificial intelligence (AI)-supported smart optical biosensors combining nanotechnology, AI, and optical sensing to enable rapid, sensitive, and real-time point-of-care diagnostics in AMR may prove beneficial (Taha et al. 2024). These advancements reflect the impact of nanotechnology in the prevention, detection, and treatment of microbial diseases. To achieve a deep understanding, factors and mechanisms driving the spread of ARGs and ARB in the environment are highlighted, along with the potential of advancements in nanoscience and nanotechnology to combat environmental AMR. A comprehensive overview of emerging nanotechnology-based methods, examining their effectiveness and underlying mechanisms in addressing AMR is also presented.

## 5.2 Environmental Dissemination of AMR: Prime Sources and Mechanism

AMR is the ability of a microorganism to resist the effect of an antibiotic to which it previously renders susceptible. It is a natural phenomenon acquired by a microorganism through genetic manipulation as a medium to replicate and survive. Antibiotics disrupt critical survival processes in bacteria, such as blocking cell wall synthesis and interfering with the production of essential proteins, DNA, and RNA. However, bacteria can quickly adapt through mutations and the horizontal transfer of DNA, allowing them to resist the effects of these antibacterial agents (Gupta et al. 2019). The same microbe can acquire multiple drug resistance genes from different organisms, resulting in the rise of multidrug-resistant (MDR) "superbugs." Exploration of resistant bacteria has discovered a "super-resistance" gene (NDM-1) that allows for the enzymatic degradation of β-lactam antibiotics, making these bacteria resistant to a broad spectrum of antibiotics (Wang et al. 2017). Well, it is not specific to a bacteria or a group of bacteria to be equally susceptible or resistant to any given antimicrobial agent. Even among related bacterial families, resistance levels might differ significantly. The resistance is usually classified under two main categories:

- **Intrinsic**: A characteristic that is common to all members of bacterial species, where the species acquired resistant genes genetically from the family and unaffected by prior antibiotic exposure is known as intrinsic resistance.
- **Acquired**: All three of the primary ways that bacteria can acquire genetic material through horizontal gene transfer (HGT) include transduction, conjugation, and transformation. In addition, the bacteria may undergo mutations in chromosomal DNA. It could be either short-term or long-term.

### *5.2.1 AMR in Soil, Water, and Air Ecosystems*

AMR has become a vital part of the human-animal-environment cycle since it can spread through a number of environmental channels (FAO 2018; Stanley et al. 2022), as depicted in Fig. 5.1. Although pollution, inadequate local sanitation, and other factors also contribute significantly, the main causes of AMR are antibiotics and ARGs. Numerous environmental settings, including hospitals, intensive care units, outpatient clinics, animal farms, feedlots, and veterinary clinics, use antimicrobial compounds, again contributing to the rise in resistance. However, the widespread

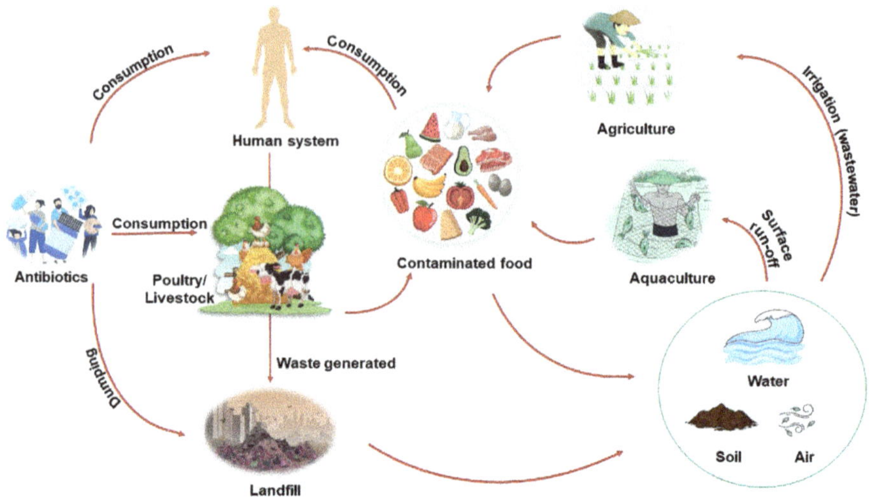


**Fig. 5.1** Crosstalk between various components leading to AMR in the environment

and negligent use of antibiotics over the past few years has caused microorganisms to become resistant to almost all of the antibiotics that have been developed to date (Peterson and Kaur 2018). Zhou et al. (2023a, b), reported that aminoglycoside resistance genes predominated in water (58.5%), whereas multidrug resistance genes were more common in soil (55.2%) and air (54.2%). Furthermore, five ARGs and 19 bacterial genera were found to be prevalent in all environmental samples, demonstrating their broad distribution (Zhou et al. 2023a, b).

#### 5.2.1.1 AMR in Soil Systems

Soil is an important reservoir for AMR due to its diverse microbial community and extensive exposure to anthropogenic causes, including agricultural practices and waste disposal. It is composed of multiple antibiotic-producing/resistant genes that can spread rapidly through HGT (Lee et al. 2018). Symochko et al. (2023) conducted extensive research on antibiotic resistance (AR) in soil microbial communities in a range of terrestrial ecosystems. 57 of the 389 bacterial strains that were isolated showed antibiotic resistance, with some strains having resistance levels above 70% (Symochko et al. 2023). These findings highlight how soil serves as a major ARG reservoir, endangering human health and the environment. Some of the primary factors that contribute to AMR in soil are agricultural inputs, waste carrying unmetabolized discharge, and the presence of persistent antibiotics in the soil and leachate.

- ***Agricultural Inputs***

Antibiotics and resistant bacteria are introduced into the soil through the overuse of antibiotics in livestock as well as the application of manure and biosolids. In sectors dealing with poultry and livestock, the use of antibiotics as growth enhancers rather than for infection control, poor sanitary conditions, and the environmental release of unmetabolized antibiotics or their residues through manure and feces, improper disposal, and ignorance are some of the main causes of antibiotic resistance. Together, these factors demonstrate the force of genetic selection that causes bacterial infections in the population to become resistant to several medications (Samreen et al. 2021; Endale et al. 2023). Neher et al. (2023) reported the presence of ermB and tetM ARGs in swine manure. Some other studies have also reported higher antibiotic resistance in dairy farm manure, linking agricultural practices to AMR spread (Sun et al. 2024; Zalewska et al. 2024). Additionally, ARB and ARGs can proliferate and spread in aquatic environments, including irrigation-purpose wastewaters, which can have a negative impact on human health and the sustainability of the environment (Tian et al. 2021). Continuous monitoring of ARB and ARG in wastewater used for irrigation is critically needed, given the serious threat posed by the growing antibiotic resistance to human health.

- ***Waste carrying unmetabolized discharge***

A large percentage of antibiotics administered to humans and animals remains unmetabolized, entering water and soil via wastewater, livestock manure, and biosludge. In nutrient-rich soils, the highly interactive soil microbiome permits the exchange of resistance genes through HGT. Through plasmid exchange at the gene level, pathogenic bacteria acquire ARGs and become highly resistant to antibiotics. HGT can occur between strains of the same bacterial species as well as different species via ARG-carrying plasmids, integrons (In), and transposons (Tn). Under the protection of deoxynucleotide enzymes, exposed DNA containing ARGs persists in the environment for a considerable amount of time even after the death of resistant strains (Lerminiaux and Cameron 2019; Jian et al. 2021).

- ***Soil and leachate containing persistent antibiotics***

Long-term soil persistence of antibiotics and resistant bacteria raises the possibility that resistance will spread to crops and make its way into the human food chain. There are certain antibiotics that do not degrade quickly and show long persistence in the soil, e.g., azithromycin (Cela-Dablanca et al. 2022). Sorption to organic matter and processes of degradation or transformation are the key factors influencing the persistence of antibiotic residues in soil. Fluoroquinolones (ciprofloxacin, norfloxacin, ofloxacin, enrofloxacin) tend to accumulate in surface soils due to their high adsorption coefficients (Kd), which limit their mobility. Similarly, macrolides strongly bind to mineral-rich soils containing iron (Fe), aluminum (Al), and manganese (Mn), further restricting their migration (Akhter et al. 2023). The broad range of DT50 (Half-life) values for antibiotic residues in soils indicates that the mechanisms controlling persistence are dependent on several variables, including the residue's physicochemical characteristics, the soil's properties, and meteorological variables (temperature, precipitation, and humidity). Further, the leaching of antibiotics depends on factors like water solubility, dissociation constants, sorption–desorption, and partitioning coefficients. Sulfonamides, with low partitioning in soil, are more likely to contaminate groundwater (Cycoń et al. 2019; Akhter et al. 2023; Reitra et al. 2023).

#### 5.2.1.2 AMR in Water Ecosystems

Rivers, lakes, and seas are examples of water bodies that serve as important reservoirs and routes for the spread of AMR. The extensive presence of ARB in soil, river surface water, irrigation wastewater, and other materials due to the dumping of antibiotics in the form of unmetabolized waste or contaminated sewage poses a risk of spreading to susceptible human groups, including children, immunocompromised individuals, and the elderly. In a study, Pärnänen et al. (2019) investigated the urban sewage influent resistome using 168 wastewater samples from seven European nations. Resistance of bacteria to several antibiotic classes, including tetracyclines, aminoglycosides, and β-lactams, was reported throughout the investigation, further proving the link between antibiotic use and ARG prevalence. Higher levels of gene transfer elements and resistance genes in wastewater were also highlighted in nations with maximum antibiotic consumption (Pärnänen et al. 2019).

Water bodies are exposed to antibiotics, ARB, and ARG through untreated or partially treated wastewater from households, businesses, and hospitals. These pollutants may impact aquatic life and human communities as they move downstream (Nappier et al. 2020; Avatsingh et al. 2023). Further, AMR may arise and proliferate in aquatic environments as a result of the use of antibiotics in aquaculture. Humans and wild fish populations may contract resistant bacteria from fish farms. Even at concentrations lower than the community's bacterial strains' minimum inhibitory concentration (MIC), antibiotic residues can favor resistant aquatic bacteria and aid in the spread of antibiotic resistance. Modified antibiotics in an aquaculture facility have a high potential to exert selective pressure and increase the frequency of antibiotic resistance in other environmental bacteria, as evidenced by the high frequencies of ARB reported in sites close to aquaculture where antibiotics have been used (Pepi and Focardi 2021). Drinking water supplies may be at risk from resistant bacteria that are introduced into surface and groundwater by sewage leaks and runoff from agricultural areas. Environmental loading of ARB and ARGs can occur from a number of points and nonpoint sources, such as contaminated wastewater effluents, runoff from animal feeding operations, and sewer overflow events.

#### 5.2.1.3 AMR in Air Ecosystems

Air can also contribute to the spread of antibiotic resistance, though there is relatively little evidence to support long-distance aerial transmission of bacterial diseases; urban air can have a high diversity of ARGs. Additionally, there are a number of ways that resistant bacteria can spread in crowded public areas, such as through the skin, coughing, breathing, and human airways (Larsson and Flach 2022). Huijbers et al. (2015) reviewed 157 studies on environmental AMR bacteria up to 2014, finding only 3% focused on bioaerosols or settled dust, compared to 25 on soil, 56 on water, and 71 on wildlife (Huijbers et al. 2015). However, due to their critical role in microbial transmission across a variety of environments, the number of studies on airborne ARGs has since increased. Owing to their long-range ability, bioaerosols from landfills, wastewater treatment facilities, and livestock operations can also spread resistant genes and bacteria into the atmosphere in both urban and rural settings (Lee and Yoo 2022). According to a study, poultry layer farms contributed 14.6% of ARGs in Beijing's urban air, with approximately 46% of target genes having the potential to enter the respiratory tract of the residents (Gao et al. 2022). In densely populated areas, pollution and human activity increase the risk of exposure to these resistant bacteria, which can then settle on mucous membranes, the digestive and respiratory systems, and within the lungs after exposure (Zhou et al. 2023a, b). In places with poor air quality, particularly industrial and farm areas, dust particles rich in particulate matter contribute as a major source of spreading resistant bacteria and genes in the environment. Further, hospital air conditioners have the ability to acquire ARGs and can disperse the resistome over time (Li et al. 2021). Zhou et al. (2021) reported the occurrence of *Campylobacter* spp. and *Staphylococcus* spp. in airborne particulate matter and dust, quantified from an urban hospital, responsible for fluoroquinolone (*vat*C-02, *mex*D) and β-lactams (*bla*Z, *mec*A) resistance genes (Zhou et al. 2021). Some other studies also reported the airborne transmission of ARGs and ARB through dust, particulate matter, and human airways (Li et al. 2018; Jin et al. 2021).

### 5.2.2 Major Factors Driving Environmental AMR

Environmental AMR is primarily driven by the ability of various bacteria, fungi, and microorganisms to acquire and transmit resistance traits. Among the primary contributors is ARB, which, through the presence of mobile genetic elements such as plasmids, transposons, and integrons, facilitates the HGT of ARGs across microbial populations. These mobile genetic elements serve as efficient genetic vehicles, enabling the rapid spread of resistance traits within and across microbial communities. Additionally, bacteriophages, which can carry ARGs, play a significant role in AMR dissemination through the process of transduction, wherein genetic material is transferred from one bacterial cell to another (Goryluk-Salmonowicz and Popowska 2022). Further, there is a documented correlation between high levels of heavy metals in the environment and increased antibiotic resistance, either phenotypic or genotypic. This relationship is attributed to co-selection mechanisms, which allow resistance traits to emerge and persist under environmental stressors. Co-selection occurs in two forms: co-resistance, where different resistance genes are located on the same mobile genetic element, enabling their simultaneous selection, and cross-resistance, where a single genetic determinant provides resistance to multiple stressors, including both heavy metals and antibiotics. These mechanisms illustrate how heavy metals inadvertently promote antimicrobial resistance, compounding the AMR crisis and complicating its management (Imran et al. 2019).

- **ARB and ARGs**

Antibiotic resistance was first observed in the 1930s and 1940s with the emergence of sulfonamide-resistant *Streptococcus pyogenes* and penicillin-resistant *Staphylococcus aureus* in hospitals (Barber and Rozwadowska-Dowzenko 1948). Shortly after, *Mycobacterium tuberculosis* developed resistance to streptomycin (Crofton and Mitchison 1948). Over time, multidrug resistance has become prevalent in various pathogens, including *Escherichia coli, Shigella, Salmonella,* and *Klebsiella pneumoniae,* which are resistant to broad-spectrum antibiotics like cephalosporins and carbapenems (Havenga et al. 2019). Environmental pathogens such as *Pseudomonas aeruginosa*, common in soil and water, exhibit resistance to aminoglycosides and beta-lactams, while *Vibrio* species predominant in aquatic settings are resistant to quinolones and tetracyclines. Methicillin- and vancomycin-resistant *S. aureus* pose a serious public health threat due to its ability to cause severe infections (Denissen et al. 2022). Overuse of antifungals in healthcare and agriculture has led to resistance in species like *Aspergillus* and *Candida albicans* to multiple antifungals, including azoles and echinocandins. The widespread use of azole fungicides has likely contributed to developing azole resistance in *Aspergillus fumigatus*, a common bioaerosol component associated with lung infections (Verweij et al. 2009; Revie et al. 2018). Research has also identified ARGs in bacteriophages from various sources. Phages isolated from human stools carry genes such as blaTEM, blaCTX–M, armA, and qnrA. Meanwhile, soilborne and waterborne phages harbor genes including aadA, strA, strB, sul1, blaTEM, and mecA, highlighting their role in the environmental spread of resistance (Quirós et al. 2014; Ross and Topp 2015). Some of the common microbial species and their associated ARGs contributing to environmental AMR are depicted in Table 5.1.

- **Heavy metals and biocides**

Heavy metals and biocides are also contributing to the development and persistence of ARGs in the environment. Heavy metals enter soil, wastewater, and aquaculture settings through agricultural inputs such as fertilizers, livestock manure, and pesticides, resulting in metal toxicity that creates selection pressure on microorganisms. This pressure drives the development of resistance genes. Studies show that bacteria with metal resistance genes are more likely to carry ARGs than those without, and these genes often reside on mobile genetic elements like plasmids, facilitating their transfer between organisms (Pal et al. 2015; Di Cesare et al. 2016). Similarly, biocides, including disinfectants and surfactants, contribute to antimicrobial resistance through co-selection. For instance, biocides like triclocarban and triclosan accumulate in sludge, creating reservoirs that promote the persistence of resistance

**Table 5.1** Key microbial agents and associated genes responsible for AMR

| Antibiotic family | Genes involved | Associated microorganisms | Diseases | Source in environment | References |
|---|---|---|---|---|---|
| Carbapenem-resistance genes | *bla*KPC, *bla*NDM, *bla*VIM, *bla*OXA blaIMP, blaVIM, blaOXA-51, blaIMP-1, *bla*OXA | *Klebsiella pneumoniae, Acinetobacter* spp., *Pseudomonas, Rheinheimera, Stenotrophomonas, Shewanella, Raoultella, Vibrio, Pseudoalteromonas, Algoriphagus, Bowmanella* and*Thalassospira* | Urinary tract infections (UTIs), pneumonia, bloodstream infections (sepsis), gastroenteritis, wound infections | Wastewater, seawater | Mills and Lee (2019); Xin et al. (2019); Dewi et al. (2020); Hubeny et al. (2022) |
| Polymyxin-resistance genes | *mcr-1, mcr-1 to 8, arnA and pmrF, pmrE and pmrF, pmrE1, pmrE2, pmrE3, pmrE4* | *Escherichia coli, Yersinia enterocolitica, Burkholderia cepacia, Klebsiella pneumoniae* | Gastroenteritis (diarrhea), neonatal meningitis, liver abscesses | Groundwater, seawater | Liu et al. (2016); Joo et al. (2023); Zhou et al. (2025) |
| Tetracycline-resistance genes | *tet*B, *tet*G, *tet*L (efflux genes), *tet*M, *tet*Q, tetW (ribosomal protection genes), *tet*X and *tet*37 (an activation gene), *tet*W, *tet*S, *tet*O, *tet*L, *tet*44 | Firmicutes and Proteobacteria *Citrobacter, Bhargavaea, Chryseobacterium, Paenibacillus, Streptomyces, Bacillus, Serratia, Streptomyces, Burkholderia, Microlunatus, Solirubrobacter, and Escherichia coli* | Skin infections, sepsis, and respiratory infections | Livestock (manure), Soil (farmland), Air (poultry farm) Agricultural soil | Gao et al. (2017); Lu et al. (2021); Nogrado et al. (2021); Ali et al. (2024) |
| Sulfonamide-resistance genes | sul1, sul2, sul3 | *Achro-mobacter, Arthrobacter, Bacillus, Brevibac-terium, Chryseobacterium, Citrobacter, Cupri-avidus, Escherichia, Flavobac-terium, Hydrogenophaga, Klebsiella, Lysini-bacillus, Massilia, Microbac-terium, Microvirga, Pseudomonas, Pseudoxan-thomonas, Rhizo-bium, Rhodococcus, Shigella, Sphingobac-terium, Sphingopyxis, Staphylo-coccus, Stenotrophomonas, Streptococcus, Streptomyces* | Endocarditis (rare), bloodstream infections (sepsis), hemolytic uremic syndrome, bone and joint infections, skin and soft tissue infections | Groundwater and surface water, animal manure | Wang et al. (2014); Wilson et al. (2021) |

traits. These substances apply selective pressure, enabling bacteria with both metal and antibiotic resistance genes to thrive (Buffet-Bataillon et al. 2016). Research has highlighted their role in driving AMR, emphasizing how environmental contaminants act as catalysts for the proliferation of resistance.

Together, these factors highlight the complex interplay between environmental contaminants and the emergence of antimicrobial resistance, underscoring the need for strategies to mitigate their impact.

## 5.3 Nano-Based Therapeutics in Combating AMR

Technological advancements have led to significant challenges like antibiotic pollution, and environmental contamination by toxic substances. These challenges emphasize the importance of developing innovative solutions to reduce ecological damage while maximizing benefits. NPs, with their unique properties like better surface area and reactivity, present promising strategies to address these challenges. They facilitate across diverse fields, including medicine, energy storage, and environmental restoration, facilitating targeted drug delivery, enhancing solar cell efficiency, and creating advanced materials with tailored characteristics.

There are several methods employed for synthesizing NPs. In addition to traditional chemical synthesis methods, green synthesis presents a more environment-friendly approach. This approach employs natural, biologically active compounds as reducing, stabilizing, or capping agents in a single step process for NP synthesis. Microbes or other biological materials such as bacteria, fungi, yeast, algae, actinomycetes, and plant extracts facilitate the intra- or extracellular biosynthesis of NPs. The use of renewable, non-toxic materials in green synthesis contributes to reducing the potential hazards associated with nanoparticle production for both human health and the environment (Bhardwaj et al. 2020). The diverse types of NPs exhibiting different properties and characteristics, particularly in addressing AMR are discussed.

### *5.3.1 Type of Nanomaterials*

#### 5.3.1.1 Carbon-Based Nanomaterials

Carbon-based nanomaterials represent a significant category of materials encompassing various nanoscale forms of carbon. Their unique properties make them suitable for multiple applications across different fields. Research on carbon-based nanomaterials has experienced sharp exponential growth recently. Some main types of carbon-based NPs include carbon nanotubes, carbon nanohorns, and carbon dots.

#### 5.3.1.2 Carbon Nanotubes (CNTs)

CNTs are tubular nanostructures composed of carbon atoms arranged in a hexagonal pattern. CNTs can be synthesized by several methods, with chemical vapor deposition (CVD), laser ablation, and arc discharge being the most common (Singh et al. 2017; Hasnain et al. 2019; Toleukhanuly et al. 2020). They are considered one of the ideal materials due to their high surface area, high aspect ratio, impressive small size, and chemical, thermal, biological, mechanical, and electrical properties (Ito et al. 2001). Their large surface area and the ability to modify their surfaces make them ideal for transporting therapeutic agents like peptides, genes, and chemotherapeutic drugs to specific target sites in the body (Yao et al. 2020). CNTs have shown potential in enhancing the effectiveness of cancer treatments, and offer novel therapeutic possibilities for addressing AMR when combined with bioactive molecules (Son et al. 2016; Saleemi et al. 2022). They are efficient in inhibiting bacterial adhesion and biofilm formation (Teixeira-Santos et al. 2021). They effectively neutralize *E. coli*, *P. aeruginosa*, *Bacillus subtilis*, *Staphylococcus epidermidis,* and viruses in wastewater effluent and river water when incorporated into membranes (Kang et al. 2009). Single-walled nanotubes with oxidizing antimicrobial agents effectively deactivate *Bacillus anthracis* spores (Lilly et al. 2012). The vast applications of CNTs in the biomedical field are due to chemical inertia, fast electron transfer kinetics, and antifungal and antibacterial properties (Anzar et al. 2020). CNTs may offer greater effectiveness and cost-efficiency than traditional antibiotic treatments, showing significant potential in addressing the rise of MDR.

#### 5.3.1.3 Carbon Nanohorns (CNHs)

CNHs are one-dimensional carbon nanostructures of conical shape and have exceptional properties compared to other carbon nanostructures. These are also known as "nanocones" and range from 40 to 50 nm in length and 2–5 nm in diameter (Dutta et al. 2023). Among different carbon-based nanomaterials, CNHs are one of the best candidates with their high drug-loading capability due to their high surface area (Chechetka et al. 2016). CNHs composite effectively demonstrates the antimicrobial potential toward peri-implantitis treatments (Hirata et al. 2023; Konishi et al. 2024). They are also employed in various other fields, such as electrochemical sensing, biofuel cells, and biomedical applications (Andrews et al. 2019).

#### 5.3.1.4 Carbon Dots (CDs)

CDs are carbon-based NPs classified into different subgroups based on crystallinity and morphology. They have attracted considerable attention due to their excellent and tunable photoluminescence, small size, high quantum yield, appreciable biocompatibility, and abundant low-cost sources providing applications in many fields (Zhu et al. 2015; Liu et al. 2019; Zhang et al. 2020). CDs inhibit bacterial growth or eliminate bacteria by generating reactive oxygen species (ROS), disrupting cell structures, and causing fragmentation and condensation of genomic DNA (Hussen et al. 2024). They exhibit strong antimicrobial activity against both Gram-positive and Gram-negative bacteria, with MICs of 64, 128, 64, and 128 μg/mL against *E. coli*, *P. aeruginosa*, *S. aureus*, and *B. subtilis*, respectively (Liang et al. 2021). CDs are highly effective in treating wounds and skin infections due to their excellent water solubility, biocompatibility, and low cytotoxicity. A solution of CDs derived from biogenic polyamines can treat bacterial keratitis and offer a compelling alternative to traditional eye drops for addressing ocular diseases caused by bacterial infections (Li et al. 2022). The non-toxic nature of CDs makes them highly promising for a range of biological and medical applications (Liu et al. 2018).

#### 5.3.1.5 Polymeric Nanomaterials

Polymeric nanomaterials range from 1 to 1000 nm and exhibit diverse shapes, sizes, and structures. They are biodegradable and biocompatible polymers, showing low toxicity, increased therapeutic effect, high drug loading, improved drug penetration, controlled drug release, and enhanced chemical and physical stability (Soppimath et al. 2001; Bhardwaj and Jangde 2023). Polymeric NPs can be formed from various natural and synthetic polymers, including cellulose, pectin, dextran, agarose, chitosan, xanthan gum, gelatin, albumin, starch, and alginate. They have vast applications in agriculture, wastewater treatment, environmental protection, regenerative technology, drug delivery, gene therapy, cancer therapy, and antimicrobial treatments (Lei et al. 2014; Reshad et al. 2021; Meng et al. 2022).

#### 5.3.1.6 Metal-Based Nanoparticles (MNPs)

MNPs range from 1 to 100 nm, and the most familiar ones are derived from Ag, Au, Fe, Al, Cu, Zn, Co, Pb, Ni, Bi, Si, Sn, Ti, Mo, and Cd (Mujahid et al. 2022). Physical methods involve the top-down approach, while chemical and biological methods involve the bottom-up approach (Nyabadza et al. 2023). In biomedical sciences and engineering, MNPs such as silver, gold, copper, and zinc are widely used for various medical and pharmaceutical applications viz. antiviral, antibacterial, antifungal, antiprotozoal, anti-inflammatory, anticancer, and anti-angiogenic (Wahab et al. 2023). MNPs have been proposed as an alternative to traditional antibiotics to overcome bacterial resistance because of their well-described antimicrobial activity against Gram-negative and Gram-positive bacteria (Diaz-Puertas et al. 2023).

### *5.3.2 Emerging Nano-Based Antimicrobial Agents*

Nanomaterials exhibit unique properties and can kill microbes through different mechanisms (Wang et al. 2017; Borehalli et al. 2023; Mondal et al. 2024). The composition of antimicrobial NPs varies in physical and chemical properties. In the last two decades, efforts have been made to develop various antimicrobial NPs like silver NPs (AgNPs), zinc NPs (ZnNPs), copper NPs (CuNPs) (metal-based); zinc oxide NPs (ZnO-NPs), titanium oxide NPs ($TiO_2$NPs), magnesium oxide NPs, iron oxide NPs ($Fe_3O_4$/$Fe_2O_3$ NPs) (metal oxide-based); liposomes (LipoNPs), micelles (MNPs), solid–liquid NPs (SLNPs) (lipid-based); carbon nanotubes (CNTs), graphene oxide (GO) (carbon-based); chitosan (ChitNPs), poly-lactic-co-glycolic acid (PLGA NPs), polycaprolactone NPs, polyelectrolyte NPs (composite-based); plant-based NPs; microorganism-based NPs and so on. Some of these examples are summarized in Table 5.2.

## 5.4 Nanoparticles–Mediated AMR Management: Mechanistic Insights

NPs offer alternatives in combating AMR not only as drug delivery systems but also for their intrinsic antimicrobial properties (Vallet-Regi et al. 2019). They can significantly contribute to preventing antibiotic resistance through multiple mechanisms providing innovative strategies to tackle the challenges of bacterial resistance. They can help combat antibiotic resistance by disrupting cell walls and membranes, affecting enzymes and proteins, damaging efflux pumps, inhibiting the electron transport chain and biofilm formation, generating ROS, etc. (Balkrishna et al. 2023). This production of ROS induces oxidative stress, ultimately leading to cell death.

### *5.4.1 Disruption of Cell Wall and Cell Membrane*

The positive charge on the surface of MNPs is necessary to cling to the bacterial outer surface (Wang et al. 2017). AgNPs on exposure to bacteria attach to the cell wall and membranes resulting in their antimicrobial potential (Salleh et al. 2020). They significantly impair membrane permeability and disrupt cellular respiration by leveraging electrostatic attraction and their affinity for sulfur proteins, ultimately disrupting the bacterial envelope (Salleh et al. 2020; Yin et al. 2020) as depicted in Fig. 5.2. Studies demonstrate that AgNPs are effective against Gram-positive and Gram-negative pathogens, with smaller dimensions exhibiting superior penetration abilities (Mukundan et al. 2017; More et al. 2023). The adsorption of NPs depolarizes the cell wall by altering the usual negative charge and increasing permeability (Mukha

**Table 5.2** Types of nanoparticles with their application against multidrug-resistant (MDR) bacterial strains

| Type of NPs | Plants/Microorganisms utilized | Antimicrobial activity | Mechanism | References |
|---|---|---|---|---|
| Plant-based NPs | | | | |
| Ag | *Ocimum. gratissimum* | MIC and MBC: *Escherichia coli*—4 μg/ML; MIC: *Staphylococcus aureus*—8 μg/mL<br>MBC: S. *aureus*—16 μg/mL | Cell membrane damage by ROS | Das et al. (2017) |
| | *Cymbopogon citratus* | MIC: *E. coli*—3 μg/mL<br>MIC: *E. coli* CTXM strain* -12.5 μg/mL | Cell membrane penetration, followed by accumulation and cell death | Ranjani et al. (2020) |
| | *Alphonsea sclerocarpa* | MIC: *Streptococcus mutans*- 8.66 mg/100 mL; *Pseudomonas aeruginosa*- 44.33 mg/100 μL | Penetration of silver ions into cells and intercalating between purine and pyrimidine bases leads to denaturation of DNA | Doddapaneni et al. (2018) |
| | *Curcuma longa* | MBC: *E.coli* BL-21—50 mg/L | Outer membrane destabilization and plasma membrane rupture lead to ATP depletion | Kumar et al. (2010) |
| | *Acalypha indica* | MIC: *E. coli* and *Vibrio cholera*—10 μg/ml | Alteration in membrane permeability and respiration | Krishanraj et al. (2010) |
| | *Garcinia mangostana* | $AgNO_3$ NPs<br>ZOI at 20 μg/mL: *E. coli*- 15 mm; *S. aureus*- 20 mm | $Ag^+$ exhibits a bactericidal effect, while polymer subunits disrupt membranes | Veerasamy et al. (2011) |

(continued)

**Table 5.2** (continued)

| Type of NPs | Plants/Microorganisms utilized | Antimicrobial activity | Mechanism | References |
|---|---|---|---|---|
| $TiO_2$ | *Psidium guajava* | ZOI at 20 μg/mL: *E. coli* -23 mm; *S. aureus*-25 mm | Exhibit strong oxidizing properties when exposed to sunlight or UV light | Kumar et al. (2014) |
| | *Luffa acutangula* | $TiO_2$ NPs—(5–100 μg/ml)<br>*MIC: E. coli* and *Pseudomonas aeruginosa*- higher than 5 μg/ml<br>*Klebsiella pneumoniae and S. aureus*- 5 μg/ml<br>*Enterococcus faecalis and Bacillus subtilis*- Lower than 5 μg/ml | Disruption of cell membranes by inducing ROS, and hydroxyl radicals, leads to cell death | Anbumani et al. (2022) |
| Au | *Syzygium cumini* | ZOI at 200 μg/mL<br>*S. aureus*- 17.87 mm, *Acinetobacter baumannii*- 17.69 mm, *E. coli* -17.65 mm, *P. aeruginosa*- 17.79 mm, *E. faecalis*- 17.65 mm, *K. pneumoniae*- 16.7 mm, and *Proteus vulgaris*- 20.05 mm; MIC- 450 μg/mL—250 μg/mL | Disruption of MDR bacteria wall and membrane, lipid peroxidation and oxidative damage of DNA and proteins, interference with the process of cell signal transduction, kills the cells | Franzolin et al. (2022); Diksha et al. (2023) |
| | *Nigella arvensis* | MIC: *E. coli* and *B. subtilis*- 62.5 μg/mL<br>MIC: *S. aureus and Serratia marcescens*- 125 μg/mL | Synergic effect of plant extract and AuNPs. AuNPs disrupt the membrane, and inhibit the transcription of DNA strains, causing growth inhibition and cellular damage | Chahardoli et al. (2018) |
| CuO | *Ocimum lamiifolium* and *Withana somnifera* | MIC: *O. lamiifolium*-CuO-NPs—0.625—2.5 μg/mL against *S. aureus*, *E. coli*, *P. aeruginosa*<br>MIC: *W. somnifera*-CuO-NPs—0.236—5 μg/mL against *S. aureus*, *E. coli*, and *P. aeruginosa* | Synergistic effect of plant extract and CuO-NPs | Mengesha et al. (2024) |

(continued)

**Table 5.2** (continued)

| Type of NPs | Plants/Microorganisms utilized | Antimicrobial activity | Mechanism | References |
|---|---|---|---|---|
| Microorganism-based NPs | | | | |
| Ag | *Bacillus cereus* (Endophytic) | ZOI: *Salmonella* Typhi-14 mm, *K. pneumoniae*- 15 mm, *P. aeruginosa* -15 mm, *E. coli* -18 mm, and *S. aureus*-22 mm | Disrupts membrane permeability and respiration, interacting with sulfur and phosphorus compounds like DNA, leading to cell death | Sunkar et al. (2012) |
| | *Pseudomonas alloputida* B003 UAM | MIC: *B. subtilis*, *E. coli*, *K. pneumoniae*, *P. aeruginosa*, *S. aureus*, and *S. epidermidis*—0.24—9.68 µg/mL | ROS produced by the AgNPs damages cellular components, leading to cell envelope disruption and cell death | Pernas-Pleite et al. (2022) |
| Au | *Trichoderma saturnisporum* | MIC: MRSA, MSSA, *K. pneumoniae*, and *P. aeruginosa*- 500–1000 µg/mL; reduction of biofilm formation against *P. aeruginosa* | Inhibition of respiratory enzymes of bacterial cells stops ATP production, leading to cell death | Soliman et al. (2023) |
| | *Brevibacillus formosus* | AuNPs<br>ZOI: *E. coli* -10 mm, *S. aureus*- 26 mm at 150 µg/mL<br>*E. coli*- 11 mm, *S. aureus*-28 mm at 200 µg/mL | Disrupts bacterial cell membrane permeability and inhibits DNA replication | Srinath et al. (2017) |
| Chemical-based NPs | | | | |
| Type of NPs | Bacterial strains | Antimicrobial activity | Mechanism | References |
| ZnO | *E. coli* | 50 µg/mL | Release of zinc ions from nano-ZnO causes nanotoxicity | Li et al. (2011) |
| $TiO_2$ | *E.coli* MG 1655 | 100 µg/mL | Degradation of cell wall through ROS | Simon-Deckers et al. (2009) |

(continued)

**Table 5.2** (continued)

| Type of NPs | Plants/Microorganisms utilized | Antimicrobial activity | Mechanism | References |
|---|---|---|---|---|
| $Al_2O_3$ | *E.coli* MG 1655 | 106 µg/mL | Nano alumina formation affects bacterial membrane fluidity and integrity | Simon-Deckers et al. (2009) |
| $CeO_2$ | *E. coli* | 500 µg/mL | NPs absorption causes oxidative stress to bacterial membranes | Thill et al. (2006) |
| $Fe_3O_4$ | *Bacillus brevis, E. coli, S. aureus, S. epidermidis, B. subtilis, Bacillus licheniformis* | ZOI at 50 mg/mL: *B. brevis* -9 mm, *E. coli* -11 mm, *S. aureus*- 12 mm, *S. epidermidis*-14 mm, *B. subtilis*- 20 mm, *B. licheniformis*- 22 mm | Produces ROS like superoxide radicals ($O_2^-$), hydroxyl radicals (–OH), hydrogen peroxide ($H_2O_2$), and singlet oxygen ($^1O_2$) damage bacterial proteins and DNA | Behera et al. (2012) |

* CTXM strains are highly resistant to β-lactams, like penicillins and cephalosporins; MIC: minimum inhibitory concentration; MBC: minimum bactericidal concentration; MRSA: methicillin-resistant *Staphylococcus aureus;* MSSA: methicillin-sensitive *Staphylococcus aureus;* ZOI: zone of inhibition

et al. 2013). A similar action of cell damage is also displayed by green NPs (Sardar et al. 2022).

In addition to electrostatic attraction, MNPs interact with sulfur-containing proteins in the cell wall, causing irreversible structural alterations that lead to the destruction of cell wall. This destruction affects the permeability of cell membrane and lipid bilayer's integrity. As the cell morphology changes, their ability to regulate transport activities through the plasma membrane is compromised. Additionally, NPs can influence the transport and release of potassium ($K^+$) ions from microbial cells (Girma 2023). Moreover, impacting transport activity, increased membrane permeability can lead to more significant consequences, such as the leakage of cellular contents, including ions, proteins, reducing sugars, and occasionally the cellular energy reservoir, ATP (Dakal et al. 2016). After penetrating the cell wall, many NPs induce oxidative stress and generate ROS as they diffuse into the cell. Another mechanism of MNP absorption involves the release of ions into the surrounding environment, which then bind to the carboxyl or phosphate groups on bacterial cell membranes. After getting into the cell, these ions interfere with enzyme activity, disrupting crucial intracellular processes. For example, ZnO-NPs release $Zn^{2+}$ ions, which attach to –SH groups on membrane proteins, therefore disrupting the membrane permeability and inactivating thiol-containing enzymes, ultimately leading to bacterial cell death (Padmavathy and Vijayaraghavan 2011). Proteomic analysis revealed that exposure to AgNPs led to the upregulation or downregulation of various membrane proteins involved in ion binding, transport, pore formation, and membrane stability (Yan et al. 2018).

NPs exert their distinctive toxicity by penetrating bacterial membranes and disrupting critical cellular pathways (Lee et al. 2019). Research has shown that Ag and AuNPs can enter cells, with the released $Ag^+$ and $Au^+$ ions inhibiting cytochrome function in the electron transport chain, thereby damaging the respiratory chain. A

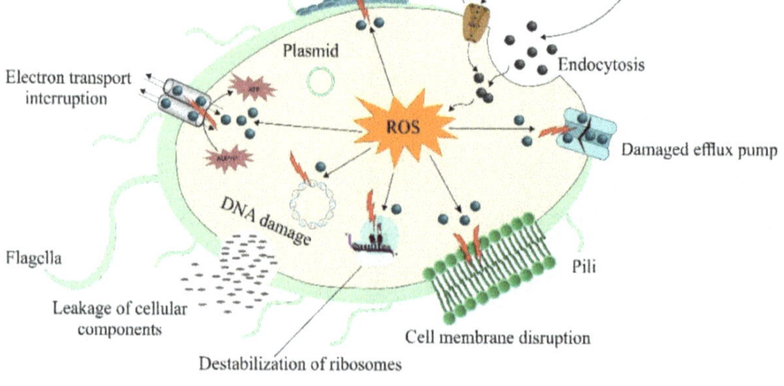


**Fig. 5.2** Disintegration of bacterial cell wall and membrane through NPs

similar effect has been reported with copper (Cu) and nickel oxide (NiO) NPs (Hemeg 2017). Additionally, NPs destabilize plasma membrane potential by damaging the efflux pump system which in turn is not able to expel linezolid and weakens the biological activity of membrane-bound respiratory chain dehydrogenase, and DNA integrity, so that linezolid could reach a concentration that effectively inhibits protein synthesis, thus mitigating the behavior of bacterial resistance (Raffi et al. 2008; Boakye et al. 2019).

### *5.4.2 Production of Reactive Oxygen Species (ROS) and Reactive Nitrogen Oxide Species (RNOS)*

NPs induce genotoxicity through various mechanisms which include direct contact and lead to cellular deformation. Alternatively, NPs may cause indirect DNA damage by altering replication or repair processes and generating ROS. These cellular changes can impact bacterial resistance (Shukla et al. 2021; Xu et al. 2023). ROS disrupts diverse cellular substances viz. nucleic acids, proteins, and lipids crucial for proper cell function (Gao et al. 2021). For example, nitric oxide-releasing NPs (NONPs) utilize multiple antimicrobial mechanisms by which microbes will develop a minimal resistance (Pelgrift and Friedman 2013). The NONPs exert antimicrobial effects through the formation of RNOS, resulting in the interaction of NO with superoxide ($O_2$). These RNOS can cause bacterial cell death by reacting with amino acid residues of bacterial cells, including cys, met, tyr, and others (Qasim et al. 2014). RNOS leads to an increased production of hydrogen peroxide ($H_2O_2$) and alkylating agents, which can damage DNA. Additionally, they inhibit DNA repair enzymes, such as DNA alkyl transferases, by S-nitrosylating their cysteine residues. RNOS also interacts with the prosthetic groups of proteins, including iron-sulfur (Fe–S) clusters and heme. Heme is found in various enzymes, such as nitric oxide synthase (NOS), cytochrome P450 (CYP450) enzymes, and guanylate cyclase. At sufficiently high concentrations, RNOS can irreversibly bind to the Fe (II) in heme, which removes heme from the protein and results in iron depletion within the bacterial cell (Schairer et al. 2012). Studies have demonstrated that NONPs can effectively inhibit the growth of antibiotic-resistant strains of *P. aeruginosa*, *E. faecalis*, *K. pneumoniae*, and *E. coli* (Qasim et al. 2014).

### *5.4.3 Binding to and Damaging Intracellular Components*

The antibacterial mechanism of MNPs includes protein dysfunction and enzyme inactivation (Godoy-Gallardo et al. 2021). For example, $Ag^+$ is believed to exert its antibacterial effects through various actions, such as denaturing the 30S ribosomal subunit and inhibiting the synthesis of essential proteins and enzymes involved in

ATP production by oxidizing amino acid side chains (More et al. 2023). Protein inactivation occurs when $Ag^{+}$ ions form stable bonds with thiol groups on proteins in the cell membrane, resulting in changes to the protein's three-dimensional structure, disruption of disulfide bonds, and blockage of active sites, all of which impair the microorganism's functions. Additionally, the inhibition of protein phosphorylation can suppress enzymatic activity, further hindering bacterial growth (Gordon et al. 2010; Shukla et al. 2021; Girma 2023). NPs were found to enforce the variations in genomic and proteomic profiles that suppress vital cellular functions and lead to cell necrosis and death (Slavin et al. 2017). The studies suggested the interaction of NPs with proteins through the SH group of mannose phosphate isomerase halts the cell wall synthesis. This inhibition causes leakage of internal components, resulting in cell death (Mukha et al. 2013). Moreover, the interaction of AgNPs with genetic material causes the denaturation and tearing of DNA as well as inhibition and alteration in cell division (Radzig et al. 2013). NP-induced genotoxicity includes chromosomal aberrations such as mutations, DNA strand breaks, and oxidative DNA base damage (Trouiller et al. 2009). In *E. coli*, AgNPs cause DNA damage and mutations in DNA containing oxidative lesions (*mutY*, *mutS*, *mutM*, *mutT*, and *nth*), making mutant strains more susceptible to AgNP-mediated antimicrobial treatment compared to wild-type strains (Radzig et al. 2013).

### 5.4.4 Destruction of Biofilm Architecture

NPs disrupt biofilms due to their high penetration ability and quorum-sensing mechanism. They penetrate the condensed extracellular polysaccharide matrix of the biofilm and get through its protective barrier. After removing biofilm, NPs interact with bacteria and display a bactericidal effect (Kumar et al. 2023). The charge on the NPs plays a key role in both penetration and interaction with the biofilm. Positively charged (cationic) NPs specifically interact with the anionic components of the biofilm matrix (Hosnedlova et al. 2022). Ag, Au, and ZnNPs have been extensively used as powerful antibacterial agents against a range of bacterial pathogens. Histidine-functionalized AgNPs have been used to inhibit *K. pneumoniae* biofilms and showed marked effects (Chhibber et al. 2017).

A scientific study demonstrated biofilm inhibition of *E. coli* and *P. aeruginosa* using AgNPs. The AgNPs exhibited MIC of 6.25 μg/mL and 3.12 μg/mL, for *E. coli* and *P. aeruginosa*, respectively (Singh et al. 2020). Goda et al. (2022) studied the effect of AgNPs synthesized from pomegranate rind (*Punica granatum*) extract to prevent bacterial colonization and biofilm formation on urinary tract catheters. The AgNPs-coated catheters significantly inhibited the colonization of antibiotic-resistant clinical strains of both Gram-positive (*S. epidermidis* and *S. aureus*) and Gram-negative bacteria (*E. coli*, *K. pneumoniae*, *Proteus mirabilis*, and *P. aeruginosa*). The AgNPs varnish proved particularly effective against Gram-negative bacteria (Goda et al. 2022). Furthermore, pyelonephritis is a severe infection that can lead to mortality. Currently, there are no effective treatments for managing this infection, as its pathogenesis is driven by quorum sensing-regulated virulence and persistent biofilm formation. In a novel approach, NPs loaded with quorum-sensing inhibitors were developed using biopolymers, specifically chitosan and heparin sulfate, and were intravesically delivered to treat biofilm-linked murine pyelonephritis. The results demonstrated significant antivirulence effects, including reduced biofilm formation, by interfering with the extracellular polymeric substance (EPS) matrix and disrupting quorum-sensing mechanisms (Vadekeetil et al. 2019).

### 5.4.5 Synergistic Effects of NPs with Conventional Antibiotics

Using antibiotics with NPs raises their efficacy compared to using them alone. The combination halts the development of bacterial resistance, minimizes the treatment duration, and lowers the antibiotic dose requirements (Hwang et al 2012). AgNPs are the most studied MNPs due to their synergistic effects. However, Cu, Au, CuO, copper sulfide CuS, Fe, $Fe_3O_4$, $Fe_2O_3$, Zn, ZnO, and platinum (Pt) are also reported for their synergistic potential (Ribeiro et al. 2022). The preparation of MNPs conjugates with antimicrobial drugs is usually carried out through one of the following methods demonstrated in Fig. 5.3.

Method A describes that the MNPs are synthesized separately and mixed with the combining agents, and characterized. In method B, MNPs are created by combining/

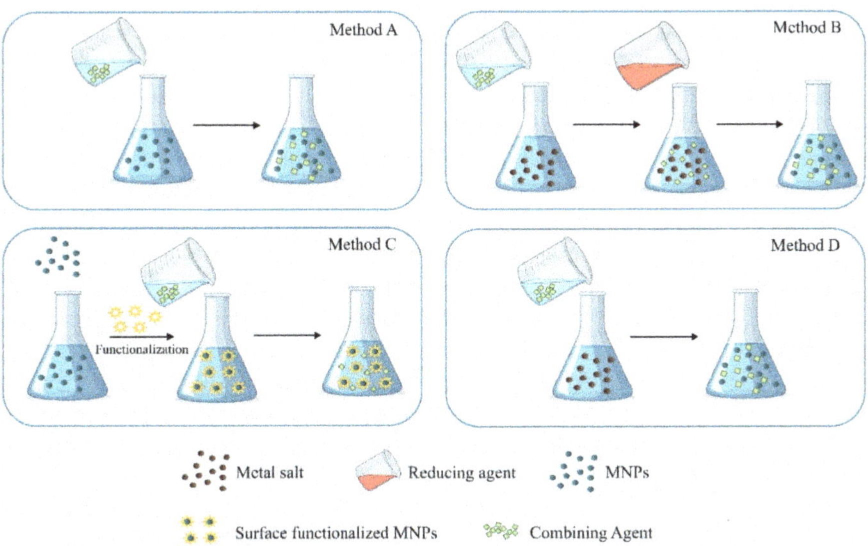


**Fig. 5.3** Common methods to conjugate MNPs and antibiotics, antifungals, or antivirals. Adapted with permission (Ribeiro et al. 2022)

conjugating agents (may or not act as a reducing agent). However, it is always associated with another reducing agent during the synthesis (Ribeiro et al. 2022). Method C demonstrated the steps like MNP synthesis, subsequent functionalization, and posterior conjugation step. In this method, MNPs can be prepared separately by any of the methods and afterward lead to surface functionalization using phosphonates or silanes. This creates strong multivalent interactions with organic molecules and biological cells, raising the concentration of MNPs in specific targets (Chen et al. 2015). In method D, the reaction time is long, but MNPs preparation is achieved in a single step using fewer chemicals (Ribeiro et al. 2022).

Wan et al. (2016) reported synergistic effects of AgNPs with the antibiotics polymixin B and rifampicin, and an additive effect of tigecyclines in combatting *Acinetobacter baumannii* infection was observed. In vivo studies resulted in the survival of mice treated with the AgNP-antibiotic combination. The synergistic mechanism was demonstrated by using antisense RNA-induced gene silencing to target specific genes in *E. coli.* The findings indicated that silencing genes such as rpoD, kdsA, and B, lpxC, and mutL led to increased sensitivity to both AgNO3 and AgNPs (Wan et al. 2016). The synergistic effect of AgNPs-antibiotics has also been found effective against antibiotic-resistant veterinary bacteria such as *Salmonella enterica* LT2, *S. aureus* GP0004, *E. coli* eae+ GN2514, *Actinobacillus pleuropneumoniae* 17/06L, *Pasteurella multocida* P-813, and *Streptococcus uberis* GP1037. Moreover, *A. pleuropneumoniae* and *P. multocida*, resistant to amoxicillin, gentamicin, and colistin, were found to be sensitive to these antibiotics when combined with AgNPs (Smekalova et al. 2016). Cephalexin NPs exert synergistic properties in the presence of AgNPs against *S. aureus* (Salarian et al. 2017). Green-synthesized NPs, in combination with other materials, have also been shown to exhibit synergistic effects in disease resistance (Begum and Jayawardana 2023). From the research, it has been confirmed that antibiotics synergistically halt the multiplication of *Salmonella typhimurium* when combined with the AgNPs. Four different classes of antibiotics, β-lactam (ampicillin and penicillin), quinolone (enoxacin), aminoglycoside (kanamycin and neomycin), and polypeptide (tetracycline) were explored for their synergistic mechanism. Enoxacin, kanamycin, neomycin, and tetracycline showed synergistic growth inhibition against the investigated microorganism when combined with AgNPs, while ampicillin and penicillin do not (Deng et al. 2016). The synergistic effect of ZnO-NPs was enhanced when combined with the antibiotics against *E. coli.* A similar effect of ZnO-NPs was observed against *S. aureus* when combined with azithromycin, cefotaxime, cefuroxime, fosfomycin, chloramphenicol, and oxytetracycline (Abo-Shama et al. 2020). Although the exact mechanism of the antibacterial action of ZnO-NPs is obscure. However, previous research suggests that generating ROS, such as hydrogen peroxide, contributes to disrupting bacterial cell walls and membranes (Xie et al. 2011).

There have been few reports on bimetallic NPs and their use in conjunction with antibiotics. Fakhri et al. (2017) synthesized Ag-Au NPs using a core–shell method and characterized them using different modern techniques. These Ag-Au NPs were combined with doxycycline and tested against *P. aeruginosa*, *E. coli*, *S. aureus*, and *Micrococcus luteus.* Combining bimetallic NPs with antibiotics showed promising results, demonstrating synergistic properties beneficial for burn healing therapy (Fakhri et al. 2017). An amalgam of polyol-coated CuO, ZnO, and CuZn NPs with meropenem and ciprofloxacin acts as an efflux pump inhibitor against multidrug-resistant strains of *P. aeruginosa*. These NPs act synergistically in a concentration-dependent manner, enhancing the efficacy of antibiotics affected by efflux mechanisms. At their lowest tested concentrations, all NPs demonstrated an increased antimicrobial activity of ciprofloxacin. This concluded the effectiveness of NPs as an effective source of efflux pump inhibitors to combat MDR (Eleftheriadou et al. 2021). Ranpariya et al. (2021) investigated bimetallic Ag-PtNPs in combination with antibiotics such as streptomycin, rifampicin, chloramphenicol, novobiocin, and ampicillin against *E. coli*, *P. aeruginosa*, and *S. aureus*. The study found that Ag-PtNPs exhibited greater inhibitory effects on all tested pathogens compared to individual AgNPs or PtNPs. Notably, combining rifampicin and novobiocin with Ag-PtNPs demonstrated significant activity against *S. aureus* (Ranpariya et al. 2021). These findings highlight the promising antimicrobial properties of bimetallic MNPs-conjugated antibiotics, suggesting their potential as novel therapeutic agents. The integration of vancomycin-carrying polydopamine with AgNPs boosts antibiotic efficacy against *S. aureus* and *E. coli* and reduces the required dosage, thereby minimizing the risk of developing drug resistance. In vivo testing demonstrates its bacteria-killing effectiveness and potential for wound healing in mouse skin defect models (Ma et al. 2020).

The functionalization of the MNP surface is typically a posterior step. The only reported MNPs using this method are AuNPs, AgNPs, and ZnO-NPs, with AuNPs being the most commonly used due to the easy functionalization with thiol groups. Brown et al. (2012) synthesized AgNPs and AuNPs stabilized with citrate, which were then functionalized with ampicillin. These ampicillin-functionalized NPs displayed bactericidal activity against both Gram-negative and Gram-positive bacteria. The resulting conjugates are emerging as powerful bactericidal agents with distinct properties that target the antibiotic resistance mechanisms of multidrug-resistant strains (Brown et al. 2012). De Oliveira et al. (2017) developed a customized protocol for synthesizing nano-antibiotics, optimizing the binding of the antibiotic to the surface of silver-silica core–shell NPs to enhance biological efficacy against MDR bacteria. The approach is quite promising for designing non-toxic functionalized NPs that can be used in antimicrobial, antiviral, or anti-cancerous therapies (De Oliveira et al. 2017).

Hur et al. (2014) described the one-step functionalization of Ag and Au NPs using ampicillin (Amp). Here, ampicillin eliminates the requirement of the conjugate, reducing agent, and stabilizing agent. Interestingly, the newly prepared NPs with the method exert remarkable antibacterial activity against *S. pyogenes* (Hur et al. 2014). A second-generation beta-lactam antibiotic in combination with NPs synthesized using ampicillin (reducing agent) with dual properties of antibiotic and NPs. The antibacterial activity of Amp-AgNPs was evaluated against both sensitive and antibiotic-resistant bacteria. The MIC values for Amp-AgNPs ranged from 3 to 28 μg/mL for six different bacterial strains, much lower than those for ampicillin (12–720 μg/mL) and AgNPs (280–640 μg/mL) alone. Notably, no bacterial resistance developed even after 15 consecutive exposure cycles. The MTT assay confirmed that

Amp-AgNPs were non-toxic, indicating their potential as a treatment for multidrug-resistant bacteria. (Khatoon et al. 2019). These advances and innovations in nano methodologies and research hold significant promise for addressing the urgent challenge of multidrug-resistant microorganisms, concreting the way for more effective and sustainable therapeutic options in the future.

## 5.5 Challenges and Limitations

The use of engineered nanomaterials in cosmetics, food packaging, drug delivery systems, therapeutics, and other applications has increased dramatically in recent years. They are widely used in a variety of commercial products, including wound dressings, detergents, and antimicrobial coatings for their antibacterial and odor-fighting qualities (Ray et al. 2009). However, engineered NPs' due to their large surface area and high particle density per unit mass, exhibit significant surface reactivity, potentially generating elevated levels of ROS, which can cause cytotoxicity and genotoxicity. Further, the use of inorganic/carbon-based NPs raises concerns about organ toxicity (Kumar et al. 2023). As these toxicity effects are unique and not typically observed in larger particles, the specialized field of nano-toxicology has emerged to address these concerns.

In nanomedicine research, the toxicity of nano-based treatments is a crucial concern, although they are very good at destroying particular cells or infections. These substances can have unpredictable interactions with biological systems (Xuan et al. 2023). The global health community is paying more attention to the ethical issues surrounding the development and use of nanomedicine as it advances. Assessing and controlling the risks associated with engineered NPs, as well as making sure that these risks are communicated clearly throughout clinical trials, are important issues. For a better evaluation of the toxicity, the following factors should be considered:

- Surface Area and Size
- Material Composition and Functionalization
- Administration Route and Dosage
- Activation of the Immune System
- Oxidative Stress and DNA Damage
- Biodegradability and Clearance
- Hazards of Environmental and Secondary Exposure

### 5.5.1 *Toxicity and Safety Concerns Associated with NPs*

NPs can infiltrate the human body via various routes, including inhalation through the lungs, ingestion through the gastrointestinal tract, or absorption through the skin (Lu et al. 2014; Singh et al. 2024). Once inside, these tiny particles have the potential to pose significant health risks owing to their unique size, high surface area-to-volume ratio, and chemical composition. It has been observed that certain NPs can trigger adverse effects such as lung inflammation, cardiac dysfunction, and even neurotoxicity (Xuan et al. 2023). Furthermore, their ability to penetrate biological barriers enables them to reach sensitive organs, where they may induce oxidative stress and cellular damage (Oberdoster et al. 2005, Kumah et al. 2023). Prolonged exposure to these particles can lead to permanent damage at the cellular and organ levels, exacerbating risks of chronic conditions and systemic toxicity. This highlights the need to understand their biocompatibility and potential health effects, particularly in industrial, environmental, and medical applications.

#### 5.5.1.1 Biocompatibility and Toxicity (Cellular and Organ)

For nano-based treatments, ensuring biocompatibility and reducing toxicity are crucial obstacles. NPs' small size allows them to enter tissues and cells that larger particles cannot, which raises concerns about cellular toxicity, including DNA damage, inflammation, and oxidative stress. The liver, kidneys, and lungs, organs that process or eliminate NPs, are especially susceptible to possible damage (Kyriakides et al. 2021; Kus-Liśkiewicz et al. 2021). To ensure they do not damage cells or cause unexpected immunological reactions, NPs must undergo extensive testing to prevent unfavorable biological reactions.

#### 5.5.1.2 Distribution and Accumulation in the Body

Slow-degrading or non-biodegradable NPs can build up in organs over time and cause chronic toxicity because they stay in tissues. Furthermore, NPs' retention and distribution within the body can be erratic, potentially reaching unexpected places like the brain, which can cause neural toxicity. Conventional MNPs exhibit rapid uptake and preferential accumulation in the liver and can be limited by the application of PEGylation (Hunt et al. 2022; Zeng et al. 2024). These elements emphasize the necessity of meticulous observation and planning to reduce the hazards connected to exposure to NPs.

#### 5.5.1.3 Dosage and Overexposure

Two essential aspects of the safety of nano-based treatments are exposure duration and dose-dependent toxicity. Higher doses of NPs can cause toxicity because of particle accumulation in the body, even though smaller doses may be safe. Because therapeutic and toxic doses can differ significantly depending on the type of NPs and their target, it can be difficult to establish safe dosing guidelines. Long-term exposure raises the risk of chronic toxicity, especially when administered repeatedly or cleared ineffectively, highlighting the necessity of cautious dosing and monitoring to reduce long-term negative effects (Damasco et al. 2020; Xuan et al. 2023).

#### 5.5.1.4 Clearance and Degradation Pathways

Particularly for non-biodegradable NPs, poor clearance and persistency can cause accumulation in organs like the liver and spleen, posing long-term safety concerns. While non-biodegradable options need careful consideration and lengthy safety studies, biodegradable NPs are frequently preferred because they decompose into non-toxic metabolites naturally eliminated from the body. Several studies revealed that the degradation process of some NPs can take a few days, and the excretion of mesoporous silica nanoparticles (MSNs) takes several weeks from the body (Liu et al. 2011; MacCuaig et al. 2022). Consequently, such a substantial number of NPs reside in the tissues/organs and cannot undergo further degradation or excretion, which may cause cytotoxicity and other side effects (Zhang et al. 2024).

### *5.5.2 Potential Environmental Impacts of Nanomaterials*

There are various ways that nanomaterials can appear in the environment: (i) purposefully created artificial nanomaterials that allow for exact control over size and characteristics; (ii) accidental nanomaterials produced by industrial processes like heat engine combustion and natural events like fires; and (iii) naturally occurring nanomaterials, such as viral NPs (Arora et al. 2024). Unquestionably, the environment and living things are impacted by the increasing production of NPs in many different sectors, which has an effect on people's health and well-being. Applications of nanotechnology in healthcare will impact disease diagnosis, prevention, and treatment over the course of the next few decades.

The potential environmental risks associated with the increased use of nanomedical products have become a concern, as pharmaceutical products (PPs) have been detected in various environmental compartments. Prolonged, low-level exposure to PPs leads to negative, and sometimes unexpected, effects on non-target organisms. Despite the potential benefits and usefulness of nanomedicine, there is a chance that its use could lead to toxicities through various mechanisms of action, such as different pharmacodynamics or toxicodynamics, different organ and cellular distributions, or different clearance rates (toxicokinetics or pharmacokinetics). Furthermore, the variety of nanomaterial compositions, sizes, surface coatings, and functionalities may lead to a range of toxic potentials (for example, in different cell types and organisms) (Bottini et al. 2011; Joseph et al. 2023). As NPs have dual characteristics, necessitating careful measures to minimize their adverse effects is crucial.

### 5.5.3 Regulatory and Ethical Considerations Surrounding Nanotherapeutics

Nanotherapeutics, which are characterized by their nanoscale size (ranging from a few nanometers to 250 nm), encompasses a wide range of constructs such as polymeric micelles, dendrimers, nanoparticles, and liposomes. These nanosystems are transforming drug delivery and diagnostics by enabling targeted treatment of diseased cells and minimizing systemic side effects, as demonstrated in chemotherapy (Ebbesen and Jensen 2006; Tewari et al. 2022). Their potential to improve treatments for conditions like cancer, diabetes, chronic infections, and autoimmune diseases highlights their importance, particularly in addressing challenges like antibiotic resistance (Ho et al. 2019). However, advancements in nanomedicine have also sparked ethical concerns, such as uncertainties about potential risks to human health and the environment, issues with the integrity of industry-funded research, and differing social acceptance of emerging technologies (Sheng et al. 2022).

In 2000, the US Government introduced the National Nanotechnology Initiative to support the responsible research and development of nanotechnology (National Nanotechnology Initiative 2021). India's New Drugs and Clinical Trials Rules, 2019, outline specific requirements for nanopharmaceuticals, including chemical, pharmaceutical, and clinical data related to nanotechnology (Ninawe et al. 2019). The rules emphasize case-by-case evaluations to address the complexities of nanoproducts, focusing on their unique manufacturing processes, controls, and waste disposal methods, as well as assessing the environmental impact of nanomaterial waste disposal. Specific US-FDA guidelines for "Liposomal Drug Products" (2018) complement these regulations, particularly for liposomal formulations. The European Union (2012) highlights the importance of tailored assessments for ensuring the safety, efficacy, and quality of nanoformulations, which often differ significantly from conventional drugs.

While a framework convention was proposed as an effective approach for governance and global collaboration, it lacks clear structure and mandatory participation requirements, limiting its practical application in addressing potential health risks (Wasti et al. 2023). Beyond safety, issues such as data security, patient privacy, and equitable access to care remain significant ethical concerns (Gagliardi et al. 2021). As regulatory frameworks struggle to keep pace with rapid advancements in nanomedicine, it is crucial for researchers and practitioners to address these challenges. Product descriptions for nanoformulations should include details on manufacturing processes, quality controls, and waste disposal methods in line with regulatory standards, highlighting the environmental impact of disposing of nanomaterial-containing waste.

## 5.6 Conclusion

Antibiotics have long been the cornerstone of modern medicine, playing a pivotal role in advancing healthcare and significantly reducing mortality rates from infectious diseases, but the rise of AMR has created an unprecedented global health threat. AMR arises from the ability of bacteria, fungi, and microorganisms to acquire and transmit resistance traits, primarily via ARB and mobile genetic elements like plasmids, transposons, and integrons. These elements facilitate the HGT of ARGs, accelerating the spread of resistance. NPs, with their unique physicochemical properties, hold great potential in enhancing the efficacy of antimicrobial treatments. Their nano size, surface area, and functionalization capabilities enable targeted delivery, and controlled, sustained release of antimicrobial agents. Their multifaceted antimicrobial actions include disrupting bacterial membranes, generating ROS, inhibiting efflux pumps, and interfering with metabolic pathways, which make them effective against MDR pathogens. Synthesized through eco-friendly methods using plant extracts, green nano colloids further added a rapid, scalable, and economically viable approach that eliminates the need for toxic chemicals. These NPs are not only effective against MDR pathogens but may also prove safer for humans, animals, and the environment.

Despite the promising potential of nanotechnology in antimicrobial treatments, there remain critical challenges that need to be addressed before widespread clinical applications can be realized. Key concerns include the environmental impact of NPs, as their long-term effects on ecosystems are not yet fully understood, even with the use of eco-friendly synthesis methods. Biocompatibility and toxicity are also critical, as NPs may cause oxidative stress or inflammation in humans and animals, necessitating thorough safety evaluations. Scalability and stability present additional hurdles, requiring methods to ensure consistent quality, prevent aggregation, and maintain functional properties during large-scale production and storage. Furthermore, cost-effective production and the establishment of regulatory frameworks are essential to make NP-based treatments accessible and viable in healthcare settings. Overcoming these challenges demands a multidisciplinary effort to ensure nanotechnology's safe and effective integration into medical practice.

In conclusion, while nanotechnology holds tremendous promise in addressing the growing threat of AMR, continued research is crucial to fully understand how NPs can impact bacterial viability and ensure their safe integration into clinical practice. With ongoing advancements in nanomedicine, there is a strong potential for NPs to revolutionize the treatment of infectious diseases, offering new solutions to one of the most pressing challenges in modern medicine.

## References


Abo-Shama UH, El-Gendy H, Mousa WS, Hamouda RA, Yousuf WE, Hetta HF, et al (2020) Synergistic and antagonistic effects of metal nanoparticles in combination with antibiotics against some reference strains of pathogenic microorganisms. Infect Drug Resist 351–362

Ahmadi S, Fazilati M, Mousavi SM, Nazem H (2020) Anti-bacterial/fungal and anti-cancer performance of green synthesized Ag nanoparticles using summer savory extract. J Exp Nanosci 15(1):363–380

Akhter S, Bhat MA, Hashem A, Abd_Allah EF, Ahmed S, Siddiqi WA, et al (2023) Profiling of antibiotic residues in soil and vegetables irrigated using pharmaceutical-contaminated water in the Delhi stretch of the Yamuna river, India. Water 15(23):4197

Ali I, Naz B, Liu Z, Chen J, Yang Z, Attia K et al (2024) Interplay among manures, vegetable types, and tetracycline resistance genes in rhizosphere microbiome. Front Microbiol 15:1392789

Anbumani D, vizhi Dhandapani K, Manoharan J, Babujanarthanam R, Bashir AKH, Muthusamy K, et al (2022) Green synthesis and antimicrobial efficacy of titanium dioxide nanoparticles using *Luffa acutangula* leaf extract. J King Saud Univ-Sci 34(3):101896

Andrews JB (2019) Printed carbon nanotube thin films for electronic sensing (Doctoral dissertation, Duke University)

Anzar N, Hasan R, Tyagi M, Yadav N, Narang J (2020) Carbon nanotube-A review on Synthesis, Properties and plethora of applications in the field of biomedical science. SensS Int 1:100003

Arora R, Roy T, Adak P (2024) A review of the impact of nanoparticles on environmental processes. In: BIO web of conferences, vol 86. EDP Sciences, p 01001

Avatsingh AU, Sharma S, Kour S, Arora Y, Sharma S, Joshi D et al (2023) Prevalence of antibiotic-resistant Gram-negative bacteria having extended-spectrum β-lactamase phenotypes in polluted irrigation-purpose wastewaters from Indian agro-ecosystems. Front Microbiol 14:1227132

Balkrishna A, Thakur N, Patial B, Sharma S, Kumar A, Arya V, Amarowicz R (2023) Synthesis, characterization and antibacterial efficacy of *Catharanthus roseus* and *Ocimum tenuiflorum*-mediated silver nanoparticles: phytonanotechnology in disease management. Processes 11(5):1479

Barber M, Rozwadowska-Dowzenko M (1948) Infection by penicillin-resistant staphylococci. Lancet 252:641–644

Behera SS, Patra JK, Pramanik K, Panda N, Thatoi H (2012) Characterization and evaluation of antibacterial activities of chemically synthesized iron oxide nanoparticles. World J Nanosci Eng 2(4)

Bhardwaj H, Jangde RK (2023) Current updated review on preparation of polymeric nanoparticles for drug delivery and biomedical applications. Next Nanotechnol 2:100013

Bhardwaj B, Singh P, Kumar A, Kumar S, Budhwar V (2020) Eco-friendly greener synthesis of nanoparticles. Adv Pharm Bull 10(4):566–576

Boakye YD, Osafo N, Danquah CA, Adu F, Agyare C (2019) Antimicrobial agents: antibacterial agents, anti-biofilm agents, antibacterial natural compounds, and antibacterial chemicals. Antimicrob, Antibiot Resist, Antibiofilm Strat Act Methods 13:75

Borehalli Mayegowda S, Roy A, NG M, Pandit S, Alghamdi S, Almehmadi M, et al (2023) Eco-friendly synthesized nanoparticles as antimicrobial agents: an updated review. Front Cell Infect Microbiol 13:1224778

Bottini M, Rosato N, Gloria F, Adanti S, Corradino N, Bergamaschi A, et al (2011) Public optimism towards nanomedicine. Int J Nanomedicine 3473–3485

Brown AN, Smith K, Samuels TA, Lu J, Obare SO, Scott ME (2012) Nanoparticles functionalized with ampicillin destroy multiple-antibiotic-resistant isolates of *Pseudomonas aeruginosa* and *Enterobacter aerogenes* and methicillin-resistant *Staphylococcus aureus*. Appl Environ Microbiol 78(8):2768–2774

Buffet-Bataillon S, Tattevin P, Bonnaure-Mallet M, Jolivet-Gougeon A (2012) Emergence of resistance to antibacterial agents: the role of quaternary ammonium compounds–a critical review. Int J Antimicrob Agents 39:381–389

Cela-Dablanca R, Barreiro A, Rodríguez-López L, Pérez-Rodríguez P, Arias-Estévez M, Fernández-Sanjurjo MJ et al (2022) Azithromycin adsorption onto different soils. Processes 10(12):2565
Chahardoli A, Karimi N, Sadeghi F, Fattahi A (2018) Green approach for synthesis of gold nanoparticles from *Nigella arvensis* leaf extract and evaluation of their antibacterial, antioxidant, cytotoxicity and catalytic activities. Artif Cells, Nanomedicine, Biotechnol 46(3):579–588
Chaudhary P, Singh P (2014) Isolation, identification and molecular characterization of microflora obtained from spices and spice mixes. World J Pharm Res 3(2):2020–2030
Chaudhary P, Singh P (2015) Fungicidal potency of *Cinnamomum tamala* leaves against common food borne pathogens. Int Res J Pharm 6(9):644–648
Chechetka SA, Zhang M, Yudasaka M, Miyako E (2016) Physicochemically functionalized carbon nanohorns for multi-dimensional cancer elimination. Carbon 97:45–53
Chen WY, Chang HY, Lu JK, Huang YC, Harroun SG, Tseng YT et al (2015) Self-assembly of antimicrobial peptides on gold nanodots: against multidrug-resistant bacteria and wound-healing application. Adv Func Mater 25(46):7189–7199
Chhibber S, Gondil VS, Sharma S, Kumar M, Wangoo N, Sharma RK (2017) A novel approach for combating *Klebsiella pneumoniae* biofilm using histidine functionalized silver nanoparticles. Front Microbiol 8:1104
Crofton J, Mitchison D (1948) Streptomycin resistance in pulmonary tuberculosis. BMJ 2:1009
Cycoń M, Mrozik A, Piotrowska-Seget Z (2019) Antibiotics in the soil environment—degradation and their impact on microbial activity and diversity. Front Microbiol 10:338
Dakal TC, Kumar A, Majumdar RS, Yadav V (2016) Mechanistic basis of antimicrobial actions of silver nanoparticles. Front Microbiol 7:1831
Damasco JA, Ravi S, Perez JD, Hagaman DE, Melancon MP (2020) Understanding nanoparticle toxicity to direct a safe-by-design approach in cancer nanomedicine. Nanomaterials 10(11):2186
Das B, Dash SK, Mandal D, Ghosh T, Chattopadhyay S, Tripathy S et al (2017) Green Synthesized silver nanoparticles destroy multidrug-resistant bacteria via reactive oxygen species mediated membrane damage. Arab J Chem 10(6):862–876
De Oliveira JFA, Saito Â, Bido AT, Kobarg J, Stassen HK, Cardoso MB (2017) Defeating bacterial resistance and preventing mammalian cells toxicity through rational design of antibiotic-functionalized nanoparticles. Sci Rep 7(1):1326
Deng H, McShan D, Zhang Y, Sinha SS, Arslan Z, Ray PC et al (2016) Mechanistic study of the synergistic antibacterial activity of combined silver nanoparticles and common antibiotics. Environ Sci Technol 50(16):8840–8848
Denissen J, Reyneke B, Waso-Reyneke M, Havenga B, Barnard T, Khan S et al (2022) Prevalence of ESKAPE pathogens in the environment: antibiotic resistance status, community-acquired infection and risk to human health. Int J Hyg Environ Health 244:114006
Dewi DAR, Götz B, Thomas T (2020) Diversity and genetic basis for carbapenem resistance in a coastal marine environment. Appl Environ Microbiol 86(10):e02939-e3019
Di Cesare A, Eckert EM, D'Urso S, Bertoni R, Gillan DC, Wattiez R et al (2016) Co-occurrence of integrase 1, antibiotic and heavy metal resistance genes in municipal wastewater treatment plants. Water Resour 94:208–214
Díaz-Puertas R, Álvarez-Martínez FJ, Falco A, Barrajón-Catalán E, Mallavia R (2023) Phytochemical-based nanomaterials against antibiotic-resistant bacteria: an updated review. Polymers 15(6):1392
Diksha D, Gupta SK, Gupta P, Banerjee UC, Kalita D (2023) Antibacterial potential of gold nanoparticles synthesized from leaf extract of *Syzygium cumini* against multidrug-resistant urinary tract pathogens. Cureus 15(2)
Doddapaneni SJDS, Amgoth C, Kalle AM, Suryadevara SN, Alapati KS (2018) Antimicrobial and anticancer activity of AgNPs coated with *Alphonsea sclerocarpa* extract. 3 Biotech 8(3):156
Dutta G, Guha N, Sugumaran A, Kamaruz Zaman M (2023) Carbon nanohorns in drug delivery and medical applications. Carbon Nanostructures in Biomedical Applications. Springer International Publishing, Cham, pp 95–121

Ebbesen M, Jensen TG (2006) Nanomedicine: techniques, potentials, and ethical implications. Biomed Res Int 2006(1):051516

Eleftheriadou I, Giannousi K, Protonotariou E, Skoura L, Arsenakis M, Dendrinou-Samara C et al (2021) Cocktail of CuO, ZnO, or CuZn nanoparticles and antibiotics for combating multidrug-resistant *Pseudomonas aeruginosa* via efflux pump inhibition. ACS Appl Nano Mater 4(9):9799–9810

Endale H, Mathewos M, Abdeta D (2023) Potential causes of spread of antimicrobial resistance and preventive measures in one health perspective-a review. Infect Drug Resist 7515–7545

Fakhri A, Tahami S, Naji M (2017) Synthesis and characterization of core-shell bimetallic nanoparticles for synergistic antimicrobial effect studies in combination with doxycycline on burn specific pathogens. J Photochem Photobiol, B 169:21–26

FAO (2018) Antimicrobial resistance in the environment summary report of an FAO meeting of experts. FAO Antimicrobial Resistance Working Group, Rome

Franzolin MR, Courrol DDS, Silva FRO, Courrol LC (2022) Antimicrobial activity of silver and gold nanoparticles prepared by photoreduction process with leaves and fruit extracts of *Plinia cauliflora* and *Punica granatum*. Molecules (Basel, Switzerland) 27(20):6860

Gagliardi A, Giuliano E, Venkateswararao E, Fresta M, Bulotta S, Awasthi V et al (2021) Biodegradable polymeric nanoparticles for drug delivery to solid tumors. Front Pharmacol 12:601626

Gao M, Jia R, Qiu T, Han M, Wang X (2017) Size-related bacterial diversity and tetracycline resistance gene abundance in the air of concentrated poultry feeding operations. Environ Pollut 220:1342–1348

Gao F, Shao T, Yu Y, Xiong Y, Yang L (2021) Surface-bound reactive oxygen species generating nanozymes for selective antibacterial action. Nat Commun 12(1):745

Gao M, Zhang X, Yue Y, Qiu T, Wang J, Wang X (2022) Air path of antimicrobial resistance related genes from layer farms: Emission inventory, atmospheric transport, and human exposure. J Hazard Mater 430:128417

Gdoura-Ben Amor M, Siala M, Zayani M, Grosset N, Smaoui S, Messadi-Akrout F et al (2018) Isolation, identification, prevalence, and genetic diversity of *Bacillus cereus* group bacteria from different foodstuffs in Tunisia. Front Microbiol 9:447

Girma A (2023) Alternative mechanisms of action of metallic nanoparticles to mitigate the global spread of antibiotic-resistant bacteria. The Cell Surface 10:100112

Goda RM, El-Baz AM, Khalaf EM, Alharbi NK, Elkhooly TA, Shohayeb MM (2022) Combating bacterial biofilm formation in urinary catheter by green silver nanoparticle. Antibiotics 11(4):495

Godoy-Gallardo M, Eckhard U, Delgado LM, de Roo Puente YJ, Hoyos-Nogués M, Gil FJ et al (2021) Antibacterial approaches in tissue engineering using metal ions and nanoparticles: from mechanisms to applications. Bioact Mater 6(12):4470–4490

Gordon O, Vig Slenters T, Brunetto PS, Villaruz AE, Sturdevant DE, Otto M et al (2010) Silver coordination polymers for prevention of implant infection: thiol interaction, impact on respiratory chain enzymes, and hydroxyl radical induction. Antimicrob Agents Chemother 54(10):4208–4218

Goryluk-Salmonowicz A, Popowska M (2022) Factors promoting and limiting antimicrobial resistance in the environment—existing knowledge gaps. Front Microbiol 13:992268

Gupta A, Mumtaz S, Li CH, Hussain I, Rotello VM (2019) Combatting antibiotic-resistant bacteria using nanomaterials. Chem Soc Rev 48(2):415–427

Hasnain MS, Nayak AK, Hasnain MS, Nayak AK (2019) Synthesis of carbon nanotubes. Carbon Nanotub Target Drug Deliv 17–20

Havenga B, Ndlovu T, Clements T, Reyneke B, Waso M, Khan W (2019) Exploring the antimicrobial resistance profiles of WHO critical priority list bacterial strains. BMC Microbiol 19:1–16

Hemeg HA (2017) Nanomaterials for alternative antibacterial therapy. Int J Nanomedicine 8211–8225

Hernando-Amado S, Coque TM, Baquero F, Martínez JL (2019) Defining and combating antibiotic resistance from one health and global health perspectives. Nat Microbiol 4(9):1432–1442

Hirata E, Takano Y, Konishi D, Maeda Y, Ushijima N, Yudasaka M et al (2023) An antibacterial conjugate of carbon nanohorns for NIR-light mediated peri-implantitis treatment. Chem Commun 59(73):11000–11003

Ho D, Wang P, Kee T (2019) Artificial intelligence in nanomedicine. Nanoscale Horiz 4(2):365–377

Hosnedlova B, Kabanov D, Kepinska M, Narayanan VHB, Parikesit AA, Fernandez C, et al (2022) Effect of biosynthesized silver nanoparticles on bacterial biofilm changes in *S. aureus* and *E. coli*. Nanomaterials 12:2183

Hubeny J, Korzeniewska E, Buta-Hubeny M, Zieliński W, Rolbiecki D, Harnisz M (2022) Characterization of carbapenem resistance in environmental samples and Acinetobacter spp. isolates from wastewater and river water in Poland. Sci Total Environ 822:153437

Huijbers PM, Blaak H, de Jong MC, Graat EA, Vandenbroucke-Grauls CM, de Roda Husman AM (2015) Role of the environment in the transmission of antimicrobial resistance to humans: a review. Environ Sci Technol 49(20):11993–12004

Hunt NJ, McCourt PA, Kuncic Z, Le Couteur DG, Cogger VC (2022) Opportunities and challenges for nanotherapeutics for the aging population. Front Nanotechnol 4:832524

Hur YE, Kim S, Kim JH, Cha SH, Choi MJ, Cho S et al (2014) One-step functionalization of gold and silver nanoparticles by ampicillin. Mater Lett 129:185–190

Hussen NH, Hasan AH, FaqiKhedr YM, Bogoyavlenskiy A, Bhat AR, Jamalis J (2024) Carbon dot based carbon nanoparticles as potent antimicrobial, antiviral, and anticancer agents. ACS Omega 9(9):9849–9864

Hwang IS, Hwang JH, Choi H, Kim KJ, Lee DG (2012) Synergistic effects between silver nanoparticles and antibiotics and the mechanisms involved. J Med Microbiol 61(12):1719–1726

Imran M, Das KR, Naik MM (2019) Co-selection of multi-antibiotic resistance in bacterial pathogens in metal and microplastic contaminated environments: an emerging health threat. Chemosphere 215:846–857

Ito F, Konuma K, Okamoto A (2001) Electron emission from single-walled carbon nanotubes with sharpened bundles. J Appl Phys 89(12):8141–8145

Jeevanandam J, Chan YS, Danquah MK (2020) Cytotoxicity and insulin resistance reversal ability of biofunctional phytosynthesized MgO nanoparticles. 3 Biotech 10(11):489

Jian Z, Zeng L, Xu T, Sun S, Yan S, Yang L et al (2021) Antibiotic resistance genes in bacteria: occurrence, spread, and control. J Basic Microbiol 61(12):1049–1070

Jin L, Xie J, He T, Wu D, Li X (2021) Airborne transmission as an integral environmental dimension of antimicrobial resistance through the "One Health" lens. Crit Rev Environ Sci Technol 52(23):4172–4193

Joo H, Eom H, Cho Y, Rho M, Song WJ (2023) Discovery and characterization of polymyxin-resistance genes pmrE and pmrF from sediment and seawater microbiome. Microbiol Spectr 11(1):e02736-e2822

Joseph TM, Kar Mahapatra D, Esmaeili A, Piszczyk Ł, Hasanin MS, Kattali M et al (2023) Nanoparticles: taking a unique position in medicine. Nanomaterials 13(3):574

Kang S, Mauter MS, Elimelech M (2009) Microbial cytotoxicity of carbon-based nanomaterials: implications for river water and wastewater effluent. Environ Sci Technol 43(7):2648–2653

Khatoon N, Alam H, Khan A, Raza K, Sardar M (2019) Ampicillin silver nanoformulations against multidrug resistant bacteria. Sci Rep 9(1):6848

Konishi D, Hirata E, Takano Y, Maeda Y, Ushijima N, Yudasaka M et al (2024) Near-infrared light-boosted antimicrobial activity of minocycline/hyaluronan/carbon nanohorn composite toward peri-implantitis treatments. Nanoscale 16(28):13425–13434

Krishnaraj C, Jagan EG, Rajasekar S, Selvakumar P, Kalaichelvan PT, Mohan NJCSBB (2010) Synthesis of silver nanoparticles using *Acalypha indica* leaf extracts and its antibacterial activity against water borne pathogens. Colloids Surf, B 76(1):50–56

Kumah EA, Fopa RD, Harati S, Boadu P, Zohoori FV, Pak T (2023) Human and environmental impacts of nanoparticles: a scoping review of the current literature. BMC Public Health 23:1059

Kumar MS, Sneha K, Yun YS (2010) Immobilization of silver nanoparticles synthesized using *Curcuma longa* tuber powder and extract on cotton cloth for bactericidal activity. Biores Technol 101(20):7958–7965
Kumar TS, Rahuman AA, Jayaseelan C, Rajakumar G, Marimuthu S, Kirthi AV et al (2014) Green synthesis of titanium dioxide nanoparticles using *Psidium guajava* extract and its antibacterial and antioxidant properties. Asian Pac J Trop Med 7(12):968–976
Kumar L, Bisen M, Harjai K, Chhibber S, Azizov S, Lalhlenmawia H et al (2023) Advances in nanotechnology for biofilm inhibition. ACS Omega 8(24):21391–21409
Kumar K, Kumar R, Jasrotia R, Kalia S, Arya V, Kumar A, Khargotra R, Singh T, Thakur N (2024) In-vitro bactericidal and anti-oxidant efficacy of biosynthesized $CuO/Cu_2O$-NiO nanocomposites against the pathogenic bacteria and DPPH free radical. Discov Appl Sci 6(3):87
Kus-Liśkiewicz M, Fickers P, Ben Tahar I (2021) Biocompatibility and cytotoxicity of gold nanoparticles: recent advances in methodologies and regulations. Int J Mol Sci 22(20):10952
Kyriakides TR, Raj A, Tseng TH, Xiao H, Nguyen R, Mohammed FS et al (2021) Biocompatibility of nanomaterials and their immunological properties. Biomed Mater 16(4):042005
Larsson DG, Flach CF (2022) Antibiotic resistance in the environment. Nat Rev Microbiol 20(5):257–269
Lee G, Yoo K (2022) A review of the emergence of antibiotic resistance in bioaerosols and its monitoring methods. Rev Environ Sci Bio/Technol 21(3):799–827
Lee JH, Park KS, Jeon JH, Lee SH (2018) Antibiotic resistance in soil. Lancet Infect Dis 18(12):1306–1307
Lee NY, Ko WC, Hsueh PR (2019) Nanoparticles in the treatment of infections caused by multidrug-resistant organisms. Front Pharmacol 10:1153
Lei Y, Nosoudi N, Vyavahare N (2014) Targeted chelation therapy with EDTA-loaded albumin nanoparticles regresses arterial calcification without causing systemic side effects. J Control Release 196:79–86
Lerminiaux NA, Cameron AD (2019) Horizontal transfer of antibiotic resistance genes in clinical environments. Can J Microbiol 65(1):34–44
Li M, Zhu L, Lin D (2011) Toxicity of ZnO nanoparticles to *Escherichia coli*: mechanism and the influence of medium components. Environ Sci Technol 45(5):1977–1983
Li LG, Yin X, Zhang T (2018) Tracking antibiotic resistance gene pollution from different sources using machine-learning classification. Microbiome 6:93
Li X, Wu Z, Dang C, Zhang M, Zhao B, Cheng Z et al (2021) A metagenomic-based method to study hospital air dust resistome. Chem Eng J 406:126854
Li P, Sun L, Xue S, Qu D, An L, Wang X, Sun Z (2022) Recent advances of carbon dots as new antimicrobial agents. SmartMat 3(2):226–248
Liang J, Li W, Chen J, Huang X, Liu Y, Zhang X et al (2021) Antibacterial activity and synergetic mechanism of carbon dots against Gram-positive and-negative bacteria. ACS Appl Bio Mater 4(9):6937–6945
Lilly M, Dong X, McCoy E, Yang L (2012) Inactivation of *Bacillus anthracis* spores by single-walled carbon nanotubes coupled with oxidizing antimicrobial chemicals. Environ Sci Technol 46(24):13417–13424
Liposome Drug Products: Chemistry, Manufacturing, and Controls; Human Pharmacokinetics and Bioavailability; and Labeling Documentation, Guidance for Industry: U.S. Department of Health and Human Services Food and Drug Administration, Center for Drug Evaluation and Research (CDER): April 2018 Pharmaceutical Quality/CMC. https://www.fda.gov/downloads/drugs/guidances/ucm070570.pdf
Liu T, Li L, Teng X, Huang X, Liu H, Chen D et al (2011) Single and repeated dose toxicity of mesoporous hollow silica nanoparticles in intravenously exposed mice. Biomaterials 32(6):1657–1668
Liu YY, Wang Y, Walsh TR, Yi LX, Zhang R, Spencer J et al (2016) Emergence of plasmid-mediated colistin resistance mechanism MCR-1 in animals and human beings in China: a microbiological and molecular biological study. Lancet Infect Dis 16(2):161–168

Liu Z, Zou H, Wang N, Yang T, Peng Z, Wang J et al (2018) Photoluminescence of carbon quantum dots: coarsely adjusted by quantum confinement effects and finely by surface trap states. Science China Chem 61:490–496

Liu ML, Chen BB, Li CM, Huang CZ (2019) Carbon dots: synthesis, formation mechanism, fluorescence origin and sensing applications. Green Chem 21(3):449–471

Lu X, Zhu T, Chen C, Liu Y (2014) Right or left: the role of nanoparticles in pulmonary diseases. Int J Mol Sci 15(10):17577–17600

Lu W, Wang M, Wu J, Jiang Q, Jin J, Jin Q et al (2020) Spread of chloramphenicol and tetracycline resistance genes by plasmid mobilization in agricultural soil. Environ Pollut 260:113998

Ma K, Dong P, Liang M, Yu S, Chen Y, Wang F (2020) Facile assembly of multifunctional antibacterial nanoplatform leveraging synergistic sensitization between silver nanostructure and vancomycin. ACS Appl Mater Interfaces 12(6):6955–6965

MacCuaig WM, Samykutty A, Foote J, Luo W, Filatenkov A, Li M et al (2022) Toxicity assessment of mesoporous silica nanoparticles upon intravenous injection in mice: implications for drug delivery. Pharmaceutics 14(5):969

Meng R, Zhu H, Wang Z, Hao S, Wang B (2022) Preparation of drug-loaded albumin nanoparticles and its application in cancer therapy. J Nanomater 2022(1):3052175

Mengesha SM, Abebe GM, Habtemariam TH (2024) Biosynthesis of CuO nanoparticle using leaf extracts of *Ocimum lamiifolium* Hochst. ex Benth and *Withana somnifera* (L) Dunal for antibacterial activity. Sci Rep 14(1):1–12

Mills MC, Lee J (2019) The threat of carbapenem-resistant bacteria in the environment: evidence of widespread contamination of reservoirs at a global scale. Environ Pollut 255:113143

Mondal SK, Chakraborty S, Manna S, Mandal SM (2024) Antimicrobial nanoparticles: current landscape and future challenges. RSC Pharm

More PR, Pandit S, Filippis AD, Franci G, Mijakovic I, Galdiero M (2023) Silver nanoparticles: bactericidal and mechanistic approach against drug resistant pathogens. Microorganisms 11(2):369

Mujahid MH, Upadhyay TK, Khan F, Pandey P, Park MN, Sharangi AB et al (2022) Metallic and metal oxide-derived nanohybrid as a tool for biomedical applications. Biomed Pharmacother 155:113791

Mukha IP, Eremenko AM, Smirnova NP, Mikhienkova AI, Korchak GI, Gorchev VF et al (2013) Antimicrobial activity of stable silver nanoparticles of a certain size. Appl Biochem Microbiol 49:199–206

Mukundan D, Mohankumar R, Vasanthakumari R (2017) Comparative study of synthesized silver and gold nanoparticles using leaves extract of *Bauhinia tomentosa* Linn and their anticancer efficacy. Bull Mater Sci 40:335–344

Murray CJ, Ikuta KS, Sharara F, Swetschinski L, Aguilar GR, Gray A et al (2022) Global burden of bacterial antimicrobial resistance in 2019: a systematic analysis. The Lancet 399(10325):629–655

Naghavi M, Vollset SE, Ikuta KS, Swetschinski LR, Gray AP, Wool EE et al (2024) Global burden of bacterial antimicrobial resistance 1990–2021: a systematic analysis with forecasts to 2050. The Lancet 404(10459):1199–1226

Nanomaterials: Commission proposes 'case by case approach' to assessment.—Second Regulatory Review on Nanopharmaceuticals, European Union, 2012. http://europa.eu/rapid/press-release_MEMO 12–732_en.htm

Nappier SP, Liguori K, Ichida AM, Stewart JR, Jones KR (2020) Antibiotic resistance in recreational waters: state of the science. Int J Environ Res Public Health 17(21):8034

National Nanotechnology Initiative (2021) Nano.gov. (Accessed April 12, 2021)

Neher TP, Soupir ML, Andersen DS, O'Neill ML, Howe A (2023) Comparison of antibiotic resistance genes in swine manure storage pits of Iowa, USA. Front Antibiot 2:1116785

New Drugs and Clinical Trials Rules 2019 G.S.R. 227(E), Government of India Gazette Notification dated 19th March 2019. https://cdsco.gov.in/opencms/opencms/system/modules/CDSCO.WEB/elements/download_file_division.jsp?num_id=NDI2MQ _ assessed on 21st June 2019.

Ninawe S, Tiwary D, Krishna AV (2019) Guidelines for Evaluation of Nanopharmaceuticals in India. In: Gupta YK, Dinda AK, Pradhan AK, Dhawan A, Ninawe S (eds) New Delhi: department of biotechnology government of india. retrieved from department of biotechnology government of India website: https://dbtindia.gov.in/sites/default/files/uploadfiles/Guidelines_For_Evaluation_of_Nanopharmaceuticals_in_India_24.10.19.pdf

Nogrado K, Unno T, Hur HG, Lee JH (2021) Tetracycline-resistant bacteria and ribosomal protection protein genes in soils from selected agricultural fields and livestock farms. Appl Biol Chem 64(1):42

Nyabadza A, McCarthy É, Makhesana M, Heidarinassab S, Plouze A, Vazquez M, Brabazon D (2023) A review of physical, chemical and biological synthesis methods of bimetallic nanoparticles and applications in sensing, water treatment, biomedicine, catalysis and hydrogen storage. Adv Colloid Interface Sci 103010

Oberdorster G, Oberdorster E, Oberdorster J (2005) Nanotoxicology: an emerging discipline evolving from studies of ultrafine particles. Environ Health Perspect 113:823–839

Padmavathy N, Vijayaraghavan R (2011) Interaction of ZnO nanoparticles with microbes—a physio and biochemical assay. J Biomed Nanotechnol 7(6):813–822

Pal C, Bengtsson-Palme J, Kristiansson E, Larsson DGJ (2015) Co-occurrence of resistance genes to antibiotics, biocides and metals reveals novel insights into their co-selection potential. BMC Genomics 16:964

Pärnänen KM, Narciso-da-Rocha C, Kneis D, Berendonk TU, Cacace D, Do T, et al (2019) Antibiotic resistance in European wastewater treatment plants mirrors the pattern of clinical antibiotic resistance prevalence. Sci Adv 5(3):eaau9124

Pelgrift RY, Friedman AJ (2013) Nanotechnology as a therapeutic tool to combat microbial resistance. Adv Drug Deliv Rev 65(13–14):1803–1815

Pepi M, Focardi S (2021) Antibiotic-resistant bacteria in aquaculture and climate change: a challenge for health in the Mediterranean area. Int J Environ Res Public Health 18(11):5723

Pernas-Pleite C, Conejo-Martínez AM, Marín I, Abad JP (2022) Green extracellular synthesis of silver nanoparticles by *Pseudomonas alloputida*, their growth and biofilm-formation inhibitory activities and synergic behavior with three classical antibiotics. Molecules (Basel, Switzerland) 27(21):7589

Peterson E, Kaur P (2018) Antibiotic resistance mechanisms in bacteria: relationships between resistance determinants of antibiotic producers, environmental bacteria, and clinical pathogens. Front Microbiol 9:2928

Qasim M, Lim DJ, Park H, Na D (2014) Nanotechnology for diagnosis and treatment of infectious diseases. J Nanosci Nanotechnol 14(10):7374–7387

Quirós P, Colomer-Lluch M, Martínez-Castillo A, Miró E, Argente M, Jofre J et al (2014) Antibiotic resistance genes in the bacteriophage DNA fraction of human fecal samples. Antimicrob Agents Chemother 58:606–609

Radzig MA, Nadtochenko VA, Koksharova OA, Kiwi J, Lipasova VA, Khmel IA (2013) Antibacterial effects of silver nanoparticles on Gram-negative bacteria: influence on the growth and biofilms formation, mechanisms of action. Colloids Surf, B 102:300–306

Raffi M, Hussain F, Bhatti TM, Akhter JI, Hameed A, Hasan MM (2008) Antibacterial characterization of silver nanoparticles against *E. coli* ATCC-15224. J Mater Sci Technol 24(2):192–196

Ramakrishna D, Rao P (2011) Nanoparticles: is toxicity a concern? Electron J Int Fed Clin Chem Lab Med 22(4):92

Ranjani S, Shariq Ahmed M, Ruckmani K, Hemalatha S (2020) Green nanocolloids control multi drug resistant pathogenic bacteria. J Cluster Sci 31:861–866

Ranpariya B, Salunke G, Karmakar S, Babiya K, Sutar S, Kadoo N et al (2021) Antimicrobial synergy of silver-platinum nanohybrids with antibiotics. Front Microbiol 11:610968

Ray PC, Yu H, Fu PP (2009) Toxicity and environmental risks of nanomaterials: challenges and future needs. J Environ Sci Health C 27(1):1–35

Reshad RAI, Jishan TA, Chowdhury NN (2021) Chitosan and its broad applications: a brief review. J Clin Exp Investig. SSRN 3842055

Revie NM, Iyer KR, Robbins N, Cowen LE (2018) Antifungal drug resistance: evolution, mechanisms and impact. Curr Opin Microbiol 45:70–76
Ribeiro AI, Dias AM, Zille A (2022) Synergistic effects between metal nanoparticles and commercial antimicrobial agents: a review. ACS Applied Nano Materials 5(3):3030–3064
Rietra RPJJ, Berendsen BJA, Mi-Gegotek Y, Römkens PFAM, Pustjens AM (2023) Prediction of the mobility and persistence of eight antibiotics based on soil characteristics. Heliyon 10(1)
Ross J, Topp E (2015) Abundance of antibiotic resistance genes in bacteriophage following soil fertilization with dairy manure or municipal biosolids, and evidence for potential transduction. Appl Environ Microbiol 81:7905–7913
Salarian AA, Mollamahale YB, Hami Z, Soltani-Rezaee-Rad M (2017) Cephalexin nanoparticles: Synthesis, cytotoxicity and their synergistic antibacterial study in combination with silver nanoparticles. Mater Chem Phys 198:125–130
Saleemi MA, Kong YL, Yong PVC, Wong EH (2022) An overview of antimicrobial properties of carbon nanotubes-based nanocomposites. Adv Pharm Bull 12(3):449–465
Salleh A, Naomi R, Utami ND, Mohammad AW, Mahmoudi E, Mustafa N et al (2020) The potential of silver nanoparticles for antiviral and antibacterial applications: a mechanism of action. Nanomaterials 10(8):1566
Samreen, Ahmad I, Malak HA, Abulreesh HH (2021) Environmental antimicrobial resistance and its drivers: a potential threat to public health. J Glob Antimicrob Resist 27:101–111
Santos-Lopez A, Marshall CW, Haas AL, Turner C, Rasero J, Cooper VS (2021) The roles of history, chance, and natural selection in the evolution of antibiotic resistance. Elife 10:e70676
Sardar M, Ahmed W, Al Ayoubi S, Nisa S, Bibi Y, Sabir M et al (2022) Fungicidal synergistic effect of biogenically synthesized zinc oxide and copper oxide nanoparticles against *Alternaria citri* causing citrus black rot disease. Saudi J Biol Sci 29(1):88–95
Schairer DO, Chouake JS, Nosanchuk JD, Friedman AJ (2012) The potential of nitric oxide releasing therapies as antimicrobial agents. Virulence 3(3):271–279
Shahriari M, Hemmati S, Zangeneh A, Zangeneh MM (2019) Biosynthesis of gold nanoparticles using *Allium noeanum* Reut. ex Regel leaves aqueous extract; characterization and analysis of their cytotoxicity, antioxidant, and antibacterial properties. Appl Organomet Chem 33(11):e5189
Sharma S, Devi M, Kumar P, Thakur N, Kumar K, Jeet K, Sharma A, Kumar S, Kumar A, Thakur N (2025) Photocatalytic, antibacterial and antioxidant study of *Vitex negundo* mediated green synthesized nickel and neodymium doped zinc oxide nanoparticles. Toxicol Environ Chem 107(1):178–206
Sheng J, Wu L, Ding H, Zhang Y, Gu N (2022) Ethics of nanomedicine. In: Gu N (ed) Nanomedicine. Micro/Nano Technologies. Springer, Singapore
Shukla RK, Badiye A, Vajpayee K, Kapoor N (2021) Genotoxic potential of nanoparticles: structural and functional modifications in DNA. Front Genet 12:728250
Simon-Deckers A, Loo S, Mayne-L'hermite M, Herlin-Boime N, Menguy N, Reynaud C, et al (2009) Size-, composition-and shape-dependent toxicological impact of metal oxide nanoparticles and carbon nanotubes toward bacteria. Environ Sci Technol 43(21):8423–8429
Singh KK, Chaudhary SK, Venugopal R, Gaurav A (2017) Bulk synthesis of multi-walled carbon nanotubes by AC arc discharge method. Proceedings of the Institution of Mechanical Engineers, Part N: J Nanomater, Nanoeng Nanosyst 231(3):141–151
Singh P, Pandit S, Mokkapati VRSS, Garnæs J, Mijakovic I (2020) A sustainable approach for the green synthesis of silver nanoparticles from *Solibacillus isronensis* sp. and their application in biofilm inhibition. Molecules 25(12):2783
Singh G, Thakur N, Kumar R (2024) Nanoparticles in drinking water: assessing health risks and regulatory challenges. Sci Total Environ 174940
Slavin YN, Asnis J, Hńfeli UO, Bach H (2017) Metal nanoparticles: understanding the mechanisms behind antibacterial activity. J Nanobiotechnology 15:1–20
Smekalova M, Aragon V, Panacek A, Prucek R, Zboril R, Kvitek L (2016) Enhanced antibacterial effect of antibiotics in combination with silver nanoparticles against animal pathogens. Vet J 209:174–179

Soliman MK, Salem SS, Abu-Elghait M, Azab MS (2023) Biosynthesis of silver and gold nanoparticles and their efficacy towards antibacterial, antibiofilm, cytotoxicity, and antioxidant activities. Appl Biochem Biotechnol 195(2):1158–1183

Son KH, Hong JH, Lee JW (2016) Carbon nanotubes as cancer therapeutic carriers and mediators. Int J Nanomedicine 5163–5185

Soppimath KS, Aminabhavi TM, Kulkarni AR, Rudzinski WE (2001) Biodegradable polymeric nanoparticles as drug delivery devices. J Control Release 70(1–2):1–20

Srinath BS, Namratha K, Byrappa K (2017) Eco-friendly synthesis of gold nanoparticles by gold mine bacteria Brevibacillus formosus and their antibacterial and biocompatible studies. IOSR J. Pharm 7(8):53–60

Stanley D, Batacan R Jr, Bajagai YS (2022) Rapid growth of antimicrobial resistance: the role of agriculture in the problem and the solutions. Appl Microbiol Biotechnol 106(21):6953–6962

Sun Z, Hong W, Xue C, Dong N (2024) A comprehensive review of antibiotic resistance gene contamination in agriculture: challenges and AI-driven solutions. Sci Total Environ 175971

Sunkar S, Nachiyar CV (2012) Biogenesis of antibacterial silver nanoparticles using the endophytic bacterium *Bacillus cereus* isolated from *Garcinia xanthochymus*. Asian Pac J Trop Biomed 2(12):953–959

Symochko L, Demyanyuk O, Symochko V, Grulova D, Fejer J, Mariychuk R (2023) The spreading of antibiotic-resistant bacteria in terrestrial ecosystems and the formation of soil resistome. Land 12(4):769

Taha BA, Ahmed NM, Talreja RK, Haider AJ, Al Mashhadany Y, Al-Jubouri et al (2024) Synergizing nanomaterials and artificial intelligence in advanced optical biosensors for precision antimicrobial resistance diagnosis. ACS Synth Biol 13(6):1600–1620

Teixeira-Santos R, Gomes M, Gomes LC, Mergulhao FJ (2021) Antimicrobial and anti-adhesive properties of carbon nanotube-based surfaces for medical applications: a systematic review. Iscience 24(1)

Tewari AK, Upadhyay SC, Kumar M, Pathak K, Kaushik D, Verma R et al (2022) Insights on development aspects of polymeric nanocarriers: the translation from bench to clinic. Polymers 14(17):3545

Thakur BK, Kumar A, Kumar D (2019) Green synthesis of titanium dioxide nanoparticles using *Azadirachta indica* leaf extract and evaluation of their antibacterial activity. S Afr J Bot 124:223–227

Thill A, Zeyons O, Spalla O, Chauvat F, Rose J, Auffan M, Flank AM (2006) Cytotoxicity of $CeO_2$ nanoparticles for Escherichia coli. Physico-chemical insight of the cytotoxicity mechanism. Environ Sci Technol 40(19):6151–6156

Tian M, He X, Feng Y, Wang W, Chen H, Gong M et al (2021) Pollution by antibiotics and antimicrobial resistance in livestock and poultry manure in China, and countermeasures. Antibiotics 10(5):539

Toleukhanuly YB, Kuttybaevna KM, Muratbekovna KZ, Turlangyzy NA (2020) Synthesis of carbon nanotubes by the electric arc-discharge method. Ser Chem Technol 5(443):126–133

Trouiller B, Reliene R, Westbrook A, Solaimani P, Schiestl RH (2009) Titanium dioxide nanoparticles induce DNA damage and genetic instability *in vivo* in mice. Can Res 69(22):8784–8789

Vadekeetil A, Chhibber S, Harjai K (2019) Efficacy of intravesical targeting of novel quorum sensing inhibitor nanoparticles against *Pseudomonas aeruginosa* biofilm-associated murine pyelonephritis. J Drug Target 27(9):995–1003

Vallet-Regí M, González B, Izquierdo-Barba I (2019) Nanomaterials as promising alternative in the infection treatment. Int J Mol Sci 20(15):3806

Veerasamy R, Xin TZ, Gunasagaran S, Xiang TFW, Yang EFC, Jeyakumar N, Dhanaraj SA (2011) Biosynthesis of silver nanoparticles using mangosteen leaf extract and evaluation of their antimicrobial activities. J Saudi Chem Soc 15(2):113–120

Verweij PE, Snelders E, Kema GHJ, Mellado E, Melchers WJG (2009) Azole resistance in *Aspergillus fumigatus*: a side-effect of environmental fungicide use? Lancet Infect Dis 9:789–795

Wahab S, Salman A, Khan Z, Khan S, Krishnaraj C, Yun SI (2023) Metallic nanoparticles: a promising arsenal against antimicrobial resistance—unraveling mechanisms and enhancing medication efficacy. Int J Mol Sci 24(19):14897

Wan G, Ruan L, Yin Y, Yang T, Ge M, Cheng X (2016) Effects of silver nanoparticles in combination with antibiotics on the resistant bacteria *Acinetobacter baumannii*. Int J Nanomedicine 3789–3800

Wang N, Yang X, Jiao S, Zhang J, Ye B, Gao S (2014) Sulfonamide-resistant bacteria and their resistance genes in soils fertilized with manures from Jiangsu Province Southeastern China. Plos One 9(11):e112626

Wang L, Hu C, Shao L (2017) The antimicrobial activity of nanoparticles: present situation and prospects for the future. Int J Nanomedicine 1227–1249

Wasti S, Lee IH, Kim S, Lee J-H, Kim H (2023) Ethical and legal challenges in nanomedical innovations: a scoping review. Front Genet 14:1163392

WHO (2024) Estimating the burden of foodborne diseases. https://www.who.int/activities/estimating-the-burden-of-foodborne-diseases

Wilson GJ, Perez-Zabaleta M, Owusu-Agyeman I, Kumar A, Ghosh A, Polya DA et al (2024) Discovery of sulfonamide resistance genes in deep groundwater below Patna. Environmental Pollution, India, p 124205

Xie Y, He Y, Irwin PL, Jin T, Shi X (2011) Antibacterial activity and mechanism of action of zinc oxide nanoparticles against *Campylobacter jejuni*. Appl Environ Microbiol 77(7):2325–2331

Xin R, Zhang K, Wu N, Zhang Y, Niu Z (2019) The pollution level of the blaOXA-58 carbapenemase gene in coastal water and its host bacteria characteristics. Environ Pollut 244:66–71

Xu Y, Li H, Li X, Liu W (2023) What happens when nanoparticles encounter bacterial antibiotic resistance? Sci Total Environ 876:162856

Xuan L, Ju Z, Skonieczna M, Zhou PK, Huang R (2023) Nanoparticles-induced potential toxicity on human health: applications, toxicity mechanisms, and evaluation models. MedComm 4(4):e327

Yan X, He B, Liu L, Qu G, Shi J, Hu L, Jiang G (2018) Antibacterial mechanism of silver nanoparticles in *Pseudomonas aeruginosa*: proteomics approach. Metallomics 10(4):557–564

Yao Y, Zhou Y, Liu L, Xu Y, Chen Q, Wang Y et al (2020) Nanoparticle-based drug delivery in cancer therapy and its role in overcoming drug resistance. Front Mol Biosci 7:193

Yin IX, Zhang J, Zhao IS, Mei ML, Li Q, Chu CH (2020) The antibacterial mechanism of silver nanoparticles and its application in dentistry. Int J Nanomedicine 2555–2562

Zalewska M, Błażejewska A, Szadziul M, Ciuchciński K, Popowska M (2024) Effect of composting and storage on the microbiome and resistome of cattle manure from a commercial dairy farm in Poland. Environ Sci Pollut Res 1–17

Zeng H, Lv Z, Sun X, Tong Y, Wu W, Dong S, Mao L (2024) Predicting bioaccumulation of nanomaterials: modeling approaches with challenges. Environ Health 2(4):189–201

Zhang Z, Yi G, Li P, Zhang X, Fan H, Zhang Y et al (2020) A minireview on doped carbon dots for photocatalytic and electrocatalytic applications. Nanoscale 12(26):13899–13906

Zhang Y, Lin X, Chen X, Fang W, Yu K, Gu W et al (2024) Strategies to regulate the degradation and clearance of mesoporous silica nanoparticles: a review. Int J Nanomed 19:5859

Zhou ZC, Liu Y, Lin ZJ, Yi Shuai X, Zhu L, Xu L et al (2021) Spread of antibiotic resistance genes and microbiota in airborne particulate matter, dust, and human airways in the urban hospital. Environ Int 153:106501

Zhou XY, Li H, Zhang YS, Su JQ (2023a) City-scale distribution of airborne antibiotic resistance genes. Sci Total Environ 856:159176

Zhou ZC, Shuai XY, Lin ZJ, Zheng J, Chen H (2023b) Comprehensive profiling and risk assessment of antibiotic resistance genes in a drinking water watershed by integrated analysis of air-water-soil. J Environ Manage 347:119092

Zhou S, Wang H, Li H, Yang Y, Shi D, Yang Z, et al (2024) Emergence of polymyxin-resistant *Yersinia enterocolitica* strains in natural aquatic environments. Environ Pollut 125341
Zhu S, Song Y, Zhao X, Shao J, Zhang J, Yang B (2015) The photoluminescence mechanism in carbon dots (graphene quantum dots, carbon nanodots, and polymer dots): current state and future perspective. Nano Res 8:355–381

# Chapter 6
# Nanocarriers for Controlled Release of Agrochemicals: Toward Cleaner and Smart Agriculture

**Ashwani Kumari, Anuj Choudhary, Ashwani Tapwal, and Anil Kumar**

**Abstract** The rising population has escalated the demand for food and land, necessitating the widespread use of agrochemicals to increase crop productivity. However, their extensive application has disrupted environmental equilibrium, led to the emergence of resistance in pests and pathogens, and posed risks to human health. In response to these issues, there is increasing interest in natural alternatives to synthetic agrochemicals. The natural products also face challenges, such as complex extraction processes and susceptibility to environmental fluctuations. Nanotechnology presents a promising solution by enabling the development of nano-carried agrochemicals, which enhance efficiency by improved targeted delivery and controlled release of natural/synthetic agrochemicals, thereby reducing the quantities of agrochemicals used. Additionally, these nanomaterials are considered comparatively safe for the environment. This innovative strategy has the potential to significantly enhance crop productivity without compromising environmental integrity and human health.

## 6.1 Introduction

The burgeoning population pressures intensify demands for land, driving agricultural expansion, deforestation, and increased resource consumption (FAO 2019). Subsequently, disrupting environmental balance further exacerbates these trends (Machado et al. 2022). The third agricultural revolution, widely known as the "Green Revolution" in the 1960s, laid the foundation for undeniable growth in agricultural productivity by incorporating high-yield varieties, modern equipment, and agrochemicals

A. Kumari (✉) · A. Tapwal
ICFRE—Himalayan Forest Research Institute, Shimla, India
e-mail: akchandel395@gmail.com

A. Choudhary
Department of Biology and Environmental Sciences, CSKHPKV, Palampur, India

A. Kumar
School of Chemical and Metallurgical Engineering, University of the Witwatersrand, Johannesburg, South Africa

 
A. Balkrishna et al. (eds.), *Nanotechnology in Environmental Science and Healthcare*,
https://doi.org/10.1007/978-981-95-6714-0_6

such as pesticides and fertilizers. Ultimately, the crop yield per unit land has been enhanced, and over the years, the global food system has relied more on the use of agrochemicals alone for improving crop productivity (Evenson And Gollin 2003; Stevenson et al. 2013). Agrochemicals refer to all chemicals required to protect plants from diseases, pests, and weeds, or to deliver fertilizers and other essential compounds. Agrochemicals are classified into fertilizers, herbicides, insecticides, and fungicides, and chemicals within each category can have various modes of action. The excessive use of these agrochemicals led to their accumulation in soil, resulting in significant losses due to runoff, pollution of adjacent water bodies, and the volatilization of gases such as NH3 and N2O into the atmosphere, thereby contributing to climate change (Park et al. 2012; Arora 2019). The long-term effects of frequent insecticide and fungicide applications have further led to resistance in fungal pathogens and insect pests, worsening the scenario (Fisher et al. 2012; Ishi and Holloman 2015). The indiscriminate use of agrochemicals has compromised environmental sustainability, leading to soil degradation, increased soil and water pollution, accumulation in the food chain, and a risk to human health (Babu et al. 2013; Lechenet et al. 2017; Brauer et al. 2019).

To mitigate the above-mentioned drawbacks of synthetic agrochemicals, the utilization of natural products and their alternatives has gained attention. However, the path of natural product-based agrochemicals has substantial challenges. Firstly, among several natural sources of agrochemicals, cyanobacteria have more potential as fertilizers, plant-derived compounds have more potential as insecticides, while fungi-based compounds are more potent as fungicides, while all of them have comparatively lesser herbicidal effect. Secondly, the process of extraction and determination of their action mechanism with particular beneficial activity to crops is difficult and rare event. Additionally, majority of bio-based agrochemicals have varied success rates in different environmental conditions, have short shelf-life period, photo instability or storage issues which make them less suitable for agriculture crop protection. Hence, although natural agrochemicals themselves might not be ideal as such, with certain novel modifications, these compounds could be broadly utilized in modern agriculture.

In this context, the emerging science of nanotechnology could provide a promising solution to the aforementioned drawbacks of agrochemicals by developing safer, more durable, and highly efficient nano-agrochemicals (Xin et al. 2020). Out of all the nanotechnological tools, encapsulation of agrochemicals as respective nanocarriers having controlled release systems could serve as adequate preparations in agriculture (Fig. 6.1). These systems improve the overall uptake and assimilation by the plants in small quantities in comparison with their counterparts (Parisi et al. 2015; Kah et al. 2019; Kumar et al. 2019). Based on the ideas presented in the following sections of this chapter, we will further explore various types of nanocarriers, their mode of action, applications as nanofungicides and nanoherbicides, and the challenges associated with their commercialization and field administration.

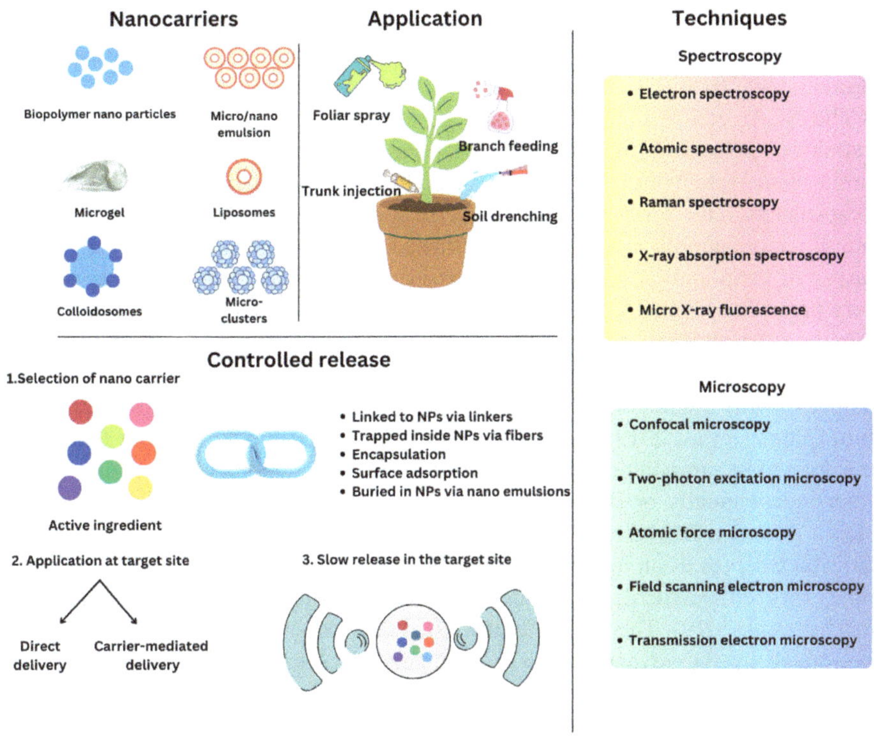


**Fig. 6.1** Schematic illustration depicting administration and characteristics of different nanocarriers

## 6.2 Nanocarriers: Controlled Release Systems in Precision Agriculture

Due to the enormous scope of nanotechnology, a diverse range of materials with at least one dimension in the nanoscale, i.e., 1–100 nm, have secured their place in boosting the agriculture sector (Kumari And Tapwal 2023). Several types of nanoparticles (NPs), including metallic NPs, metal oxide NPs, polymeric NPs, nanocarriers, and nanoemulsions, have been utilized in the agricultural sector as nanofungicides, nanofertilizers, nanopesticides, and nanoherbicides (Kumari et al. 2024a, b). Most of the structural characteristics of these nanomaterials can be determined using the spectroscopic and microscopic techniques mentioned in Fig. 6.1. According to Xin et al. (2020), three major classes of nano-agrochemicals are nanoscale agrochemicals, nanoscale additives, and nano-carried agrochemicals (NCAs). Nanoscale agrochemicals are generally nanosized metal/ metal oxide particles or mixtures containing essential nutrients or pesticides, designed to benefit crops (Kumari et al. 2024a, b). The examples under this category are nano-apatite, nano-manganese, nano-calcite, etc. Certain nanomaterials with single and multiwalled carbon tubes or titanium dioxide nanoparticles, when added to conventional fertilizers or pesticides enhance their uptake by crops, such type of nanomaterials are known as nanoscale additives (Hamdi et al. 2015; Rizwan et al. 2019).

The NCAs can be defined as nanoscale delivery systems designed to transport pesticides, herbicides, fungicides, growth regulators, fertilizers, etc. to specific sites within plant systems or environment (Choudhary et al. 2019). These can be either encapsulated in the form of emulsions or as polymeric nanoparticles, or carried by natural nanomaterials such as zeolite, chitosan, and clay. Among all the mentioned classes of nano-agrochemicals, nano-carried agrochemicals stand out due to their more precise application as controlled release systems (CRSs) (Karny et al. 2018; Choudhary et al. 2019; Grillo et al. 2021). The term CRSs refers to the efficient and slow release of active ingredients in plants and soil, allowing for targeted action against specific organisms while minimizing human exposure and ecosystem alteration (Kumar et al. 2019; Shao et al. 2022). The release process from CRSs can be triggered by environmental factors such as pH, temperature, and moisture levels, either independently or in combination (An et al. 2022). CRSs promote the slow release of active ingredients/agrochemicals due to differences in chemical potential, which serves as an instigating force for mass transfer based on diffusion or, in some cases, enzymatic degradation, thereby accelerating the release of such actives (Fischer et al. 2019). The initial motivation to develop strategies for CRSs has been accelerated by the advancements in nanomaterials for medicine delivery (Huang et al. 2018; Mahmoudpour et al. 2021). The size of NCAs may vary from 1 to 1000 nm, and their synthesis is aimed at enhancing the solubility, stability, and bioavailability of agrochemicals, i.e., the active ingredients of plant benefits (Nuruzzaman et al. 2016). Additional benefits associated with NCAs include enhanced penetration, dispersibility, increased efficacy with reduced inputs, and savings on excessive labor costs, contributing to precision agriculture (Wu et al. 2022; Fincheira et al. 2023). Furthermore, NCAs facilitate the mitigation of negative environmental impacts, which can result in a 30% increase in crop production (Rajput et al. 2021). In general, the NCAs can be classified under different categories.

### *6.2.1 Types of Nanocarriers*

NCAs can be classified by means of various criteria such as degradation nature, chemical composition, structural characteristics, form of active ingredients, methods of synthesis, functionality, and application types (Agrawal et al. 2014; Hayles et al. 2017; Usman et al. 2020). In this study, the most important types of NCAs have been discussed in the following part, and their respective applications have been elaborated in the upcoming sections of the article.

- **Nanoemulsions and Nano-lipid Carriers**

In general, emulsions are mixtures of two liquids that usually do not mix well, and one liquid (dispersed phase) remains suspended within another liquid (continuous phase). Based on the basic composition principle of emulsions, various agriculturally important macroemulsions (400–1000 nm), microemulsions (100–400 nm), and nanoemulsions (1–100 nm) can be synthesized (Gupta et al. 2024). For the preparation of nanoemulsions, oils, surfactants, co-surfactants, and gelling agents, etc., can be used as the continuous medium, while the dispersed phase is usually the agrochemical/ active ingredient of interest. The smallest size of nanoemulsions provides them with better stability and effectiveness compared to macro- and microemulsions (McClements And Rao 2011). Depending upon the properties of active ingredient, nanoemulsions can be used for management of weeds, fungal pathogens, and insect pests (Mustafa And Hussein 2020; Somala et al. 2022; Wang et al. 2022).

Nano-lipid carriers are new age formulations fabricated by use of solid lipid nanoparticles or a mixture of solid lipid nanoparticles and oils (Campos et al. 2015). They are claimed to have advantage over classical nanoemulsions as they have lower leakage of entrapped active ingredients and offers better control of the size and release process (Katouzian et al. 2017). The most common methods for the synthesis of nanoemulsions and NLCs include hot and cold homogenization, high-pressure homogenization, melting dispersion, double emulsification, and solvent emulsification-diffusion (Wani et al. 2019). Colloidosomes are a type of carrier which are used in nanoemulsions and NLCs, which have a size usually ranging from 100 to 1000 nm (Gupta et al. 2024).

- **Nanoliposomes**

Nanoliposomes are nanosized, spherical liposomal molecules having amphiphilic polar lipids arranged in a bilayer structure. In this configuration, hydrophilic heads of lipids face outward and their hydrophobic tails are oriented inward, away from the water. This arrangement creates a compartment where hydrophilic active ingredients can be contained within the aqueous core, while hydrophobic compounds are encapsulated within the inner lipid layer (Ghorbanzade et al. 2017). The most common methods employed for the synthesis of nanoliposomes are thin film hydration, the solvent injection method, the detergent removal method, and the reverse-phase method (Demirci et al. 2017).

- **Polymer-Based Nanocarriers**

Polymeric nanocarriers have been least explored in the field of crop protection; however, polymeric compounds have been utilized for various soil analyses and diagnostics in plants (Duhan et al. 2017). Among PBNs, the most utilized ones are dendrimers, which are generally synthesized by repetitive addition of monomers from a central polyfunctional core, having multiple functional groups. Addition of monomers occurs in a dendritic pattern, providing more sites for further addition (Keshawani et al. 2014; Chandra et al. 2017). With emerging applications of nanotechnology in agriculture, new terminologies have been introduced at a fast rate. Some other important PBNs include nano-coatings or surface coatings, nanocapsules and micro/nanogels, etc. Nano-coatings are nanomaterials of any type that are added to active ingredients of agriculture importance (Ghormade et al. 2011). These coatings exhibit higher surface tension compared to conventional coatings and allowed to adherence of active ingredients more effectively which helps to retain them over time and as CRSs (Riyajan 2017). Nanospheres are aggregate materials of nano-size having active agrochemicals homogenously arranged into the polymeric matrix (Perlatti et al. 2013) and similar are nanocapsules which have actives concentrated in the central core, limed by polymer matrix (Machado et al. 2022). Similarly, nanogels are other class of polymer-based hydrophilic and cross-linked nanomaterials which have the capacity of absorbing high volumes of water (Iavicoli et al. 2017).

Current trends are focused on the use of biopolymer-based NCAs, which utilize biopolymers like chitosan, alginate, cellulose, hemicellulose, lignin, polypeptides, etc., that have proved their importance as nanocarriers within precision agriculture in comparison with synthetic polymers by harboring interesting features like biocompatibility and biodegradability (Machado et al. 2022). These are the most promising CRSs, conferring high selectivity and effectiveness compared to conventional and other classes of NCAs. These NCAs can be synthesized using ionic gelation, miniemulsion, and cross-linking with surfactants, among other methods (Choudhary et al. 2017).

- **Niosomes**

Niosomes are self-assembled, nanoscale vesicles, i.e., bubble-like structures enclosed by a lipid membrane. Their synthesis involves the hydration of surfactants, along with cholesterol or similar amphiphilic molecules. They are efficient in carrying both water-soluble and fat-soluble agrochemicals, such as nanoliposomes, and contain non-ionic surfactants, making them effective for targeted delivery to specific sites with enhanced encapsulation efficiency and biocompatibility (Marianecci et al. 2014).

## 6.3 Mechanism of Action

The administration of NCAs to crops can be achieved through several methods that leverage their unique properties and enhance their specificity. Most conventional methods are soil drenching, seed coatings, and foliar spray (Kumari et al. 2024a, b). While most innovative methods involve aerial applications using drones, trunk injections, and branch feedings, among others, which are suitable for the targeted delivery of NCAs (Singh et al. 2021). While NCAs can be applied to crops using various methods discussed above, their mechanisms of action can differ significantly based on their composition and intended purpose. For instance, NCAs loaded with fertilizers and those containing pesticides can both be administered via foliar application, but they may function through distinct mechanisms. Fertilizer-based NCAs may penetrate plant tissues, enhancing physiological processes such as photosynthesis and subsequent accumulation of biomass and yield (Nile et al. 2022). On the other hand, NCAs formulated with pesticides may primarily adhere to the surface of plant leaves, forming a protective barrier against insect pests (Lucas et al. 2008; Babu et al. 2013; Beckers et al. 2019). Additionally, there are instances where NCAs with fertilizers may adhere to roots, influencing the uptake of minerals, while some pesticide-loaded NCAs can enter plant cells and induce the production of metabolites that might act as feeding deterrents for pests. Therefore, while there are general mechanisms of action that could be applied to all NCAs, such as enhanced delivery, controlled release, and increased bioavailability, the specific effects depend on the nature of the active ingredients involved. Hence, understanding the general mechanisms of action of NCAs is crucial for optimizing their specific mode of action.

The first step in the general mechanism of action is the penetration of NCAs within plant cells, which can be achieved either through natural openings, such as stomata, or by the direct transfer of smaller nanoparticles through cell walls and other membranes. Some NCAs can enter cells via the lipid exchange envelope penetration model, which enables nanocarriers to easily penetrate cell membranes and chloroplast membranes (Giraldo et al. 2014; Wong et al. 2016). When NCAs enter through stomata, they are transported via phloem. Conversely, when they enter through roots, they follow xylem transport to move through the plant system (Wang et al. 2022; Kumari et al. 2024a, b). Some NCAs may attach to cell walls, where they can play a role in enhancing structural integrity or influencing the entry of pathogens. Others may bind to cell membranes, potentially affecting membrane permeability and signaling pathways, as illustrated in Fig. 6.2. Certain NCAs can actually penetrate cells and act as modulators or inhibitors of various signaling pathways (Molnar et al. 2020). Such compounds can influence hormone signaling, stress responses, or metabolic pathways, and thus impact overall plant health and productivity (Garcia et al. 2015). Some NCAs can activate processes responsible for generating reactive oxygen species (ROS). ROS are widely recognized signaling molecules in plants that regulate various physiological processes during infection or infestation (Wang et al. 2022). Additionally, some NCAs can stimulate the function of specific organelles, such as chloroplasts, mitochondria, and ribosomes (Xu et al. 2008; Wang et al. 2022; Machado et al. 2022). Enhanced chloroplast activity can improve photosynthetic efficiency, increase energy production, and enhance yields (Xu et al. 2008). Ultimately, these enhancements contribute to the overall enhancement of production of agricultural crops, making NCAs valuable tools in modern agriculture for application as pesticides, fungicides, fertilizers, and herbicides as discussed in next sections.

## 6.4 NCAs as Nanopesticides

The major problems associated with synthetic pesticides are their effects on non-target organisms, the development of resistance in insect pests and pathogens, bioaccumulation, and soil and water pollution caused by runoff. While the limitations associated with biopesticides include short shelf life and instability. In this sense, nanotechnology-based pesticides, commonly referred to as nanopesticides, can be a viable alternative (Xin et al. 2020). One such type of nanopesticides belongs to NCAs, in which the compound, either synthetic or of biological origin, having an inhibitory effect against pests, is carried on certain nanomaterials called nanocarriers, which

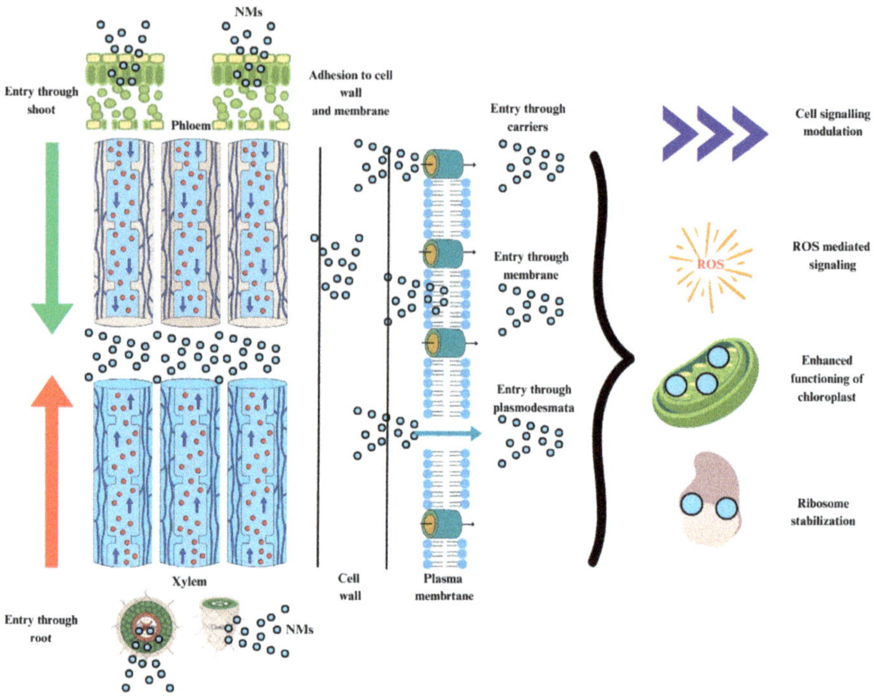


**Fig. 6.2** General mechanism of action of NCAs displaying their take up and targeted action

serve as CRSs for the over-time delivery of the desired compounds. NCAs with pesticidal activity exhibit controlled release kinetics, enhanced solubility permeability, and stability. Also, improved target affinity and enhanced dispersal range allows for reduced dose of such nanopesticides (Mishra et al. 2021). In this regard, a number of nanoformulations have been synthesized and tested in vitro against several pests as summarized in Table 6.1.

NLCs based pesticide having size less than 100 nm, carrying essential oils of *Ziziphora clinopodioides* was reported to effectively manage insect pest *Tribolium castaneum*, with a smaller $LC_{50}$ value of 30.602 μL. L $air^{-1}$ of the synthesized formulation, in comparison with pure essential oils alone ($LC_{50}$ 68.303 μL/L air) (Hosseinpour et al. 2020). Nanoliposomes can be effective management strategies against agricultural insect pests. A formulation of nanoliposomes containing the extracts of *Eleocharis dulcis* corm peels was reported to have appreciable aphidicidal activity against *Megoura crassicauda* ($LC_{50}$ 83.99 mg/L) and *Acyrthosiphon pisum* ($LC_{50}$ 154.34 mg/L) (Li et al. 2024). Similarly, β-caryophyllene nanoliposomes were found to effectively manage agricultural insect pests, such as Bemisia tabaci ($LC_{50}$ 1.51 g/L), with no short-term impact on the overall health of the plant (Gao et al. 2024).

**Table 6.1** Enumeration of different NCAs for various agricultural applications

| Category | Nanocarrier nature | Agrochemicals/ Active ingredient | Size (nm) | Target organisms/Crops/ Vegetation | References |
|---|---|---|---|---|---|
| Nanofungicides | *Aloe vera* gel | Zinc oxide | 45–60 | *A. alternata*, *Fusarium oxysporum* | Yadav et al. (2018) |
| | Cetyl trimethyl lammonium bromide (CTAB) | Copper (II) | 3–10 | *A. alternata*, *Curvularia lunata*, *F. oxysporum*, *Phoma destructiva* | Kanhed et al. (2014) |
| | Chitosan | Pepper tree essential oil | 20–100 | *Aspergillus parasiticus* | Xing et al. (2009) |
| | Chitosan | Avermectin | 175.2–245.8 | *Mgnaparthe grisea* | Chen et al. (2021) |
| | Hydroxyapatite | Copper (II) | – | *Plasmopara viticola* | Battiston et al. (2018) |
| | Lignin | Pyraclostrobin | 162.4 | *F. oxysporum* | Mustafa and Hussein (2020) |
| | Polylactide | Validamycin thifluzamide | 260 | *Rhizoctonia solani* | Saini et al. (2021) |
| | Polyvinylpyridine and polyvinylpyridine-costyrene | Tebucanazole | 100–250 | *Gloephyllum trabeum* | Liu et al. (2001) |
| | Propylene glycol | Clove and lemongrass oil | 76.73 | *F. oxysporum* | DeCastro et al. (2019) |
| Nanopesticides | Carboxymethyl chitosan | Azidobenzaldehyde | 98.6 | Armyworm larvae | Sun et al. (2014) |
| | Chitosan functionalized with β-cyclodextrin | Carvacrol linalool | 175.2–245.8 | *Tetranychus urticae* | Liang et al. (2018) |
| | Graphene oxide | Copper selenide | 20 | *Pieris rapae* | Sharma et al. (2017) |
| | Percirol ATO5 and campritol 888 | Essential oil of *Z. clinopodiodes* | 241.1 | *T. castaneum* | Keskin et al. (2021) |
| | Poly (citric acid) and poly (ethylene glycol) dendrimer | Imidacloprid | 10–20 | *Glyphodes pyloalis* | Memarizadeh et al. (2014) |

(continued)

**Table 6.1** (continued)

| Category | Nanocarrier nature | Agrochemicals/ Active ingredient | Size (nm) | Target organisms/Crops/ Vegetation | References |
|---|---|---|---|---|---|
| | Prochloraz loaded with PEG-PLGA | Essential oil of *Zanthoxylum rhoifolium* | 500 | *Bemisia tabaci* | Zhang et al. (2016) |
| | Solid lipid | Gamma cyhalothrin | – | *Dysdercus cingulatus* and *Spodoptera littoralis* | Frederiksen et al (2003) |
| | Tween 80 | Essential oil of *Mentha piperita* | 20–60 | Cotton aphid | Sharma et al. (2018) |
| | Zein | Essential oil of citronella | 142.5–172.3 | *T. urticae* | DeOliveria et al. (2019) |
| Nanofertilizers | Amorphous calcium phosphate | *Glomus mossae*, *Piriformospora indica* | 88 | *Z. mays* | Perez et al. (2020) |
| | Chitosan | Copper | 361.3 | *Z. mays* | Ha et al. (2019) |
| | Chitosan | Potassium | 39–79 | *Z. mays* | Dhlamini et al. (2020) |
| | Chitosan | Urea | – | *Solanum tuberosum* | Pereira et al. (2016) |
| | Hydroxyapatite | Urea | 15–20 | *Oryza sativa* | Pradhan et al. (2020) |
| | Hydroxyapatite | Urea, Cu NPs, Fe NPs, and Zn NPs | 39.76 | *Ablemoschus esculentus* | Yoon et al. (2020) |
| | Mesoporous silica NPs | Auxin | 40–60 | *L. usitatissimum* | Naseem et al. (2020) |
| | Nanoclays | Phosphate | 20 | *Hordeum vulgare, Triticum aestivum, Z. mays* | Figueredo et al. (2017); Bernardo et al. (2018); Songkhum et al. (2018) |

(continued)

**Table 6.1** (continued)

| Category | Nanocarrier nature | Agrochemicals/ Active ingredient | Size (nm) | Target organisms/Crops/ Vegetation | References |
|---|---|---|---|---|---|
| Nanoherbicides | Chitosan and tripolyphospahte | Paraquat | 420 | Weeds growing with *Brassica* sp. and with *Z. mays* | Grillo et al. (2014) |
| | Poly (ξ-caprolactone) | Atrazine | 483.1 | *Brassica* sp. Growing as weed in *Z. mays* field | Pereira et al. (2014) |
| | Tween 60 and Span 60 | Essential oil from *C. nardus* | 79 | *Echinichloa crus-galli* | Somala et al. (2022) |
| | Tween 80 | Essential oil of *Satureja hortensis* | < 130 | *Amaranthus retroflexus, Chenopodium album* | Hazrati et al. (2017) |

## 6.5 NCAs as Nanofungicides

Over the years, various types of nanofungicides have been synthesized and tested against several fungal pathogens (Chauhan And Tapwal 2023; Kumari And Tapwal 2023). NCAs formulated to carry actives/agrochemicals that reduce or inhibit the growth of fungi are focused in this part (Table 6.1). Sun et al. (2014) fabricated nanocapsules containing azidobenzaldehyde agrochemical, carried by carboxymethyl chitosan, with greater efficiency, and achieved 100% mortality of army larvae within a short period of 1 week. Nanoemulsion composed of pulegone was fabricated by Golden et al. (2018) to control *Sitophilus oryzae* and *T. castaneum,* which are storage pests of grains. High stability and prolonged biological activity, due to the controlled release of the pulegone nanoemulsion, effectively managed both pests, with mortality rates above 90% (Golden et al. 2018). These NCAs with fungicidal or fungistatic activity have been known to have prolonged stabilization against adverse environmental conditions. As confirmed by NLCs carrying chitosan-coated beeswax and deltamethrin, they exhibit enhanced stability against photodegradation compared to their bulk counterparts (Nguyen et al. 2012). Additionally, there are no phytotoxic effects of electrospun cellulose diacetate nanofibers on the germination rate of soybean seeds, with enhanced resistance against Alternaria alternata, providing strong evidence of the non-phytotoxic effects of NCAs (Bruneau et al. 2019).

## 6.6 NCAs as Nanofertilizers

The agricultural production must increase by 60–70% to meet the needs of increasing populations. This increase in crop productivity is not possible with conventional modes of fertilization which have low nutrient use efficiency and 50–80% of them were lost to environment. To address these losses, highly efficient nanofertilizer formulations are most promising (FAO 2019; Candido et al. 2019; Zulfiqar et al. 2019). Nanofertilizers have 20–30% higher nutrient use efficiency compared to traditional fertilizers, and they enhance nutrient uptake in roots by improving soil nutrient management, reducing nutrient immobilization, and runoff (Raliya et al. 2017; Mikula et al. 2020). Research communities worldwide have shown significant interest in nanofertilizers for precision agriculture (Table 6.1). Urea is the most commonly used fertilizer; therefore, the synthesis of NCAs based on urea has been a common interest among researchers (Pereira et al. 2016; Pradhan et al. 2020). Chitosan-urea nanocomposites have been reported to increase soil dehydrogenase activity and organic carbon content (Pereira et al. 2016). Moreover, the slow release of urea from this composite in the potato field has benefited both crop and soil microbes (Pereira et al. 2016). Furthermore, chitosan encapsulation of nitrogen, phosphorus, and potassium (NPK) has resulted in enhanced photosynthetic rates, nutrient uptake, and growth of *Coffea arabica* (Kubavat et al. 2020). Additionally, another class of fertilizer NCAs having mesoporous silica nanoparticles (MSNs) was reported to significantly boost embryogenesis, chlorophyll content, seed germination, etc. in *Zea mays*, *Linum usitatissimum*, and *Oryza sativa* (Sun et al. 2014; Rane et al. 2015; Kokina et al. 2017; Naseem et al. 2020).

## 6.7 NCAs as Nanoherbicides

The most recent application of nanomaterials is to control or kill weeds, defined as plants undesirable in agriculture, gardening, and landscaping areas, interfering with desired crops (Li et al. 2007; Kremer 2019). Various NCAs with herbicidal activity have been synthesized and tested in recent years (Table 6.1). In a study, Pereira et al. (2014) reported that encapsulating atrazine in poly(ε-caprolactone) resulted in effective herbicidal activity against *Brassica* species with no detrimental effect on *Z. mays*. Similarly, Somala et al. (2022) successfully fabricated nanoemulsions using essential oils from *Cymbopogon nardus* and surfactant mixture of Tween 60 and Span 60 with an average size of 79 nm. Appreciable herbicidal activities against *Echinichloa crus-galli* have been recorded, by lowering seed imbibition and α-amylase activity which are essential for seed germination. The fabrication of nanoherbicides presents significant challenges, as it requires careful long-term monitoring (Bombo et al. 2019). It is inevitable to test various doses to determine the optimal amounts, since a dose that is effective against weeds may inadvertently affect desirable plants once applied in the field.

## 6.8 Constraints and Challenges

Despite the numerous benefits discussed above, the commercial use of NCAs remains a highly debated topic due to their potential long-term environmental impacts and risks to human health. The toxicological implications of NCAs are complex and multifaceted, affecting various biological systems differently depending on numerous factors, including exposure route, concentration, and material properties (Meirmoeini et al. 2021; Szymanski And Dobrucka 2021). If administered without prior assessment of their potential negative impacts, NCAs can have possible detrimental effects on soil microbial communities, plants, animals, and humans, potentially impacting the entire ecosystem (Goswami et al. 2024). The introduction of NCAs into soil systems can have adverse effects on microbial populations through mechanisms such as ROS generation, DNA damage, and membrane disruption (Srikanth 2020). Consequently, the biodegradation of litter and debris may be hindered, resulting in decreased soil fertility. Furthermore, their bioaccumulation within ecosystems can result in long-term ecological consequences. The increasing use of NCAs to boost crop production raises concerns about their probable accumulation in the consumable parts of crops (Song And Lee 2016; Xiao et al. 2019; Bishta et al. 2022). Through this accumulation, they can enter the alimentary canal, circulatory system, or any other part of the human body, posing risks of long-term health effects (Qamar et al. 2024). Hence, research communities should focus on establishing proper guidelines and standard protocols for assessing all toxicological impacts discussed above to ensure their safe utilization.

## 6.9 Conclusion

NCAs offer significant benefits for delivering agrochemicals to crops by enhancing their efficacy, enabling controlled and slow release, targeted delivery, and minimizing unnecessary leaching into the environment. Encapsulating active ingredients within NCAs provides a small size, high surface-area to volume ratio, for enhanced activity, and protection from premature degradation and reducing run off losses. Additionally, NCAs also enable direct efficient application methods, such as foliar spraying or trunk injection, addressing challenges related to soil penetration and microbial degradation in the soil. Furthermore, these formulations prevent chemical runoff during irrigation and adverse weather conditions, as well as accumulation in the environment, thereby minimizing risks to soil health and human safety without compromising crop productivity.

To promote more eco-friendliness, biopolymer-based NCAs integrate agrochemicals into natural biogeochemical cycles, supporting sustainability by utilizing waste materials from biorefineries and the wood pulping industry, thus contributing to the circular economy. However, the widespread use of NCAs is currently limited by the lack of comprehensive regulatory frameworks necessary for their commercialization. Therefore, establishing robust regulatory committees and guidelines to evaluate the potential toxicity and long-term impacts of NCAs is crucial for streamlining commercialization processes and ensuring their safe and practical application.

## References

Agrawal S, Rathore P (2014) Nanotechnology pros and cons to agriculture: A review. Int J Curr Microbiol Appl Sci, 3(3), 43–55. https://doi.org/10.13140/2.1.1648.1926

An C, Sun C, Li N, Huang B et al (2022) Nanomaterials and nanotechnology for the delivery of agrochemicals: Strategies towards sustainable agriculture. Journal of Nanobiotechnology 20:11. https://doi.org/10.1186/s12951-021-01214-7

Arora NK (2019) Impact of climate change on agriculture production and its sustainable solutions. Environ Sustain 2(2):95–96. https://doi.org/10.1007/s42398-019-00078-w

Babu RP, Oconnor K, Seeram R (2013) Current progress on bio-based polymers and their future trends. Prog Biomater 2:1–16. https://doi.org/10.1186/2194-0517-2-8

Battiston E, Salvatici MC, Lavacchi A et al. (2018) Functionalization of a nanostructured hydroxyapatite with copper (II) compounds as pesticide: in situ TEM and ESEM observations of treated *Vitis vinifera* L. leaves. Pest Management. http://refhub.elsevier.com/B978-0-12-815322-2.00012-2/rf0045

Beckers SJ, Dallo IA, Del Campo I et al (2019) From compost to colloids—valorization of spent mushroom substrate. ACS Sustain Chem & Eng 7(7):6991–6998. https://doi.org/10.1021/acssuschemeng.8b06710

Bernardo MP, Guimarães GGF, Majaron VF, Ribeiro C (2018) Controlled release of phosphate from LDH structures: Dynamics in soil and application as smart fertilizer. ACS Sustain Chem & Eng 6:5152–5161. https://doi.org/10.1021/acssuschemeng.7b04806

Bishta S, Sharma V, Kumari N (2022) Biosynthesized magnetite nanoparticles from *Polyalthia longifolia* leaves improve photosynthetic performance and yield of *Trigonella foenum-graecum* under drought stress. Plant Stress 5:100090. https://doi.org/10.1016/j.stress.2022.100090

Bombo AB, Pereira AES, Lusa MG et al (2019) A mechanistic view of interactions of a nanoherbicide with target organism. J Agric Food Chem 67(16):4453–4462. https://doi.org/10.1021/acs.jafc.9b00806

Brauer VS, Rezende CP, Pessoni AM et al (2019) Antifungal agents in agriculture: friends and foes of public health. Biomolecules 9(10):521. https://doi.org/10.3390/biom9100521

Bruneau M, Bennici S, Brendle J et al (2019) Systems for stimuli-controlled release: Materials and applications. J Control Release 294:355–371. https://doi.org/10.1016/j.jconrel.2018.12.038

Campos E, Oliveira J, da Silva C et al (2015) Polymeric and solid lipid nanoparticles for sustained release of carbendazim and tebuconazole in agricultural applications. Sci Rep 5:13809. https://doi.org/10.1038/srep13809

Candido CM, Campos REV, Monteiro RA et al (2019) Development of stimuli-responsive nano-based pesticides: Emerging opportunities for agriculture. J Nanobiotechnology, 17, 100. https://doi.org/10.1186/s12951-019-0533-8

Chandra S, Mayer M, Baeumner AJ (2017) PAMAM dendrimers: A multifunctional nanomaterial for ECL biosensors. Talanta, 168, 126–129. https://doi.org/10.1016/j.talanta.2017.03.016

Chauhan P, Tapwal A (2023) Green synthesis of nanoparticles using botanicals and their application in management of fungal phytopathogens—A review. Arch Microbiol 205:94. https://doi.org/10.1007/s00203-023-03431-0

Chen H, Zhi H, Liang J et al (2021) Development of leaf-adhesive pesticide nanocapsules with pH-responsive release to enhance retention time on crop leaves and improve utilization efficiency. J Mater Chem B 9:783–792. https://doi.org/10.1039/D0TB02430A

Choudhary RC, Kumaraswamy RV, Kumari S et al (2019) Zinc encapsulated chitosan nanoparticle to promote maize crop yield. Int J Biol Macromol 127:126–135. https://doi.org/10.1016/j.ijbiomac.2018.12.274

Choudhary RC, Kumaraswamy RV, Kumari S, et al (2017) Cu-chitosan nanoparticle boost defense responses and plant growth in maize (Zea mays L.). Scientific Reports, 7(1), 9754. https://doi.org/10.1038/s41598-017-08571-0

DeCastro SP, Silva Pereira LA, de Rezende EM et al (2019) Production and efficacy of neem nanoemulsion in the control of *Aspergillus flavus* and *Penicillium citrinum* in soybean seeds. Eur J Plant Pathol 155:1105–1116. https://doi.org/10.1007/s10658-019-01838-4

Demirci M, Caglar MY, Cakir B, Gülseren İ (2017) Encapsulation by nanoliposomes. Nanoencapsulation Technol Food Nutraceutical Ind 74–113. https://doi.org/10.1016/B978-0-12-809436-5.00003-3

DeOliveira JL, Campos EVR, Germano-Costa G et al (2019) Association of zein nanoparticles with botanical compounds for effective pest control systems. Pest Manag Sci 75:1855–1865. https://doi.org/10.1002/ps.5338

Dhlamini B, Kamdem Paumo H, Katata-Seru L, Kutu FR (2020) Sulphate-supplemented NPK nanofertilizer and its effect on maize growth. Materials Research Express 7:095011. https://doi.org/10.1088/2053-1591/abb69d

Duhan JS, Kumar R, Kumar N et al (2017) Nanotechnology: The new perspective in precision agriculture. Biotechnology Reports 15:11–23. https://doi.org/10.1016/j.btre.2017.03.002

Evenson RE, Gollin D (2003) Assessing the impact of the green revolution, 1960 to 2000. Science 300:758–762. https://doi.org/10.1126/science.1078710

FAO (2019) World Fertilizer Trends and Outlook to 2022; FAO: Rome, Italy

Figueredo BB, Leopoldo CVR, Garcia Pinto F et al (2017) Layered double hydroxides: New technology in phosphate fertilizers based on nanostructured materials. ACS Sustain Chem & Eng 5:399–409. https://doi.org/10.1021/acssuschemeng.6b01784

Fincheira P, Hoffmann N, Tortella G et al (2023) Eco-efficient systems based on nanocarriers for the controlled release of fertilizers and pesticides: Toward smart agriculture. Nanomaterials 13(13):1978. https://doi.org/10.3390/nano13131978

Fischer J, Beckers SJ, Yiamsawas D et al (2019) Targeted drug delivery in plants: enzyme-responsive lignin nanocarriers for the curative treatment of the worldwide grapevine trunk disease Esca. Adv Sci 6(15):1802315. https://doi.org/10.1002/advs.201802315

Fisher MC, Henk DA, Briggs CJ et al (2012) Emerging fungal threats to animal, plant and ecosystem health. Nature 484(7393):186–194. https://doi.org/10.1038/nature10947

Frederiksen, H.K., Kristensen, H.G., and Pedersen, M. (2003). Solid lipid microparticle formulations of the pyrethroid gamma–cyhalothrin–incompatibility of the lipid and the pyrethroid and biological properties of the formulations. J Control Release, 86, 243–252. http://refhub.elsevier.com/B978-0-12-815322-2.00012-2/rf0120.

Gao L, Li K, Zhang Z, Chen J et al (2024) β-Caryophyllene wrapped by nanoliposomes efficiently increases the control effect on *Bemisia tabaci* MED. Pestic Biochem Physiol 204:106082. https://doi.org/10.1016/j.pestbp.2024.106082

Garcia SS, Bernales I, Cristobal S (2015) Early response to nanoparticles in the *Arabidopsis* transcriptome compromises plant defence and root-hair development through salicylic acid signalling. BMC Genomics 16:1–17. https://doi.org/10.1186/s12864-015-1530-4

Ghorbanzade T, Jafari SM, Akhavan S, Hadavi R (2017) Nano-encapsulation of fish oil in nano-liposomes and its application in fortification of yogurt. Food Chem 216:146–152. https://doi.org/10.1016/j.foodchem.2016.08.022

Ghormade V, Deshpande MV, Paknikar KM (2011) Perspectives for nano-biotechnology enabled protection and nutrition of plants. Biotechnol Adv 29:792–803. https://doi.org/10.1016/j.biotechadv.2011.06.007

Giraldo JP, Landry MP, Faltermeier SM et al (2014) Plant nanobionics approach to augment photosynthesis and biochemical sensing. Nat Mater 13(4):400–408. https://doi.org/10.1038/nmat3890

Golden G, Quinn E, Shaaya E, Kostyukovsky M, Poverenov E (2018) Coarse and nano emulsions for effective delivery of the natural pest control agent pulegone for stored grain protection. Pest Manag Sci 74:820–827. https://doi.org/10.1002/ps.4787

Goswami P, Kumar A, Arya V, Arya B, Kumar J, Thakur AK, Rao GP (2024) Dynamic applied interactions amid nanoparticles, beneficial soil microorganisms and phytopathogens. In Nanotechnol: Agric, Environ Health (pp. 53–74). Singapore: Springer Nature Singapore. https://doi.org/10.1007/978-981-97-6814-1_2

Grillo R, Mattos BD, Antunes DR, Forini MM (2021) Foliage adhesion and interactions with particulate delivery systems for plant nanobionics and intelligent agriculture. Nano Today 37:101078. https://doi.org/10.1016/j.nantod.2021.101078

Grillo R, Pereira AES, Nishisaka CS, de Lima R et al (2014) Chitosan/tripoly phosphate nanoparticles loaded with paraquatherbicide: an environmentally safer alternative for weed control. J Hazard Mater, 278, 163–171. http://refhub.elsevier.com/B978-0-12-815322-2.00012-2/rf0155

Gupta R, Malik P, Rani R, Solanki R et al (2024) Recent progress on nanoemulsions mediated pesticides delivery: Insights for agricultural sustainability. Plant Nano Biology 100073. https://doi.org/10.1016/j.plana.2024.100073

Ha NMC, Nguyen TH, Wang SL, Nguyen AD (2019) Preparation of NPK nanofertilizer based on chitosan nanoparticles and its effect on biophysical characteristics and growth of coffee in green house. Res Chem Intermed 45:51–63. https://doi.org/10.1007/s11164-018-3630-7

Hamdi H, De LaTorre-Roche R, Hawthorne J, White JC (2015) Impact of non-functionalized and amino-functionalized multiwall carbon nano tubes on pesticide uptake by lettuce (*Lactuca sativa* L.). Nanotoxicology 9, 172–180. https://doi.org/10.3109/17435390.2014.907456.

Hayles J, Johnson L, Worthley C, Losic D (2017) Nanopesticides: a review of current research and perspectives. New Pestic Soil Sens 193–225. https://doi.org/10.1016/b978-0-12-804299-1.00006-0

Hazrati H, Saharkhiz MJ, Niakousari M, Moein M (2017) Natural herbicide activity of *Satureja hortensis* L. essential oil nanoemulsion on the seed germination and morphophysiological features of two important weed species. Ecotoxicol Environ Saf, 142, 423–430. http://refhub.elsevier.com/B978-0-12-815322-2.00012-2/rf0175.

Hosseinpour JF, Moravvej G, Modarres Awal M, Golmohammadzadeh S (2020) Insecticidal activity of solid lipid nanoparticle loaded by *Ziziphora clinopodioides* Lam. against *Tribolium castaneum* (Herbst, 1797) (Coleoptera: Tenebrionidae). Int J Pest Manag, 67(2), 147–154. https://doi.org/10.1080/09670874.2020.1713420

Huang B, Chen F, Shen Y, Qian et al. (2018) Advances in targeted pesticides with environmentally responsive controlled release by nanotechnology. Nanomaterials, 8(2), 102. https://doi.org/10.3390/nano8020102.

Iavicoli I, Leso V, Beezhold DH, Shvedova AA (2017) Nanotechnology in agriculture: Opportunities, toxicological implications, and occupational risks. Toxicol Appl Pharmacol 329:96–111. https://doi.org/10.1016/j.taap.2017.05.025

Ishii H, Holloman DW (2015) Fungicide resistance in plant pathogens. Springer. https://doi.org/10.1007/978-4-431-55642-8

Kah M, Tufenkji N, White JC (2019) Nano-enabled strategies to enhance crop nutrition and protection. Nat Nanotechnol 14(6):532–540. https://doi.org/10.1038/s41565-019-0439-5

Kanhed P, Birla S, Gaikwad S, Gade A, Seabra AB, Rubilar O, Duran N, Rai M (2014) *In vitro* antifungal efficacy of copper nanoparticles against selected crop pathogenic fungi. Materials Letters 115, 13–17. http://refhub.elsevier.com/B978-0-12-815322-2.00012-2/rf0210

Karny A, Zinger A, Kajal A, Shainsky-Roitman J, Schroeder A (2018) Therapeutic nanoparticles penetrate leaves and deliver nutrients to agricultural crops. Sci Rep 8(1):1–10. https://doi.org/10.1038/s41598-018-25197-y

Katouzian I, Esfanjani AF, Jafari SM, Akhavan S (2017) Formulation and application of a new generation of lipid nano–carriers for the food bioactive ingredients. Trends Food Sci Technol 68, 14–25. http://refhub.elsevier.com/B978-0-12-815322-2.00012-2/rf0220.

Kesharwani P, Jain K, Jain NK (2014) Dendrimer as nanocarrier for drug delivery. Prog Polym Sci 39(2):268–307. https://doi.org/10.1016/j.progpolymsci.2013.07.005

Keskin D, Zu G, Forson AM et al (2021) Nanogels. A novel approach in antimicrobial delivery systems and antimicrobial coatings. Bioactive Materials 6:3634–3657. https://doi.org/10.1016/j.bioactmat.2021.03.004

Kokina I, Jahundovica I, Mickevica I, et al. (2017) Target transportation of auxin on mesoporous Au/$SiO_2$ nanoparticles as a method for somaclonal variation increasingin flax (*L. usitatissimum* L.). Journal of Nanomaterials, 7143269. https://doi.org/10.1155/2017/7143269

Kremer RJ (2019) Bioherbicides and nanotechnology: Current status and future trends. Nano-Biopestic Today Futur Perspect 353–366. https://doi.org/10.1016/B978-0-12-815829-6.00015-2

Kubavat D, Trivedi K, Vaghela P, Prasad K et al (2020) Characterization of a chitosan-based sustained release nanofertilizer formulation used as a soil conditioner while simultaneously improving biomass production of *Zea mays* L. Land Degrad Dev 31:2734–2746. https://doi.org/10.1002/ldr.3629

Kumar S, Nehra M, Dilbaghi N, Marrazza G et al (2019) Nano-based smart pesticide formulations: Emerging opportunities for agriculture. J Control Release 294:131–153. https://doi.org/10.1016/j.jconrel.2018.12.012

Kumari A, Kumar A, Seth A, Guleria I, Arya V, Shahi SK (2024) Nanotechnology intervention for sustainable agriculture: challenges and possibilities. In *Nanotechnology:* Agriculture, Environment and Health (pp. 1–51). Singapore: Springer Nature Singapore. https://doi.org/10.1007/978-981-97-6814-1_1

Kumari A, Tapwal A, Kumar A (2024). Influence of Nanoparticles in Orchestrating Plant Growth and Development. In: Nanotechnology: Agriculture, Environment and Health (pp. 75–92). Singapore: Springer Nature Singapore. https://doi.org/10.1007/978-981-97-6814-1_3

Kumari A, Tapwal A (2023) *Trichoderma*: A beneficial antimycotic agent and insights on its utilization in nanoparticle synthesis. BioNanoScience. https://doi.org/10.1007/s12668-023-011 70-6

Lechenet M, Dessaint F, Py G, Makowski D, Munier-Jolain N (2017) Reducing pesticide use while preserving crop productivity and profitability on arable farms. Nature Plants 3(3):1–6. https://doi.org/10.1038/nplants.2017.8

Li M, Li J, Meng Y, Wang Y, Gao M, Dong J et al (2024) Preparation of nanoliposomes containing the extracts of Eleocharis dulcis corm-peels and ascertaining their aphidicidal activity against *Megoura crassicauda* and *Acyrthosiphon pisum*. Ind Crops Prod 207:117746. https://doi.org/10.1016/j.indcrop.2023.117746

Li ZZ, Chen JF, Liu F, Liu AQ, Wang Q, Sun HY, Wen LX (2007) Study of UV–shielding properties of novel porous hollow silica nanoparticle carriers for avermectin. Pest Manag Sci 63, 241–246. http://refhub.elsevier.com/B978-0-12-815322-2.00012-2/rf0235

Liang W, Yu A, Wang G, Zheng F, Hu P, Jia J, Xu H (2018) A novel water-based chitosan-La pesticide nanocarrier enhancing defense responses in rice (*Oryza sativa* L.) growth. Carbohyd Polym 199:437–444. https://doi.org/10.1016/j.carbpol.2018.07.042

Liu Y, Yan I, Heiden P, Laks P (2001) Use of nanoparticles for controlled release of biocides in solid wood. J Appl Polym Sci, 79, 458–465. http://refhub.elsevier.com/B978-0-12-815322-2.00012-2/rf0245

Lucas N, Bienaime C, Belloy C, Queneudec M, Silvestre F, Nava-Saucedo JE (2008) Polymer biodegradation: Mechanisms and estimation techniques–A review. Chemosphere 73(4):429–442. https://doi.org/10.1016/j.chemosphere.2008.06.064

Machado TO, Grabow J, Sayer C, de Araújo PH et al (2022) Biopolymer-based nanocarriers for sustained release of agrochemicals: A review on materials and social science perspectives for a sustainable future of agri-and horticulture. Adv Coll Interface Sci 303:102645. https://doi.org/10.1016/j.cis.2022.102645

Mahmoudpour M, Karimzadeh Z, Ebrahimi G, Hasanzadeh M, Dolatabadi ENJ (2021) Synergizing functional nanomaterials with aptamers based on electrochemical strategies for pesticide detection: Current status and perspectives. CritAl Rev Anal Chem, 52(8), 1818–1845. 1080/10408347.2021.1919987

Marianecci C, Di Marzio L, Rinaldi F, Celia C et al (2014) Niosomes from 80s to present: the state of the art. Adv Coll Interface Sci 205:187–206. https://doi.org/10.1016/j.cis.2013.11.018

McClements DJ, Rao J (2011) Food-grade nanoemulsions: Formulation, fabrication, properties, performance, biological fate, and potential toxicity. Crit Rev Food Sci Nutr 51(4):285–330. https://doi.org/10.1080/10408398.2011.559558

Meirmoeini T, Pishkar L, Kahrizi D, Barzin G, Karimi N (2021) Phytotoxicity of green synthesized silver nanoparticles on *Camelina sativa* L. Physiol Mol Biol Plants 27(2):417–427. https://doi.org/10.1007/s12298-021-00946-y

Memarizadeh N, Ghadamyari M, Adeli M, Talebi K (2014) Preparation, characterization and efficiency of nano encapsulated imidacloprid under laboratory conditions. Ecotoxicol Environ Saf, 107, 77–83. http://refhub.elsevier.com/B978-0-12-815322-2.00012-2/rf0280

Mikula K, Izydorczyk G, Skrzypczak D, Mironiuk M et al (2020) Controlled release micronutrient fertilizers for precision agriculture—A review. Sci Total Environ 712:136365. https://doi.org/10.1016/j.scitotenv.2019.136365

Mishra D, Khare P (2021) Emerging nano-agrochemicals for sustainable agriculture: Benefits, challenges, and risk mitigation. In: Sustainable agriculture reviews 50; Kumar Singh V, Singh R, Lichtfouse E, Eds. Springer: Cham, Switzerland. 50. https://doi.org/10.1007/978-3-030-632 49-6_9

Molnar A, Papp M, Kovacs DZ, Belteky P et al (2020) Nitro-oxidative signalling induced by chemically synthetized zinc oxide nanoparticles (ZnO NPs) in *Brassica* species. Chemosphere 251:126419. https://doi.org/10.1016/j.chemosphere.2020.126419

Mustafa IF, Hussein MZ (2020) Synthesis and technology of nanoemulsion-based pesticide formulation. Nanomaterials (Basel) 10(8):1608. https://doi.org/10.3390/nano10081608

Naseem F, Zhi Y, Akhyar Farrukh M, Hussain F, Yin Z (2020) Mesoporous $ZnAl_2Si_{10}O_{24}$ nanofertilizers enable high yield of *Oryza sativa* L. Sci Rep 10:10841. https://doi.org/10.1038/s41598-020-67611-4

Nguyen HM, Hwang IC, Park JW, Park HJ (2012) Photoprotection for deltamethrin using chitosan-coated beeswax solid lipid nanoparticles. Pest Manag Sci 68:1062–1068. https://doi.org/10.1002/ps.3268

Nile SH, Thiruvengadam M, Wang Y, Samynathan R, Shariati MA, Rebezov M et al (2022) Nano-priming as emerging seed priming technology for sustainable agriculture—recent developments and future perspectives. J Nanobiotechnology 20(1):254. https://doi.org/10.1186/s12951-021-01214-7

Nuruzzaman M, Mahmudur Rahman M, Liu Y, Naidu R (2016) Nanoencapsulation, nano-guard for pesticides: A new window for safe application. J Agric Food Chem 64:1447–1483. https://doi.org/10.1021/acs.jafc.5b05214

Parisi C, Vigani M, Rodríguez-Cerezo E (2015) Agricultural nanotechnologies: what are the current possibilities? Nano Today 10(2):124–127. https://doi.org/10.1016/j.nantod.2014.09.009

Park S, Croteau P, Boering KA, Etheridge DM et al (2012) Trends and seasonal cycles in the isotopic composition of nitrous oxide since 1940. Nat Geosci 5(4):261–265. https://doi.org/10.1038/ngeo1421

Pereira AES, Sandoval-Herrera IE, Zavala-Betancourt SA, Oliveira HC, Ledezma-Pérez AS, Romero J, Fraceto LF (2016) Polyglutamic acid/chitosan nanoparticles for the plant growth regulator gibberellic acid: Characterization and evaluation of biological activity. Carbohyd Polym 157:1862–1873. https://doi.org/10.1016/j.carbpol.2016.11.073

Pereira AES, Grillo R, Mello NFS, Rosa AH, Fraceto LF (2014) Application of poly(epsilon–caprolactone) nanoparticles containing atrazine herbicide as an alternative technique to control weeds and reduce damage to the environment. J Hazard Mater, 268, 207–215. http://refhub.elsevier.com/B978-0-12-815322-2.00012-2/rf0295

Perez-Álvarez EP, Ramírez-Rodríguez GB, Carmona FJ et al (2020) Towards a more sustainable viticulture: Foliar application of N-doped calcium phosphate nanoparticles on Tempranillo grapes. J Sci Food Agric 101:1307–1313. https://doi.org/10.1002/jsfa.10738

Perlatti, B., de Souza Bergo, P.L., Fernandes, J.B., and Forim, M.R. (2013). Polymeric nanoparticle-based insecticides: A controlled release purpose for agrochemicals. In: Insectic-Dev Safer More Eff Technol. IntechOpen. https://doi.org/10.5772/53355.

Pradhan S, Durgam M, Rao Mailapalli D (2020) Urea loaded hydroxyapatite nanocarrier for efficient delivery of plant nutrients in rice. Arch Agron Soil Sci 67:371–382. https://doi.org/10.1080/03650340.2020.1732940

Qamar W, Gulia S, Athar M, Ahmad R et al (2024) An insight into impact of nanomaterials toxicity on human health. PeerJ 12:e17807. https://doi.org/10.7717/peerj.17807

Rajput VD, Singh A, Minkina T, Rawat S et al (2021) Nano-enabled products: Challenges and opportunities for sustainable agriculture. Plants 10:2727. https://doi.org/10.3390/plants10122727

Raliya R, Saharan V, Dimkpa C, Biswas P (2017) Nanofertilizer for precision and sustainable agriculture: current state and future perspectives. J Agric Food Chem 66(26):6487–6503. https://doi.org/10.1021/acs.jafc.7b02178

Rane M, Bawskar M, Rathod D, Nagaonkar D, Rai M (2015) Influence of calcium phosphate nanoparticles, *Piriformospora indica* and *Glomus mosseae* on growth of *Zea mays*. Adv Nat Sci: Nanosci Nanotechnol 6:045014. https://doi.org/10.1088/2043-6262/6/4/045014

Riyajan RA (2017) Physical property testing of a novel hybrid natural rubber–graft–cassava starch/sodium alginate bead for encapsulating herbicide. Polymer Testing, 58, 300–307. http://refhub.elsevier.com/B978-0-12-815322-2.00012-2/rf0320

Rizwan M, Ali S, urRehman MZ, Malik S, et al. (2019) Effect of foliar applications of silicon and titanium dioxide nanoparticles on growth, oxidative stress, and cadmium accumulation by rice (*Oryza sativa*). Acta Physiologiae Plantarum, 41, 35. https://doi.org/10.1007/S11738-019-2828-7

Saini A, Panwar D, Singh Panesar P, Bandhu Bera M (2021) Encapsulation of functional ingredients in lipidic nanocarriers and antimicrobial applications: A review. Environ Chem Lett 19:1107–1134. https://doi.org/10.1007/s10311-020-01109-3

Shao C, Zhao H, Wang P (2022) Recent development in functional nanomaterials for sustainable and smart agricultural chemical technologies. Nano Convergence 9:11. https://doi.org/10.1186/s40580-022-00302-0

Sharma A, Kumar N, Srivastava A, Kataria A et al (2018) Clove and lemongrass oil based non-ionic nanoemulsion for suppressing the growth of plant pathogenic *Fusarium oxysporum* f. sp. *lycopersici*. Ind Crops Prod 123:353–362. https://doi.org/10.1016/j.indcrop.2018.06.077

Sharma S, Singh S, Ganguli AK, Shanmugam V (2017) Anti–drift nano–stickers made of graphene oxide for targeted pesticide delivery and crop pest control. Carbon, 115, 781–790. http://refhub.elsevier.com/B978-0-12-815322-2.00012-2/rf0330

Singh H, Sharma A, Bhardwaj SK, Arya SK et al (2021) Recent advances in the applications of nano-agrochemicals for sustainable agricultural development. Environ Sci Process Impacts 23(2):213–239. https://doi.org/10.1039/D0EM00404A

Somala N, Laosinwattana C, Teerarak M (2022) Formulation process, physical stability and herbicidal activities of *Cymbopogon nardus* essential oil-based nanoemulsion. Scientific Reports, 12, 10280. https://doi.org/10.1038/s41598-022-14591-2

Song U, Lee S (2016) Phytotoxicity and accumulation of zinc oxide nanoparticles on the aquatic plants *Hydrilla verticillata* and *Phragmites australis*: leaf-type-dependent responses. Environ Sci Pollut Res 23(9):8539–8545. https://doi.org/10.1007/s11356-015-5982-5

Songkhum P, Wuttikhun T, Chanlek N, Khemthong P, Laohhasurayotin K (2018) Controlled release studies of boron and zinc from layered double hydroxides as the micronutrient hosts for agricultural application. Appl Clay Sci 152:311–322. https://doi.org/10.1016/j.clay.2017.11.028

Srikanth K (2020) Routes of Exposures and Toxicity of Nanoparticles. In: Siddhardha B, Dyavaiah M, Kasinathan K (eds) Model organisms to study biological activities and toxicity of nanoparticles. Springer, Singapore. https://doi.org/10.1007/978-981-15-1702-0_13

Stevenson JR, Villoria N, Byerlee D et al (2013) Green Revolution research saved an estimated 18 to 27 million hectares from being brought into agricultural production. PNAS 110:8363–8368. https://doi.org/10.1073/pnas.1208065110

Sun C, Shu K, Wang W, Ye Z, Liu T et al (2014). Encapsulation and controlled release of hydrophilic pesticide in shell cross–linked nanocapsules containing aqueous core. Int J Pharm, 463, 108–114. http://refhub.elsevier.com/B978-0-12-815322-2.00012-2/rf0345

Szymanski M, Dobrucka R (2021) Evaluation of phytotoxicity of bimetallic Ag/Au nanoparticles synthesized using *Geum urbanum* L. J Inorg Organomet Polym Mater 31(6):2459–2470. https://doi.org/10.1007/s10904-020-01814-0

Usman M, Farooq M, Wakeel A, Nawaz A et al (2020) Nanotechnology in agriculture: Current status, challenges and future opportunities. Sci Total Environ 721:137778. https://doi.org/10.1016/j.scitotenv.2020.137778

Wang D, Saleh NB, Byro A, Zepp R et al (2022) Nano-enabled pesticides for sustainable agriculture and global food security. Nat Nanotechnol 17(4):347–360. https://doi.org/10.1038/s41565-022-01082-8

Wani TA, Masoodi FA, Baba WN, Ahmad M, Rahmanian N, Jafari SM (2019) Nanoencapsulation of agrochemicals, fertilizers, and pesticides for improved plant production. In Advances in

Phytonanotechnology (pp 279–298). Academic Press. https://doi.org/10.1016/B978-0-12-815322-2.00012-2

Wong MH, Misra RP, Giraldo JP, Kwak SY et al (2016) Lipid exchange envelope penetration (LEEP) of nanoparticles for plant engineering: a universal localization mechanism. Nano Lett 16(2):1161–1172. https://doi.org/10.1021/acs.nanolett.5b04467

Wu L, Pan H, Huang W, Hu Z, Wang M, Zhang F (2022) PH and redox dual-responsive mesoporous silica nanoparticle as nanovehicle for improving fungicidal efficiency. Materials 15:2207. https://doi.org/10.3390/ma15062207

Xiao L, Wang S, Yang D, Zou Z, Li J (2019) Physiological effects of MgO and ZnO Nanoparticles on the *Citrus maxima*. J Wuhan Univ Technol-Mat Sci Ed 34:243–253. https://doi.org/10.1007/s11595-019-2042-x

Xin X, Judy JD, Sumerlin BB, He Z (2020) Nano-enabled agriculture: from nanoparticles to smart nanodelivery systems. Environ Chem 17(6):413–425. https://doi.org/10.1071/EN19254

Xing K, Chen XG, Liu CS, Cha DS, Park HJ (2009) Oleoyl-chitosan nanoparticles inhibits *Escherichia coli* and *Staphylococcus aureus* by damaging the cell membrane and putative binding to extracellular or intracellular targets. Int J Food Microbiol 132:127–133. https://doi.org/10.1016/j.ijfoodmicro.2009.04.013

Xu ZP, Niebert M, Porazik K, Walker TL et al (2008) Subcellular compartment targeting of layered double hydroxide nanoparticles. J Control Release 130(1):86–94. https://doi.org/10.1016/j.jconrel.2008.05.021

Yadav HJA, Eraiah B, Nagabhushana H, Prasad BD, et al (2018) Broad spectral inhibitory effects of pale green zinc oxide nanophosphor on bacterial and fungal pathogens. Arab J Chem, 11, 324–342. http://refhub.elsevier.com/B978-0-12-815322-2.00012-2/rf0420

Yoon HY, Lee JG, Esposti LD, Iafisco M et al (2020) Synergistic release of crop nutrients and stimulants from hydroxyapatite nanoparticles functionalized with humic substances: Toward a multifunctional nanofertilizer. Acsomega 5:6598–6610. https://doi.org/10.1021/acsomega.9b04354

Zhang J, Zhao C, Liu Y, Cao L et al (2016) Size-dependent effect of prochloraz-loaded PEG-PLGA micro- and nanoparticles. J Nanosci Nanotechnol 16:6231–6237. https://doi.org/10.1166/jnn.2016.10894

Zulfiqar F, Navarro M, Ashraf M, Akram NA, Munné-Bosch S (2019) Nanofertilizer use for sustainable agriculture: Advantages and limitations. Plant Sci 289:110270. https://doi.org/10.1016/j.plantsci.2019.110270

# Chapter 7
# Nanodiagnostics: Advances and Implications in Clinical Diagnosis

**Sonali Khanal, Aman Sharma, Rachna Verma, and Dinesh Kumar**

**Abstract** The incorporation of nanotechnology into clinical diagnostics has introduced novel tools and techniques at the nanoscale, which significantly improve disease detection, monitoring, and therapy. This leads to the incorporation of nanoparticles into nanodiagnostics, enhancing the sensitivity, specificity, and turnaround times of diagnostic tests. This implies that diseases can now be detected molecularly, enabling prompt treatment and improved patient outcomes. New technologies, in particular biosensors and nanomaterials for imaging, have expanded the range of diagnostics in oncology, infectious diseases, cardiovascular, and neurological conditions. The use of nanotechnology for point-of-care testing and personalized medicine has numerous possibilities for fast, non-invasive, and accurate clinical diagnosis. Nonetheless, factors such as the need for standardization, safety, cost-effectiveness, and regulatory approvals still pose significant challenges. Key emerging trends and innovations in nanodiagnostics are expected to enhance the accuracy of diagnosis and the effectiveness of treatment, ultimately leading to improved healthcare outcomes.

## 7.1 Introduction

Nanodiagnostics, defined as the application of nanotechnology to medical diagnostics, has been evolving to meet the increasing demand for clinical diagnostics that offer enhanced sensitivity and early disease detection (Baptista 2014). This field has been utilized in clinical diagnosis for over 60 years, bringing unique opportunities for more successful and efficient identification of infectious diseases (Pallares

S. Khanal · A. Sharma · D. Kumar (✉)
School of Bioengineering and Food Technology, Shoolini University of Biotechnology and Management Sciences, Solan, India
e-mail: dineshkumar@shooliniuniversity.com

R. Verma
School of Biological and Environmental Science, Shoolini University of Biotechnology and Management Sciences, Solan, India

Centre of Advanced Innovation Technologies, VSB – Technical University of Ostrava, Ostrava-Poruba 70800, Czech Republic



A. Balkrishna et al. (eds.), *Nanotechnology in Environmental Science and Healthcare*,
https://doi.org/10.1007/978-981-95-6714-0_7

et al. 2023). High sensitivity and specificity should be the top priorities for modern diagnostic techniques to ensure early disease diagnosis and enable individualized therapies in an economical manner (Park et al. 2017). These instruments should ideally be able to detect minute molecular alterations, which would facilitate a better comprehension of illness conditions (Baptista 2014). Nanotechnology is one of the fastest-growing areas of research, demonstrating great potential in solving numerous biological problems, including cancer. It enables early disease detection and diagnosis, as well as precise and effective therapy (Alharbi and Al-sheikh 2014). A growing number of therapies based on nanoparticles (NPs) have been successfully introduced in clinical settings to treat conditions such as cancer, pain, and infectious diseases (Woodman et al. 2021). The accurate and early diagnosis of hereditary diseases is crucial for advancing therapeutic opportunities. Nanodiagnostics is anticipated to streamline healthcare delivery by reducing costs and enhancing the effectiveness of personalized medicine (Falzarano et al. 2018). A novel class of biosensors, called nanobiosensors, is being greatly influenced by nanotechnology. These biosensors typically consist of a biological recognition molecule immobilized onto a signal transducer. The performance of the biosensor is primarily dependent on the interaction between the biorecognition molecule and the analyte. Biosensing research has advanced significantly with the application of novel nanomaterials, particularly in the molecular detection of disease-associated biomarkers (Prasad 2014). Overall, the integration of nanotechnology in diagnostics presents new paradigms in medicine, offering the potential for significant advancements in disease detection, treatment, and management across various fields, including human health and agriculture. The advances in nanodiagnostics not only promise to revolutionize how we detect and monitor health conditions, but they also have far-reaching implications for healthcare accessibility and cost-effectiveness. As these technologies become integrated into clinical practice, they hold the potential to simplify diagnostic processes, reduce the burden of disease, and empower healthcare providers to deliver more precise and effective treatments. By prioritizing early detection and tailored therapies, we pave the way for a healthier future, where diseases can be managed more effectively and with greater compassion.

## 7.2 Nanoparticles Used for Clinical Diagnosis

### *7.2.1 Metallic Nanoparticles*

In recent years, the phrase "metallic nanoparticles" (MNPs) has gained prominence in the domain of nanotechnology. These nanoparticles, commonly composed of noble metals such as gold, silver, and platinum, are recognized for their health benefits and are increasingly being synthesized for various purposes (Jamkhande et al. 2019). Figure 7.1 illustrates various types of nanoparticles utilized in biomedical applications. Researchers are actively focusing on the production of metal nanoparticles, nanostructures, and nanomaterials due to their significant properties, which are beneficial for composite and catalytic applications, such as polymer formulations, disease diagnosis and treatment, and sensor technologies (Shaikh et al. 2016; Banerjee et al. 2017; Moura et al. 2017).

MNPs offer considerable benefits in cancer immunotherapy due to their configurable dimensions, morphology, charge, and surface properties (Arvizo et al. 2012; Salatin et al. 2015). Strategies for cancer vaccination can be enhanced by the increased density of MNPs, as they are more readily assimilated by cells compared to non-metallic nano-formulations of similar dimensions (Barnaby et al. 2014). Moreover, MNP-mediated tumor ablation, when combined with immunotherapy, can be enhanced by the distinctive optical properties of MNPs. These enhance vaccine delivery by augmenting dendritic cell absorption of antigens, hence improving the anti-tumor cytotoxic T cell response (Evans et al. 2018).

- **Platinum Nanoparticles**

Platinum nanoparticles (PtNPs) have garnered considerable attention due to their diverse applications in the chemical, biological, and therapeutic domains (Jeyaraj et al. 2019). In comparison with other metallic nanoparticles, platinum nanoparticles demonstrate superior catalytic activity relative to palladium nanoparticles. They exhibit superior cellular biocompatibility, similar to that of gold nanoparticles; nevertheless, they display increased cytotoxicity, which may be beneficial in cancer treatment, as platinum ions can interact with DNA to inhibit its replication. Furthermore, PtNPs demonstrate antibacterial efficacy against both Gram-positive and Gram-negative infections, comparable to silver nanoparticles (AgNPs); however,

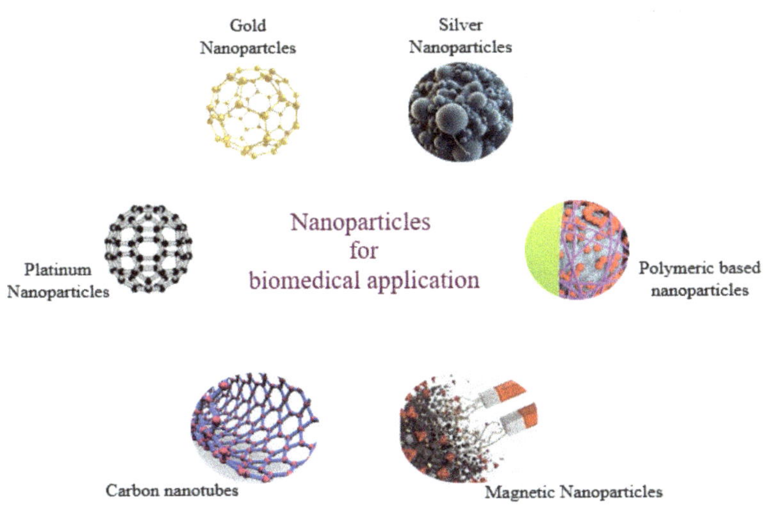


**Fig. 7.1** Different types of nanoparticles for biomedical applications

they exhibit lower toxicity, rendering them more suitable for therapeutic applications (Hashimoto et al. 2016; Martins et al. 2017; Nishanthi et al. 2019). Moreover, PtNPs may serve as coatings for sensors designed to detect glucose or other biomolecules, as reported in certain studies. Moreover, PtNPs can stimulate T cells and enhance antigen presentation, thereby bolstering the host's immune response against the tumor. PtNPs exhibit anticancer efficacy comparable to that of platinum-based medicines, although they differ in behavior. PtNPs induce cytotoxicity contingent upon their size, concentration, and incubation length following cellular entry through passive diffusion or endocytosis (Johnstone et al. 2016). This primarily results from the formation of strand breaks in the chromosomal DNA. Additionally, PtNPs may serve as coatings for sensors designed to detect glucose or other biomolecules, as shown by several reports. Moreover, PtNPs can stimulate T cells and enhance antigen presentation, thereby bolstering the host's immune response against tumors (Yu et al. 2022).

- **Gold Nanoparticles**

Gold nanoparticles (AuNPs) have been thoroughly investigated and utilized in tumor diagnostics and therapy due to their distinctive intrinsic properties. To improve their efficacy for various applications, it is crucial to methodically and comprehensively understand their inherent qualities and the interrelationships among these properties (Bai et al. 2020). AuNPs have demonstrated efficacy in identifying several cell types, including neoplastic and bacterial cells. A disposable immunosensor is engineered for the detection of *Salmonella enterica* subsp. *enterica* serovar Typhimurium LT2 (S) using an electrochemical detection approach that incorporates magneto-immunoassay and AuNPs as a labeling agent (Afonso et al. 2013). The immunosensor operates by utilizing a screen-printed carbon electrode positioned above a stable magnet. Samples containing Salmonella, including skimmed milk, were analyzed with anti-Salmonella magnetic beads (MBs-pSAb) as a detection method, subsequently combined with antibodies modified with gold nanoparticles (sSAb-AuNPs), and identified through differential pulse voltammetry, achieving a detection limit of 143 cells/mL. When employed alongside magnetic field application, these AuNPs detection methods provide a limit of detection in skimmed milk samples that surpasses the typical commercial methodology utilized for comparison. The core concept of developing a biosensor using gold nanoparticles is the functionalization or capping of the particles with a thiolated biomolecule, which modifies the optical absorbance of the AuNPs upon recognition of the complementary biomolecule. Upon binding of a specific antibody to its corresponding antigen, antigen-functionalized AuNPs agglomerate, leading to a shift in plasmon absorption. Extensive research has been conducted to elucidate the bonding mechanisms of oligonucleosides to metal surfaces and colloids for various applications, including surface-enhanced Raman spectroscopy-based rapid sequencers and multiplexed DNA detection methods (Chandra et al. 2013).

The core concept of developing a biosensor utilizing gold nanoparticles is that these particles are functionalized with a thiolated biomolecule, which modifies the optical absorbance of the AuNPs upon recognition of the complementary biomolecule (Ebrahimnezhad et al. 2013). Upon binding of a specific antibody to its corresponding antigen, antigen-functionalized AuNPs agglomerate, leading to a shift in plasmon absorption. Extensive research has been conducted to elucidate the bonding mechanisms of oligonucleosides to metal surfaces and colloids for various applications, including surface-enhanced Raman spectroscopy-based rapid sequencers and multiplexed DNA detection methods (Davaran et al. 2014).

- **Silver Nanoparticles**

Silver nanoparticles (AgNPs) have been widely used due to their unique physicochemical properties, including morphological and structural characterization at the nanoscale (Malekzad et al. 2017). According to Ibrahim et al. (2021), these characteristics enabled the synthesis and modification of MNPs with different chemical functional groups for conjugation with ligands, antibodies, and DNA probes, all of which are essential for biosensor construction. However, among metallic nanoparticles with LSPRs, AgNPs are especially effective for sensing (Loiseau et al. 2019). According to studies, when AgNPs come into contact with particular molecules, they exhibit LSPR band changes that are five times stronger than those caused by AuNPs (Lismont and Dreesen 2012).

Since the presence of hydrogen peroxide can be regarded as an industrial, environmental, or even medical hazard, its detection is crucial. A biosensor that detects the presence of hydrogen peroxide in rainwater has been created by Filippo et al. Its production technique is straightforward, inexpensive, and most importantly, reusable (Filippo et al. 2009). PVA was chosen for this sensor due to its numerous potential applications in the optical, pharmaceutical, medicinal, and membrane domains, as well as the advantages of AgNPs for the LSPR-sensing technique. The hydrogel's LSPR sensing is demonstrated by a significant shift in the absorption spectra and a corresponding change in the platform's color.

- **7.2.2. Carbon nanotubes**

As a novel form of carbon material, carbon nanotubes (CNTs) have garnered significant attention for potential biosensor applications due to their unique shape and exceptional electrical, mechanical, chemical, and thermal properties (Wang et al. 2003). In terms of reaction rate, reversibility, and detection limit, CNTs outperform other carbon electrodes as biosensors for glucose and DNA detection. To improve biocompatibility, CNTs must be functionalized with conducting polymers, guest molecules, side-wall substituents, or metal nanoparticles due to their hydrophobic surface (Chu et al. 2007; Valcárcel et al. 2007; Wang et al. 2007; Zou et al. 2008). Furthermore, by reducing the effective electron transfer distance, these metal nanoparticles can facilitate charge transfer between the electrode and the enzyme, acting as tiny conducting channels to achieve direct electron transfer between the electrode surface and the enzyme's active sites (Yang et al. 2009). Additionally, the enzyme can be immobilized by adsorption if the electrode is altered by metal nanoparticles and a CNT composite. Adsorption is the most straightforward and requires the least amount of preparation of these different enzyme immobilization techniques. Since adsorption requires no chemical reagents and rarely necessitates activation, the bioactivities of the immobilized enzyme can be effectively preserved (Tsai and Tsai 2009). By pre-treating multiwalled carbon nanotubes with $NH_3$ gas, Zhang et al. were able to coat them with small, well-dispersed Pt nanoparticles. The PtNP-MWNT-coated electrode demonstrated the best electrocatalytic property for $H_2O_2$ detection (Zhang and Gao 2010).

### *7.2.2 Magnetic Nanoparticles*

Magnetic nanoparticles (MNPs) are a prominent class of nanoparticles, generally composed of pure metals such as iron, cobalt, and nickel, or a blend of metals and polymers. Their applicability has broadened across multiple medical fields, including hyperthermia cancer therapy, regulated medication delivery, magnetic resonance imaging (MRI), and biosensing (Farzin et al. 2020). Studies have demonstrated encouraging outcomes regarding the application of magnetic nanoparticles in magnetite-based cationic liposomes for treating brain gliomas in F344 rats. In a research study, treatment resulted in a substantial decrease in average tumor volume, decreasing from 30,377 $mm^3$ to 2,684 $mm^3$ following three treatment sessions. Nonetheless, one rat in the therapy group had no response, presumably attributable to the tumor's atypical form and its metastatic dissemination. This instance highlights the challenges of utilizing magnetic hyperthermia for the treatment of multi-site and metastatic malignancies, suggesting that although the technology holds significant potential, its efficacy may vary depending on tumor characteristics (Balivada et al. 2010; Williams and Price 2017). Recently, Zhao et al. (2020) reported a technique for collecting SARS-CoV-2 viral RNA via magnetic nanoparticles (MNPs). The researchers developed iron oxide nanoparticles coated with poly(amino ester) featuring carboxyl groups (pcMNPs), which were used for effective RNA extraction. This technique optimizes the lysis and binding processes, facilitating direct PCR amplification and detection of MNP-RNA complexes. The pcMNPs demonstrate significant dispersity in water, allowing the complexes to be utilized directly in PCR, as the associated RNA is readily available for primer binding and the creation of complementary strands. Furthermore, as the process eliminates the need for precipitation and centrifugation of nucleic acids, it has been effectively automated for the high-throughput extraction of numerous samples.

Magnetic nanoparticles (MNPs) have become essential in the development of biosensors, significantly enhancing detection capabilities and improving sensor efficiency. Their applications are particularly prominent in electrochemical transducers, which dominate the field. For instance, a capacitive sensor modified with MNPs was developed to detect cancer markers, demonstrating the potential of MNPs in cancer diagnostics (Haun et al. 2010). Graphene/$Fe_3O_4$@Au nanocomposites have recently been used for electrochemiluminescence (ECL) biosensing, specifically for HeLa cells. An inventive magnetically controlled solid-state ECL sensor is produced as a result of this multifunctional nanostructure's enhanced electron transport, stability, and emission intensity. With a linear detection range of 20 to 10,000 cells/mL, this sensor demonstrated exceptional sensitivity for HeLa cells together with outstanding stability and repeatability (Gu et al. 2015). In a different application, a nanocomposite biosensor was created for the amperometric detection of dopamine using tyrosinase, $Fe_3O_4$ nanoparticles, and chitosan. Dopamine is an essential neurotransmitter involved in the cardiovascular, hormonal, renal, and central nervous systems. By acting as mediators to promote electron transmission between the electrode and the enzyme, the nanoparticles in this sensor improved the electrochemical characteristics. Even with ascorbic acid present, the sensor exhibited a wide linear response from $2.0 \times 10^{-8}$ to $7.5 \times 10^{-5}$ mol/L, with a detection limit of $6.0 \times 10^{-9}$ mol/L. This MNP-modified tyrosinase biosensor shows great promise for rapid, simple, and affordable dopamine detection in various samples, enhancing the stability of the electron transfer mediator and paving the way for cutting-edge bioelectronic devices in medical diagnostics (Wang et al. 2010). Furthermore, Chen et al. developed an electrochemical biosensor based on square wave voltammetry for detecting DNA methylation in blood, showcasing the versatility of MNPs in various biosensing applications. This sequence of advancements underscores the crucial role of MNPs in enhancing biosensor performance across various medical diagnostics (Chen et al. 2022a, b).

### *7.2.3 Polymeric-Based Nanoparticles*

Polymeric nanoparticles (PNPs) have garnered considerable interest in recent years as a drug delivery method, owing to their potential applications in medicine and pharmaceuticals. These particles exhibit stability within the gastrointestinal (GI) tract and offer numerous advantages over alternative approaches, including protection of encapsulated pharmaceuticals from adverse pH environments, enzymatic degradation, and drug efflux mechanisms. Furthermore, they can provide controlled-release characteristics and enhance medication absorption (Güncüm et al. 2018). Nanostructured polymer systems can be customized in terms of molecular weight and encapsulation parameters, facilitating controlled drug release and functioning as viable alternatives for eliciting immune responses against viral infections (Zhang et al. 2017). PNPs offer numerous benefits in combating SARS-CoV-2, including enhanced stability, accurate drug delivery regulation, biocompatibility, and the ability to modify their physical and chemical properties. They are suitable for both hydrophilic and hydrophobic pharmaceuticals. They can be synthesized using straightforward and effective procedures, considering criteria such as polymer type, nanoparticle size, the planned route of administration, and the medication being delivered (Bhardwaj et al. 2020; Yang 2021). In light of the pandemic, PNPs have become crucial instruments, as they can be tailored for specific uses, including diagnostic sensors for identifying viral pathogens and fabrics for face masks that serve as barriers to the virus (Militky et al. 2021; Yang 2021).

Polymer-based nanoparticles are an exciting field of study with potential applications across numerous industries, including medicine, energy production, and the automotive sector. These nanoparticles are created from natural or synthetic polymers that have been modified to form sub-microscopic particles, offering unique properties for various applications (Idris et al. 2023). In biosensor technology, polymeric interfaces have gained attention for their biocompatibility and ease of fabrication. These materials, integrated into sensors such as electrochemical, optical, and field-effect transistor biosensors, offer advantages including signal amplification and enhanced detection convenience (Yilmaz-Sercinoglu et al. 2022). For example, Mahato and Adhikari developed an electronic tongue (e-tongue) to monitor water quality, using functionalized polymer membrane electrodes. The study focused on detecting mineral salts in drinking water through surface interactions facilitated by a phosphorylated EVOH copolymer, which was optimized for its swelling properties (Mahato and Adhikari 2017). In another study, Magnaghi developed a polymeric optode using an EVOH copolymer functionalized with pH-sensitive dyes to detect high-protein food spoilage, which changes color in response to different pH levels (Magnaghi et al. 2021). Lastly, Cui et al. designed a sensor for detecting copper ions ($Cu^{2+}$) in aqueous solutions by modifying EVOH nanofibers, enabling efficient detection through chemical interactions. These examples demonstrate the broad applicability and transformative potential of polymer-based materials in sensor technologies (Cui et al. 2019). Apart from those mentioned above, there are other nanoparticle-based diagnostics in clinical applications as illustrated in Table 7.1.

## 7.3 Application of Nanoparticles in Clinical Diagnosis

### *7.3.1 Kidney Diseases*

The kidney serves as an advantageous site for nanoparticle use in the treatment of renal disorders, owing to its susceptibility to drug toxicity and its role in excreting metabolic by-products. This sensitivity is especially pertinent in kidney transplantation, when pharmacological toxicity presents considerable hurdles. The kidney's architecture, comprising the glomeruli's basement membrane, podocytes, and fenestrated endothelium, facilitates the fast elimination of particles smaller than 10 nm via glomerular filtration. Utilizing these characteristics of nanoparticles creates new opportunities for the detection, treatment, and monitoring of renal disorders. As interest in nanoparticles for renal applications expands, novel concepts for their utilization are rapidly arising (Brede and Labhasetwar 2013). Research performed by Williams et al. In 2018, mice were administered mesoscale nanoparticles with modified doses and routes of administration to enhance renal selectivity. After intravenous injection, MNPs concentrated in the kidneys with an efficacy 5–7 times larger than in other organs, especially in the renal tubular epithelium, and persisted for up to

**Table 7.1** Detection methods and limit of detection to diagnose disease using nanoparticles

| Nanoparticles | Diagnosis feature | Detection method | Detection limit | References |
|---|---|---|---|---|
| Graphene oxide nanoparticles | Early diagnosis of breast cancer | Electrochemical biosensor | Detection limits of 0.16 nM and 0.23 nM for ERBB2 and CD24, respectively | Saeed et al. (2017) |
| Iron oxide nanoparticles | Early diagnosis of ovarian cancer cells | Optical sensor | Separation efficiency of 61.3% | Xu et al. (2011) |
| Platinum nanoparticles | Early diagnosis of thyroid cancer | Chemiluminescent biosensor | LOD of 0.005 mU/L, which is 100-fold lower than that of the commercial enzyme immunoassay | Choi et al. (2017) |
| Gold nanoparticles | Early-stage diagnosis of prostate cancer in blood Prostate cancer | Optical biosensor | At 4.0 ng/mL showed 90–95% sensitivity | Zheng et al. (2015) |
| Cuprous oxide nanoparticles | Detection of prostate-specific antigen in human serum | Electrochemical biosensor | 0.1–100 ng/ mL (LOD of 0.03 pg/ml) | Li et al. (2017) |
| Silicon dioxide nanoparticles | Detection of prohormone BNPT to acute myocardial infarction in human serum samples | Electrochemical biosensor | 0.03 ng/mL | Liang et al. (2017) |
| Cobalt ferrite nanoparticles | Detection of heart failure | Immunosensor | 0.75 fg/mL | He et al. (2016) |
| Nickel nanoparticles | Detection of glucose concentration in human serum | Fabricated biosensor | 0.14 μM | Chen et al. (2022a, b) |
| Copper nanoparticles | Detection of glucose | Electrochemical biosensor | 7.2 μM | Soganci et al. (2018) |
| Silver nanoparticles | Cholesterol | Amperometric detection | 0.25 mg/dL | Nantaphol et al. (2015) |

7 days. Significantly, renal targeting was unaffected by the type of payload, enabling mesoscale nanoparticles to transport both hydrophobic and hydrophilic substances, such as Cy5-labeled dsDNA. Zhu et al. (2018) utilized gold nanoparticles conjugated with anti-collagen-I antibodies (Co-I-AuNPs) in mice, demonstrating enhanced retention in kidneys affected by renal artery stenosis, underscoring the significance

of evaluating renal fibrosis in disease progression. No notable alterations in hematological, electrolyte levels, or liver and kidney function were detected 24 h post-gold nanoparticle injection, indicating short-term safety.

### *7.3.2 Neurodegenerative Diseases*

- **Alzheimer's disease**

Alzheimer's disease (AD) is a prevalent age-associated disorder characterized by the accumulation of amyloid fibrils in extracellular plaques and intracellular tangles. The inability of many small-molecule inhibitors to cross the blood–brain barrier (BBB) limits their effectiveness in treating Alzheimer's disease (Long and Holtzman 2019). Curcumin, a bioactive natural polyphenol derived from *Curcuma longa*, exhibits potential applications across various sectors, including food, textiles, and pharmaceuticals, due to its unique chemical composition. Studies indicate that curcumin exhibits anti-inflammatory and antioxidant characteristics (Del Prado-Audelo et al. 2019). Nonetheless, to achieve therapeutic efficacy, an individual must ingest between 12 and 20 g daily, rendering this impractical for routine application. Consequently, curcumin has faced challenges in evolving from a culinary spice to a medicinal therapy owing to its inadequate solubility and diminished bioavailability. Nanoparticle-based drug delivery systems offer promising alternatives to enhance effectiveness and accessibility for therapeutic applications (Anand et al. 2010).

In order to mimic the binding properties of tetanus toxin, we have coupled nanoparticles with Tet-1, a 12-residue peptide that selectively activates motor neurons for retrograde trafficking within neuronal cells. Curcumin-encapsulated polylactide-co-glycolide (PLGA) nanoparticles are non-cytotoxic to cells, have antioxidative qualities, and effectively break up amyloid clumps (Mathew et al. 2012). Furthermore, curcumin-encapsulated poly(butylcyanoacrylate) (PBCA) nanoparticles markedly enhanced the half-life and concentration of curcumin in the brain relative to unencapsulated curcumin (Sun et al. 2010). In another work, curcumin-encapsulated PBCA nanoparticles with ApoE3 ligands enabled transcytosis into SH-SY5Y neuroblastoma cells via the low-density lipoprotein (LDL) receptor across the blood–brain barrier (Mulik et al. 2010). By promoting the canonical Wnt/β-catenin pathway and promoting adult neurogenesis, the results suggest that curcumin-infused PLGA nanoparticles (Cur-PLGA-NPs) may enhance neural stem cell proliferation and neuronal differentiation, thereby reducing Aβ-induced cognitive deficits. This technique may improve the brain's self-healing processes, offering a potential treatment option for neurodegenerative diseases, such as Alzheimer's (Tiwari et al. 2014).

- **Parkinson's disease**

Parkinson's disease (PD) is a chronic, progressive movement disorder primarily characterized by the degeneration of pigmented neurons in the substantia nigra, leading to a decrease in the availability of the neurotransmitter dopamine (DA) (Yang et al. 2012). Research over the past 20 years has focused on developing medication delivery methods based on nanotechnology for the targeted treatment of Parkinson's disease. One strategy was to make LDA-containing stealth liposomes that were coupled with the 36-amino acid peptide chlorotoxin (ClTx), which can direct medications to brain microvascular endothelial cells (Xiang et al. 2012). Using bone marrow-derived macrophages (BMMs), another study created catalase-loaded nanoparticles to reduce neuroinflammation and provide neuroprotection in Parkinson's disease. Catalase and poly(ethyleneimine)-poly (ethylene glycol) (PEI-PEG) were combined to create these nanoparticles, which had a core–shell structure with a PEG corona and an enzyme-polyion complex core. In C57BL/6 mice with brain inflammation brought on by MPTP, 6-OHDA, or LPS, the pharmacokinetics, biodistribution, blood–brain barrier (BBB) transit, and inflammatory targeting of the BMM-carried nanozymes were assessed. In comparison with cell-free formulations, the study demonstrated that macrophages improved nanozyme transport across the blood–brain barrier, resulting in a higher volume of distribution and a lower clearance. Additionally, these formulations were non-irritating and non-toxic, and the nanoemulsion gel formulations demonstrated higher drug absorption compared to the commercially available oral tablet (Ropitor) and traditional ropinirole gel. Biochemical studies have indicated that ropinirole nanoemulsion gel exhibits superior activity in Parkinson's-induced rats compared to conventional tablets (Zhao et al. 2011).

### *7.3.3 Cancer Diagnosis*

Biomarkers are crucial for evaluating the risk of cancer progression and tailoring treatment. BRCA1 and BRCA2 can, for example, be variants that indicate an elevated chance of recurrence in breast and ovarian cancer. Such risk markers as TP53 and APC mutations identify those at risk of developing a neoplasm; hence, early intervention can be promoted. Diagnostic markers confirm the presence of a disease, for example, HER2 in breast cancer and PSA in cancer of the prostate, and help select the best modalities for the management of the disease. Of the predictive markers, one of the most well-studied is the mutation of the EGFR in non-small cell lung cancer; therapeutically pertinent markers include CA-125, which helps track the course of therapy. However, there are still some issues to be addressed to ensure that these biomarkers are accurate, reproducible, and economical (Califf 2018; Wu and Qu 2015). Notwithstanding progress in treatment methods, cancer continues to be a predominant cause of death, chiefly owing to difficulties in the early detection of tumors. By the time cancer is diagnosed, it frequently has metastasized, complicating treatment alternatives. Therefore, it is imperative to create novel or improve current diagnostic instruments for the early or pre-symptomatic detection of cancer. The emergence of nanotechnology presents exciting opportunities for enhancing diagnostics (Chen et al. 2013). Ovarian cancer, identified as a notably fatal gynecological malignancy, is hindered by insufficient diagnostic testing, which can profoundly affect clinical outcomes and public health (Casey and Salzman 2018; Boitano et al. 2019). Electrochemical nanosensors enable the detection of minimal amounts of diverse biomarkers in urine, blood, and saliva, providing a reliable approach for identifying ovarian cancer biomarkers (Topkaya et al. 2016). Moreover, the incorporation of nanoparticles into electrochemical systems, including magnetic nanoparticles, carbon nanotubes, and gold nanoparticles, improves sensitivity and enables the multiplexing of cancer biomarkers (Wang et al. 2017). Current diagnostic methods can detect stages 1 or 2 of ovarian cancer using CA-125 in approximately 75% of patients; yet, monitoring CA-125 at reduced concentrations may enhance early identification (Wang et al. 2017). In a notable development, Gazze et al. (2018) created a screen-printed graphene nanosensor for label-free detection of CA-125. This sensor, fabricated by precipitating a polyaniline surface onto a graphene electrode and conjugating it with anti-CA-125 antibodies, demonstrated an impressive detection limit with a dynamic range of 0.92 pg/μL to 15.20 ng/μL, making it the most sensitive CA-125 detection platform published as of 2018. Additionally, the expression of e-cadherin has been shown to negatively correlate with the pathological detection of ovarian cancer, positioning it as a valuable biomarker for tumor identification.

Many new transcriptomic, proteomic, and genomic biomarkers are currently being studied. Numerous substances, such as DNA, mRNA, enzymes, metabolites, transcription factors, and cell surface receptors, are found in tumor tissues or blood as cancer biomarkers (Sawyers 2008), as shown in Fig. 7.2. The investigation of tumor molecular biomarkers, including EGFR, vascular endothelial growth factor (VEGF), interleukin 6 (IL-6), and tumor necrosis factor-alpha (TNF-α), has great promise for early cancer diagnosis and detection (Almangush et al. 2017; Ye et al. 2015). A valuable method for salivary proteomics analysis, the saliva peptide fingerprint approach, can also be used to identify potential biomarkers for cancer diagnosis. Jiang et al. (2015) and Wei et al. (2013) used matrix-assisted laser desorption ionization time-of-flight mass spectrometry (MALDI-TOF-MS) to examine the salivary peptide expression spectrum in 40 patients with OSCC and 23 healthy controls. Clinical settings use several biomarkers, such as human epididymis protein 4 (HE4) and cancer antigen-125 (CA-125) for ovarian cancer monitoring, thyroglobulin (Tg) for thyroid cancer, prostate-specific antigen (PSA) for prostate cancer, carcinoembryonic antigen (CEA) for pancreatic cancer, and CA15-3/CA27-29 and human epidermal growth factor receptor 2 (HER2/neu) for breast cancer assessment. However, these biomarkers are not explicitly associated with their particular malignancies; thus, their increased levels render them non-specific and inappropriate for screening or diagnosis (Sturgeon et al. 2010; Bevilacqua et al. 2014; Ye et al. 2018).

## 7.4 Conclusion

Nanodiagnostics is a breakthrough technique in clinical diagnosis, offering significant advances in precision, speed, and sensitivity. With the help of nanomaterials, this technology can detect diseases at a molecular level, preventing chronic and

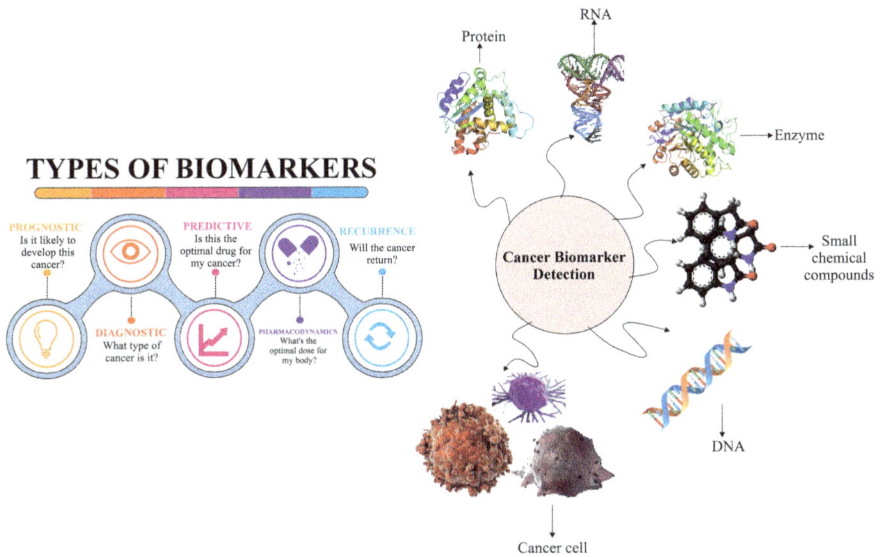


**Fig. 7.2** Different types of biomarkers used in cancer detection

severe complications as early as possible. Enhancements in nanosensors, nanodevices, and imaging technologies have overlapped the scope of the diagnostic area into several domains, including oncological, infectious diseases, and personalized medicine. In addition, the combination of nanodiagnostics with the latest trends appears promising, as it can provide more effective data analysis and interpretation to enhance diagnostic capabilities. However, the integration of these technologies into routine clinical practice is currently hindered by several challenges. Among these, the major challenges lie in safety issues, protocols and validation studies, implementation costs, and the technologies' ability to deliver on promises of mass adoption, as well as the current regulatory landscape.

## References

Afonso AS, Pérez-López B, Faria RC, Mattoso LHC, Hernández-Herrero M, Roig-Sagués AX, Maltez-da Costa M, Merkoçi A (2013) Electrochemical detection of Salmonella using gold nanoparticles. Biosens Bioelectron 40(1):121–126. https://doi.org/10.1016/J.BIOS.2012.06.054

Alharbi KK, Al-sheikh YA (2014) Role and implications of nanodiagnostics in the changing trends of clinical diagnosis. Saudi J Biol Sci 21(2):109–117. https://doi.org/10.1016/J.SJBS.2013.11.001

Almangush A, Heikkinen I, Mäkitie AA, Coletta RD, Läärä E, Leivo I, Salo T (2017) Prognostic biomarkers for oral tongue squamous cell carcinoma: a systematic review and meta-analysis. Br J Cancer 117(6):856–866. https://doi.org/10.1038/bjc.2017.244

Anand P, Nair HB, Sung B, Kunnumakkara AB, Yadav VR, Tekmal RR, Aggarwal BB (2010) RETRACTED: design of curcumin-loaded PLGA nanoparticles formulation with enhanced cellular uptake, and increased bioactivity in vitro and superior bioavailability in vivo. Biochem Pharmacol 79(3):330–338. https://doi.org/10.1016/J.BCP.2009.09.003

Arvizo RR, Bhattacharyya S, Kudgus RA, Giri K, Bhattacharya R, Mukherjee P (2012) Intrinsic therapeutic applications of noble metal nanoparticles: past, present and future. Chem Soc Rev 41(7):2943–2970. https://doi.org/10.1039/C2CS15355F

Bai X, Wang Y, Song Z, Feng Y, Chen Y, Zhang D, Feng L (2020) The basic properties of gold nanoparticles and their applications in tumor diagnosis and treatment. Int J Mol Sci 21(7):2480. https://doi.org/10.3390/IJMS21072480

Balivada S, Rachakatla RS, Wang H, Samarakoon TN, Dani RK, Pyle M, Kroh FO, Walker B, Leaym X, Koper OB, Tamura M, Chikan V, Bossmann SH, Troyer DL (2010) A/C magnetic hyperthermia of melanoma mediated by iron(0)/iron oxide core/shell magnetic nanoparticles: a mouse study. BMC Cancer 10(1):1–9. https://doi.org/10.1186/1471-2407-10-119

Banerjee K, Das S, Choudhury P, Ghosh S, Baral R, Choudhuri SK (2017) A novel approach of synthesizing and evaluating the anticancer potential of silver oxide nanoparticles in vitro. Chemotherapy 62(5):279–289. https://doi.org/10.1159/000453446

Baptista PV (2014) Nanodiagnostics: Leaving the research lab to enter the clinics? Diagnosis 1(4):305–309. https://doi.org/10.1515/DX-2014-0055/MACHINEREADABLECITATION/RIS

Barnaby SN, Lee A, Mirkin CA (2014) Probing the inherent stability of siRNA immobilized on nanoparticle constructs. Proc Natl Acad Sci USA 111(27):9739–9744. https://doi.org/10.1073/PNAS.1409431111

Bevilacqua V, Chan MK, Chen Y, Armbruster D, Schodin B, Adeli K (2014) Pediatric population reference value distributions for cancer biomarkers and covariate-stratified reference intervals in the CALIPER cohort. Clin Chem 60(12):1532–1542. https://doi.org/10.1373/CLINCHEM.2014.229799

Bhardwaj P, Bhatia E, Sharma S, Ahamad N, Banerjee R (2020) Advancements in prophylactic and therapeutic nanovaccines. Acta Biomater 108:1–21. https://doi.org/10.1016/J.ACTBIO.2020.03.020

Boitano TKL, Barrington DA, Batra S, McGwin G, Turner TB, Farmer MB, Brown AM, Straughn MJ, Leath CA (2019) Differences in referral patterns based on race for women at high-risk for ovarian cancer in the southeast: results from a gynecologic cancer risk assessment clinic. Gynecol Oncol 154(2):379–382. https://doi.org/10.1016/J.YGYNO.2019.05.031

Brede C, Labhasetwar V (2013) Applications of nanoparticles in the detection and treatment of kidney diseases. Adv Chronic Kidney Dis 20(6):454–465. https://doi.org/10.1053/J.ACKD.2013.07.006

Califf RM (2018) Biomarker definitions and their applications. Exp Biol Med 243(3):213–221. https://doi.org/10.1177/1535370217750088

Casey MJ, Salzman TA (2018) Reducing the risk of gynecologic cancer in hereditary breast ovarian cancer syndrome mutation carriers. Linacre Q 85(3):225–240. https://doi.org/10.1177/0024363918788340

Chandra P, Singh J, Singh A, Srivastava A, Goyal RN, Shim YB (2013) Gold nanoparticles and nanocomposites in clinical diagnostics using electrochemical methods. J Nanopart 2013(1):1–12. https://doi.org/10.1155/2013/535901

Chen H, Zhen Z, Todd T, Chu PK, Xie J (2013) Nanoparticles for improving cancer diagnosis. Mater Sci Eng R Rep 74(3):35–69. https://doi.org/10.1016/J.MSER.2013.03.001

Chen D, Wu Y, Tilley RD, Gooding JJ (2022a) Rapid and ultrasensitive electrochemical detection of DNA methylation for ovarian cancer diagnosis. Biosens Bioelectron 206:1–19. https://doi.org/10.1016/J.BIOS.2022.114126

Chen J, Liu G, Xiao Q, Wang C, Zhou Z, Gu C, Yin ZZ, Liu H (2022b) Glucose biosensors based on pinecone-shaped Au and Ni nanoparticle composite microelectrodes. ACS Appl Nano Mater 5(9):13319–13331. https://doi.org/10.1021/ACSANM.2C03001/SUPPL_FILE/AN2C03001_SI_001.PDF

Choi G, Kim E, Park E, Lee JH (2017) A cost-effective chemiluminescent biosensor capable of early diagnosing cancer using a combination of magnetic beads and platinum nanoparticles. Talanta 162:38–45. https://doi.org/10.1016/J.TALANTA.2016.09.061

Chu X, Duan D, Shen G, Yu R (2007) Amperometric glucose biosensor based on electrodeposition of platinum nanoparticles onto covalently immobilized carbon nanotube electrode. Talanta 71(5):2040–2047. https://doi.org/10.1016/J.TALANTA.2006.09.013

Cui X, Li T, Li J, An Y, An L, Zhang X, Zhang Z (2019) A highly selective and reversible turn-off fluorescent chemosensor for $Cu^{2+}$ based on electrospun nanofibrous membrane modified with pyrenecarboxaldehyde. Spectrochim Acta Part A Mol Biomol Spectrosc 207:173–182. https://doi.org/10.1016/J.SAA.2018.09.021

Davaran S, Alimirzalu S, Nejati-Koshki K, Nasrabadi HT, Akbarzadeh A, Khandaghi AA, Abbasian M, Alimohammadi S (2014) Physicochemical characteristics of $Fe_3O_4$ magnetic nanocomposites based on Poly(N-isopropylacrylamide) for anti-cancer drug delivery. Asian Pac J Cancer Prev 15(1):49–54. https://doi.org/10.7314/APJCP.2014.15.1.49

Del Prado-Audelo ML, Caballero-Florán IH, Meza-Toledo JA, Mendoza-Muñoz N, González-Torres M, Florán B, Cortés H, Leyva-Gómez G (2019) Formulations of curcumin nanoparticles for brain diseases. Biomolecules 9(2):1–28. https://doi.org/10.3390/BIOM9020056

Ebrahimnezhad Z, Zarghami N, Keyhani M, Amirsaadat S, Akbarzadeh A, Rahmati M, Taheri ZM, Nejati-Koshki K (2013) Inhibition of hTERT gene expression by silibinin-loaded PLGA-PEG-$Fe_3O_4$ in T47D breast cancer cell line. Bioimpacts 3(2):67. https://doi.org/10.5681/BI.2013.005

Evans ER, Bugga P, Asthana V, Drezek R (2018) Metallic nanoparticles for cancer immunotherapy. Mater Today 21(6):673–685. https://doi.org/10.1016/J.MATTOD.2017.11.022

Falzarano MS, Flesia C, Cavalli R, Guiot C, Ferlini A (2018) Nanodiagnostics and nanodelivery applications in genetic alterations. Curr Pharm des 24(15):1717–1726. https://doi.org/10.2174/1381612824666180110151318

Farzin A, Etesami SA, Quint J, Memic A, Tamayol A (2020) Magnetic nanoparticles in cancer therapy and diagnosis. Adv Healthcare Mater 9(9):1–57. https://doi.org/10.1002/ADHM.201901058

Filippo E, Serra A, Manno D (2009) Poly (vinyl alcohol) capped silver nanoparticles as localized surface plasmon resonance-based hydrogen peroxide sensor. Sens Actuators, B Chem 138(2):625–630. https://doi.org/10.1016/J.SNB.2009.02.056

Gazze A, Ademefun R, Steven Conlan R, Rodrigues Teixeira S, Sofia Rodrigues Teixeira C (2018) Electrochemical impedence spectroscopy enabled CA125 detection; toward early ovarian cancer diagnosis using graphene biosensors. J Interdisc Nanomed 3(2):82–88. https://doi.org/10.1002/JIN2.40

Gu W, Deng X, Gu X, Jia X, Lou B, Zhang X, Li J, Wang E (2015) Stabilized, superparamagnetic functionalized graphene/$Fe_3O_4$@Au nanocomposites for a magnetically-controlled solid-state electrochemiluminescence biosensing application. Anal Chem 87(3):1876–1881. https://doi.org/10.1021/AC503966U

Güncüm E, Işıklan N, Anlaş C, Ünal N, Bulut E, Bakırel T (2018) Development and characterization of polymeric-based nanoparticles for sustained release of amoxicillin—an antimicrobial drug. Artif Cells Nanomed Biotechnol 46(sup2):964–973. https://doi.org/10.1080/21691401.2018.1476371

Hashimoto M, Yamaguchi S, Sasaki JI, Kawai K, Kawakami H, Iwasaki Y, Imazato S (2016) Inhibition of matrix metalloproteinases and toxicity of gold and platinum nanoparticles in L929 fibroblast cells. Eur J Oral Sci 124(1):68–74. https://doi.org/10.1111/EOS.12235

Haun JB, Yoon TJ, Lee H, Weissleder R (2010) Magnetic nanoparticle biosensors. Wiley Interdisc Rev Nanomed Nanobiotechnol 2(3):291–304. https://doi.org/10.1002/WNAN.84

He Y, Wang Y, Yang X, Xie S, Yuan R, Chai Y (2016) Metal organic frameworks combining $CoFe_2O_4$ magnetic nanoparticles as highly efficient SERS sensing platform for ultrasensitive detection of N-terminal pro-brain natriuretic peptide. ACS Appl Mater Interfaces 8(12):7683–7690. https://doi.org/10.1021/ACSAMI.6B01112

Ibrahim N, Jamaluddin ND, Tan LL, Yusof NYM (2021) A review on the development of gold and silver nanoparticles-based biosensor as a detection strategy of emerging and pathogenic RNA virus. Sensors 21(15):1–29. https://doi.org/10.3390/S21155114

Idris DS, Roy A, Pandit S, Alghamdi S, Almehmadi M, Alsaiari AA, Abdulaziz O, Alsharif A, Khandaker MU, Faruque MRI (2023) Polymer-based nanocarriers for biomedical and environmental applications. E-Polymers 23(1):1–33. https://doi.org/10.1515/epoly-2023-0049

Jamkhande PG, Ghule NW, Bamer AH, Kalaskar MG (2019) Metal nanoparticles synthesis: an overview on methods of preparation, advantages and disadvantages, and applications. J Drug Deliv Sci Technol 53:101174. https://doi.org/10.1016/J.JDDST.2019.101174

Jeyaraj M, Gurunathan S, Qasim M, Kang MH, Kim JH (2019) A comprehensive review on the synthesis, characterization, and biomedical application of platinum nanoparticles. Nanomaterials 9(12):1–41. https://doi.org/10.3390/NANO9121719

Jiang WP, Wang Z, Xu LX, Peng X, Chen F (2015) Diagnostic model of saliva peptide finger print analysis of oral squamous cell carcinoma patients using weak cation exchange magnetic beads. Biosci Rep 35(3):1–9. https://doi.org/10.1042/BSR20150023/56796

Johnstone TC, Suntharalingam K, Lippard SJ (2016) The next generation of platinum drugs: targeted Pt(II) agents, nanoparticle delivery, and Pt(IV) prodrugs. Chem Rev 116(5):3436–3486. https://doi.org/10.1021/acs.chemrev.5b00597

Li F, Li Y, Feng J, Dong Y, Wang P, Chen L, Chen Z, Liu H, Wei Q (2017) Ultrasensitive amperometric immunosensor for PSA detection based on $Cu_2O$@$CeO_2$-Au nanocomposites as integrated triple signal amplification strategy. Biosens Bioelectron 87:630–637. https://doi.org/10.1016/J.BIOS.2016.09.018

Liang W, Zhuo Y, Xiong C, Zheng Y, Chai Y, Yuan R (2017) A sensitive immunosensor via in situ enzymatically generating efficient quencher for electrochemiluminescence of iridium complexes doped $SiO_2$ nanoparticles. Biosens Bioelectron 94:568–574. https://doi.org/10.1016/J.BIOS.2017.03.056

Lismont M, Dreesen L (2012) Comparative study of Ag and Au nanoparticles biosensors based on surface plasmon resonance phenomenon. Mater Sci Eng, C 32(6):1437–1442. https://doi.org/10.1016/J.MSEC.2012.04.023

Loiseau A, Asila V, Boitel-Aullen G, Lam M, Salmain M, Boujday S (2019) Silver-based plasmonic nanoparticles for and their use in biosensing. Biosensors 9(2):1–39. https://doi.org/10.3390/BIOS9020078

Long JM, Holtzman DM (2019) Alzheimer disease: an update on pathobiology and treatment strategies. Cell 179(2):312–339. https://doi.org/10.1016/J.CELL.2019.09.001

Magnaghi LR, Alberti G, Milanese C, Quadrelli P, Biesuz R (2021) Naked-eye food freshness detection: innovative polymeric optode for high-protein food spoilage monitoring. ACS Food Sci Technol 1(2):165–175. https://doi.org/10.1021/acsfoodscitech.0c00089

Mahato M, Adhikari B (2017) Monitoring of drinking water quality: a preliminary approach by an electronic tongue based on functionalized polymer membrane electrodes. Anal Methods 9(42):6019–6031. https://doi.org/10.1039/C7AY01756A

Malekzad H, Sahandi Zangabad P, Mirshekari H, Karimi M, Hamblin MR (2017) Noble metal nanoparticles in biosensors: recent studies and applications. Nanotechnol Rev 6(3):301–329. https://doi.org/10.1515/NTREV-2016-0014

Martins M, Mourato C, Sanches S, Noronha JP, Crespo MTB, Pereira IAC (2017) Biogenic platinum and palladium nanoparticles as new catalysts for the removal of pharmaceutical compounds. Water Res 108:160–168. https://doi.org/10.1016/J.WATRES.2016.10.071

Mathew A, Fukuda T, Nagaoka Y, Hasumura T, Morimoto H, Yoshida Y, Maekawa T, Venugopal K, Kumar DS (2012) Curcumin loaded-PLGA nanoparticles conjugated with Tet-1 peptide for potential use in Alzheimer's disease. PLoS ONE 7(3):e32616. https://doi.org/10.1371/JOURNAL.PONE.0032616

Militky J, Novak O, Kremenakova D, Wiener J, Venkataraman M, Zhu G, Yao J, Aneja A (2021) A review of impact of textile research on protective face masks. Materials 14(8):1–16. https://doi.org/10.3390/MA14081937

Moura D, Souza MT, Liverani L, Rella G, Luz GM, Mano JF, Boccaccini AR (2017) Development of a bioactive glass-polymer composite for wound healing applications. Mater Sci Eng, C 76:224–232. https://doi.org/10.1016/J.MSEC.2017.03.037

Mulik RS, M̈nkk̈nen J, Juvonen RO, Mahadik KR, Paradkar AR (2010) ApoE3 mediated poly(butyl) cyanoacrylate nanoparticles containing curcumin: study of enhanced activity of curcumin against beta amyloid induced cytotoxicity using in vitro cell culture model. Mol Pharm 7(3):815–825. https://doi.org/10.1021/MP900306X

Nantaphol S, Chailapakul O, Siangproh W (2015) A novel paper-based device coupled with a silver nanoparticle-modified boron-doped diamond electrode for cholesterol detection. Anal Chim Acta 891:136–143. https://doi.org/10.1016/J.ACA.2015.08.007

Nishanthi R, Malathi S, John Paul S, Palani P (2019) Green synthesis and characterization of bioinspired silver, gold and platinum nanoparticles and evaluation of their synergistic antibacterial activity after combining with different classes of antibiotics. Mater Sci Eng, C 96:693–707. https://doi.org/10.1016/J.MSEC.2018.11.050

Pallares RM, Kiessling F, Lammers T (2023) Bridging the gap between lab and clinic for nanodiagnostics. Nanomedicine 18(5):413–416. https://doi.org/10.2217/NNM-2023-0067

Park SM, Aalipour A, Vermesh O, Yu JH, Gambhir SS (2017) Towards clinically translatable in vivo nanodiagnostics. Nat Rev Mater 2(5):1–20. https://doi.org/10.1038/natrevmats.2017.14

Prasad S (2014) Nanobiosensors: the future for diagnosis of disease? Nanobiosens Dis Diagn 3:1–10. https://doi.org/10.2147/NDD.S39421

Saeed AA, Sánchez JLA, O'Sullivan CK, Abbas MN (2017) DNA biosensors based on gold nanoparticles-modified graphene oxide for the detection of breast cancer biomarkers for early diagnosis. Bioelectrochemistry 118:91–99. https://doi.org/10.1016/J.BIOELECHEM.2017.07.002

Salatin S, Maleki Dizaj S, Yari Khosroushahi A (2015) Effect of the surface modification, size, and shape on cellular uptake of nanoparticles. Cell Biol Int 39(8):881–890. https://doi.org/10.1002/CBIN.10459

Sawyers CL (2008) The cancer biomarker problem. Nature 452:548–552. https://doi.org/10.1038/nature06913

Shaikh SF, Mane RS, Min BK, Hwang YJ, Joo OS (2016) D-sorbitol-induced phase control of $TiO_2$ nanoparticles and its application for dye-sensitized solar cells. Sci Rep 6(1):1–10. https://doi.org/10.1038/srep20103

Soganci T, Ayranci R, Harputlu E, Ocakoglu K, Acet M, Farle M, Unlu CG, Ak M (2018) An effective non-enzymatic biosensor platform based on copper nanoparticles decorated by sputtering on CVD graphene. Sens Actuators, B Chem 273:1501–1507. https://doi.org/10.1016/J.SNB.2018.07.064

Sturgeon CM, Duffy MJ, Hofmann BR, Lamerz R, Fritsche HA, Gaarenstroom K, Bonfrer J, Ecke TH, Grossman HB, Hayes P, Hoffmann RT, Lerner SP, Löhe F, Louhimo J, Sawczuk I, Taketa K, Diamandis EP (2010) National academy of clinical biochemistry laboratory medicine practice guidelines for use of tumor markers in liver, bladder, cervical, and gastric cancers. Clin Chem 56(6):e1–e48. https://doi.org/10.1373/CLINCHEM.2009.133124

Sun M, Gao Y, Guo C, Cao F, Song Z, Xi Y, Yu A, Li A, Zhai G (2010) Enhancement of transport of curcumin to brain in mice by poly(n-butylcyanoacrylate) nanoparticle. J Nanopart Res 12(8):3111–3122. https://doi.org/10.1007/s11051-010-9907-4

Tiwari SK, Agarwal S, Seth B, Yadav A, Nair S, Bhatnagar P, Karmakar M, Kumari M, Chauhan LKS, Patel DK, Srivastava V, Singh D, Gupta SK, Tripathi A, Chaturvedi RK, Gupta KC (2014) Curcumin-loaded nanoparticles potently induce adult neurogenesis and reverse cognitive deficits in Alzheimer's disease model via canonical Wnt/β-catenin pathway. ACS Nano 8(1):76–103. https://doi.org/10.1021/nn405077y

Topkaya SN, Azimzadeh M, Ozsoz M (2016) Electrochemical biosensors for cancer biomarkers detection: recent advances and challenges. Electroanalysis 28(7):1402–1419. https://doi.org/10.1002/ELAN.201501174

Tsai MC, Tsai YC (2009) Adsorption of glucose oxidase at platinum-multiwalled carbon nanotube-alumina-coated silica nanocomposite for amperometric glucose biosensor. Sens Actuators, B Chem 141(2):592–598. https://doi.org/10.1016/J.SNB.2009.06.016
Valcárcel M, Cárdenas S, Simonet BM (2007) Role of carbon nanotubes in analytical science. Anal Chem 79(13):4788–4797. https://doi.org/10.1021/AC070196M
Wang B, Akiba U, Anzai JI (2017) Recent progress in nanomaterial-based electrochemical biosensors for cancer biomarkers: a review. Molecules 22(7):1–20. https://doi.org/10.3390/MOLECULES22071048
Wang SG, Zhang Q, Wang R, Yoon SF (2003) A novel multi-walled carbon nanotube-based biosensor for glucose detection. Biochem Biophys Res Commun 311(3):572–576. https://doi.org/10.1016/J.BBRC.2003.10.031
Wang T, Hu X, Dong S (2007) Construction of metal nanoparticle/multiwalled carbon nanotube hybrid nanostructures providing the most accessible reaction sites. J Mater Chem 17(39):4189–4195. https://doi.org/10.1039/B707481F
Wang Y, Zhang X, Chen Y, Xu H, Tan Y, Wang S, Zhang X (2010) Detection of dopamine based on tyrosinase-$Fe_3O_4$ nanoparticles-chitosan nanocomposite biosensor. Am J Biomed Sci 2(3):209–216. https://doi.org/10.5099/aj100300209
Wei P, Kuo WP, Chen F, Hua H (2013) Diagnostic model of saliva peptide finger print analysis of primary Sjogren's syndrome patients by using weak cation exchange magnetic beads. Biosci Rep 33(4):567–573. https://doi.org/10.1042/BSR20130022/56062
Williams HM, Price H (2017) The application of magnetic nanoparticles in the treatment and monitoring of cancer and infectious diseases. Biosci Horiz Int J Stud Res 10:1–10. https://doi.org/10.1093/BIOHORIZONS/HZX009
Woodman C, Vundu G, George A, Wilson CM (2021) Applications and strategies in nanodiagnosis and nanotherapy in lung cancer. Semin Cancer Biol 69:349–364. https://doi.org/10.1016/J.SEMCANCER.2020.02.009
Wu L, Qu X (2015) Cancer biomarker detection: recent achievements and challenges. Chem Soc Rev 44(10):2963–2997. https://doi.org/10.1039/C4CS00370E
Xiang Y, Wu Q, Liang L, Wang X, Wang J, Zhang X, Pu X, Zhang Q (2012) Chlorotoxin-modified stealth liposomes encapsulating levodopa for the targeting delivery against the Parkinson's disease in the MPTP-induced mice model. J Drug Target 20(1):67–75. https://doi.org/10.3109/1061186X.2011.595490
Xu H, Aguilar ZP, Yang L, Kuang M, Duan H, Xiong Y, Wei H, Wang A (2011) Antibody conjugated magnetic iron oxide nanoparticles for cancer cell separation in fresh whole blood. Biomaterials 32(36):9758–9765. https://doi.org/10.1016/J.BIOMATERIALS.2011.08.076
Yang D (2021) Application of nanotechnology in the COVID-19 pandemic. Int J Nanomed 16:623–649. https://doi.org/10.2147/IJN.S296383
Yang X, Zheng R, Cai Y, Liao M, Yuan W, Liu Z (2012) Controlled-release levodopa methyl ester/benserazide-loaded nanoparticles ameliorate levodopa-induced dyskinesia in rats. Int J Nanomed 7:2077–2086. https://doi.org/10.2147/IJN.S30463
Yang Y, Yang G, Huang Y, Bai H, Lu X (2009) A new hydrogen peroxide biosensor based on gold nanoelectrode ensembles/multiwalled carbon nanotubes/chitosan film-modified electrode. Colloids Surf, A 340(1–3):50–55. https://doi.org/10.1016/J.COLSURFA.2009.02.038
Ye F, Zhao Y, El-Sayed R, Muhammed M, Hassan M (2018) Advances in nanotechnology for cancer biomarkers. Nano Today 18:103–123. https://doi.org/10.1016/J.NANTOD.2017.12.008
Ye X, Zhang J, Tan Y, Chen G, Zhou G (2015) Meta-analysis of two computer-assisted screening methods for diagnosing oral precancer and cancer. Oral Oncol 51(11):966–975. https://doi.org/10.1016/J.ORALONCOLOGY.2015.09.002
Yilmaz-Sercinoglu Z, Kuru Cİ, Ulucan-Karnak F (2022) Polymeric-based interface for the development of COVID-19 biosensor. In: Sensing tools and techniques for COVID-19. Elsevier, pp 57–82. https://doi.org/10.1016/B978-0-323-90280-9.00013-4
Yu J, Liu S, Wang Y, He X, Zhang Q, Qi Y, Zhou D, Xie Z, Li X, Huang Y (2022) Synergistic enhancement of immunological responses triggered by hyperthermia sensitive Pt NPs via NIR

laser to inhibit cancer relapse and metastasis. Bioact Mater 7:389–400. https://doi.org/10.1016/J.BIOACTMAT.2021.05.030
Zhang J, Gao L (2010) Synthesis of highly dispersed platinum nanoparticles on multiwalled carbon nanotubes and their electrocatalytic activity toward hydrogen peroxide. J Alloy Compd 505(2):604–608. https://doi.org/10.1016/J.JALLCOM.2010.06.085
Zhang M, Hong Y, Chen W, Wang C (2017) Polymers for DNA vaccine delivery. ACS Biomater Sci Eng 3(2):108–125. https://doi.org/10.1021/ACSBIOMATERIALS.6B00418
Zhao Y, Haney MJ, Klyachko NL, Li S, Booth SL, Higginbotham SM, Jones J, Zimmerman MC, Mosley RL, Kabanov AV, Gendelman HE, Batrakova EV (2011) Polyelectrolyte complex optimization for macrophage delivery of redox enzyme nanoparticles. Nanomedicine 6(1):25–42. https://doi.org/10.2217/NNM.10.129
Zhao Z, Cui H, Song W, Ru X, Zhou W, Yu X (2020) A simple magnetic nanoparticles-based viral RNA extraction method for efficient detection of SARS-CoV-2. BioRxiv:1–18. https://doi.org/10.1101/2020.02.22.961268
Zheng T, Pierre-Pierre N, Yan X, Huo Q, Almodovar AJO, Valerio F, Rivera-Ramirez I, Griffith E, Decker DD, Chen S, Zhu N (2015) Gold nanoparticle-enabled blood test for early stage cancer detection and risk assessment. ACS Appl Mater Interfaces 7(12):6819–6827. https://doi.org/10.1021/ACSAMI.5B00371
Zhu XY, Zou X, Mukherjee R, Yu Z, Ferguson CM, Zhou W, McCollough CH, Lerman LO (2018) Targeted imaging of renal fibrosis using antibody-conjugated gold nanoparticles in renal artery stenosis. Invest Radiol 53(10):623–628. https://doi.org/10.1097/RLI.0000000000000476
Zou Y, Xiang C, Sun LX, Xu F (2008) Glucose biosensor based on electrodeposition of platinum nanoparticles onto carbon nanotubes and immobilizing enzyme with chitosan-$SiO_2$ sol–gel. Biosens Bioelectron 23(7):1010–1016. https://doi.org/10.1016/J.BIOS.2007.10.009

# Chapter 8
# Metal Nanoparticle-Assisted Disease Diagnosis: Trends and Perspectives

**Ravita Saini, Ashwani Kumar, Hemant Sharma, Razia Parveen, Dipjyoti Chakraborty, Jitender Kumar, and Ryszard Amarowicz**

**Abstract** Metal nanoparticles (MNPs) typically range in size from 1 to 100 nm, and their unique physical and chemical properties make them extremely useful in a number of applications, including disease detection. This chapter explores the essential aspects of MNPs, covering synthesis techniques and characterization methods, while emphasizing their properties, such as localized surface plasmon resonance, a high surface-to-volume ratio, as well as ferromagnetism and superparamagnetism. MNPs enhance optical, colorimetric, electrochemical sensing, and imaging contrast. Furthermore, the capability of MNPs to improve diagnostic sensitivity and precision has been demonstrated by their recent integration into advanced imaging modalities, as well as rapid and point-of-care testing, including microfluidics, lab-on-a-chip devices, and biosensor devices. Additionally, several studies demonstrated the successful application of MNPs in diagnostic settings, although issues with stability, toxicity, and regulatory barriers remain to be addressed. By addressing these trends and obstacles, this chapter aims to provide a perspective on the transformational role of MNPs in medical diagnostics and their potential to address major problems in existing diagnostic procedures.

R. Saini · A. Kumar (✉) · H. Sharma · R. Parveen
Patanjali Herbal Research Department, Patanjali Research Foundation, Haridwar, India
e-mail: dr.ashwanikumar@patanjali.res.in

D. Chakraborty
Department of Bioscience and Biotechnology, Banasthali Vidyapith, Vanasthali, Rajasthan, India
e-mail: cdipjyoti@banasthali.in

J. Kumar
Department of Plant Sciences, Central University of Himachal Pradesh, Shahpur, Kangra, India

R. Amarowicz
Department of Chemical and Physical Properties of Food, Institute of Animal Reproduction and Food Research, Polish Academy of Sciences, Olsztyn, Poland
e-mail: r.amarowicz@pan.olsztyn.pl



A. Balkrishna et al. (eds.), *Nanotechnology in Environmental Science and Healthcare*,
https://doi.org/10.1007/978-981-95-6714-0_8

## 8.1 Introduction

The advent of nanotechnology has revolutionized the field of biomedical diagnostics, particularly through the application of metal nanoparticles (MNPs). MNPs are small metal particles ranging in size from 1 to 100 nm (Vijayaram et al. 2024; Thakur et al. 2024). These NPs possess distinct physical and chemical characteristics that set them apart from their bulk counterparts, rendering them extremely valuable across various domains, including medical, electronics, and environmental sciences. MNPs exist in diverse forms and dimensions, each exhibiting its own set of specific traits and uses (Fig. 8.1). For instance, noble MNPs like gold (Au), silver (Ag), platinum (Pt), and palladium (Pd) are known for their stability and unique optical features, making them ideal for applications in drug delivery, imaging, biosensing, and as antimicrobial agents (Burlec et al. 2023; Balkrishna et al. 2023; Yashaswini et al. 2024; Kumar et al. 2024). Magnetic MNPs, which include elements such as iron (Fe), cobalt (Co), and nickel (Ni), are mainly employed in magnetic resonance imaging (MRI) and as carriers for drug delivery owing to their magnetic attributes (Caspani et al. 2020; Yashaswini et al. 2024). Metal oxide NPs (MONPs), such as zinc oxide (ZnO), titanium dioxide ($TiO_2$), and iron oxide (FeO), find extensive applications in electronics, catalysis, and as UV filters in sunscreens (Zafar et al. 2020; Haas 2021; Thakur et al. 2022). Bimetallic NPs, composed of two different metals, exhibit enhanced properties, including improved catalytic activity and stability, making them applicable in various biomedical and industrial fields (Medina-Cruz et al. 2020).

This chapter aims to investigate the multifaceted functions of MNPs within the field of disease diagnosis, highlighting their synthesis, characterization, properties, mechanisms of action, and recent innovations in MNPs-assisted diagnostics techniques. MNPs can be synthesized through various methods, including physical, chemical, and biological techniques. Each approach presents unique advantages and challenges that significantly affect the size, shape, and functionalization of the NPs. Characterization techniques are crucial for understanding these properties, as they directly impact the efficacy of MNPs in diagnostic applications. The most intriguing properties of MNPs are the ability to enhance optical sensing, surface-enhanced Raman scattering (SERS), and metal-enhanced fluorescence (MEF), calorimetric, and electrochemical sensing through key properties such as localized surface plasmon resonance (LSPR) and high surface-to-volume ratio. Furthermore, MNPs also enhance image contrast by the mechanism of ferromagnetism and superparamagnetism. This unique physical and chemical properties enable innovative applications that improve the sensitivity and specificity of diagnostic methods.

Recent advancements show that MNPs are increasingly being integrated into advanced diagnostic platforms such as imaging modalities, including X-ray, computed tomography (CT), ultrasound, MRI, and positron emission tomography (PET), which provide improved visualization of tissues and pathological conditions. Additionally, these technologies are integrated into microfluidic, lab-on-a-chip, and biosensor devices for the rapid and point-of-care diagnosis of diseases, significantly improving patient outcomes. This chapter also provides insights into several studies

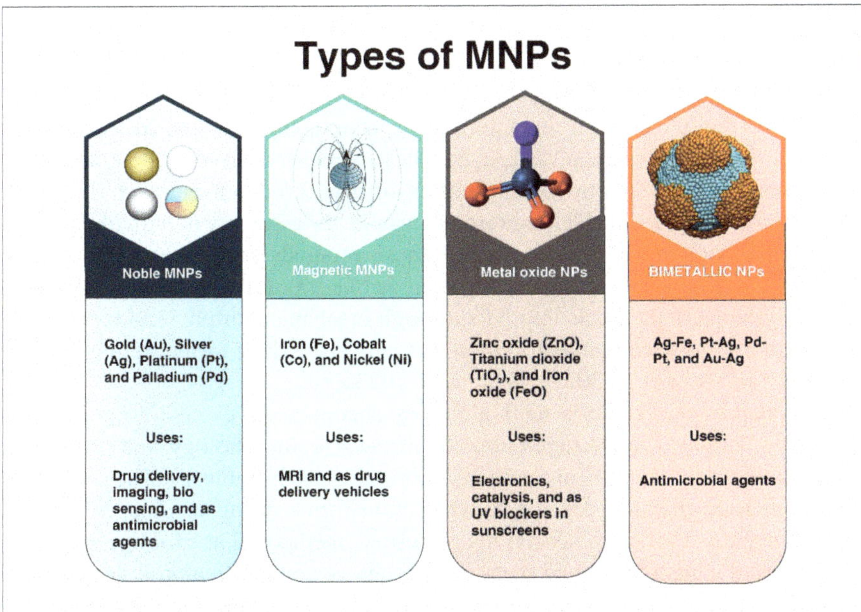


**Fig. 8.1** Various types of MNPs with their applications

that demonstrate the successful application of MNPs in diagnostic settings, including the diagnosis of cancer, cardiovascular diseases, and infectious diseases.

Despite the significant promise of MNPs in disease detection, various obstacles remain, including concerns about stability, toxicity, environmental factors, and regulatory and commercialization barriers. This chapter aims to provide a comprehensive overview of MNPs and their transformative impact on disease diagnosis. By examining the fundamental properties, recent trends, and associated challenges of MNPs, we aim to highlight the significant role they play in advancing diagnostic technologies and improving healthcare outcomes.

## 8.2 Fundamentals of Metal Nanoparticles

MNPs can be categorized by their unique properties and diverse applications (Fig. 8.1). The synthesis and characterization of MNPs are crucial areas of research, driven by their potential to revolutionize technology and industry.

### 8.2.1 Synthesis Methods of Metal Nanoparticles

The synthesis of NPs is generally classified into three primary techniques: chemical, physical, and biological, each characterized by its unique processes and applications (Zahoor et al. 2021; Singh et al. 2023a, b). These methods utilized in the synthesis of MNPs consist of two main strategies: Bottom-up and Top-down approach (Fig. 8.2). In the Bottom-up approach, NPs are formed through the assembly of individual atoms or molecules by chemical or biological synthesis methods, thereby enabling precise control over size and composition. Whereas in the top-down approach, involving breaking down bulk materials into NPs through mechanical or physical techniques, often leading to a varied size distribution that may necessitate further refinement for consistency (Zahoor et al. 2021; Singh et al. 2023a, b).

The nanoparticle synthesis methodologies encompass a diverse array of techniques that facilitate precise regulation of particle size, morphology, and properties. The principles of these synthesis methods are highlighted in Table 8.1. The prominent methods include chemical reduction, which transforms metal ions into NPs using reducing agents, and solvothermal/hydrothermal methods that promote reactions under elevated temperatures and pressures to achieve controlled morphology (Marinescu et al. 2022). The other techniques encompass electrodeposition, a process that employs electric current to transfer metal ions onto substrates (Dikusar et al. 2009;

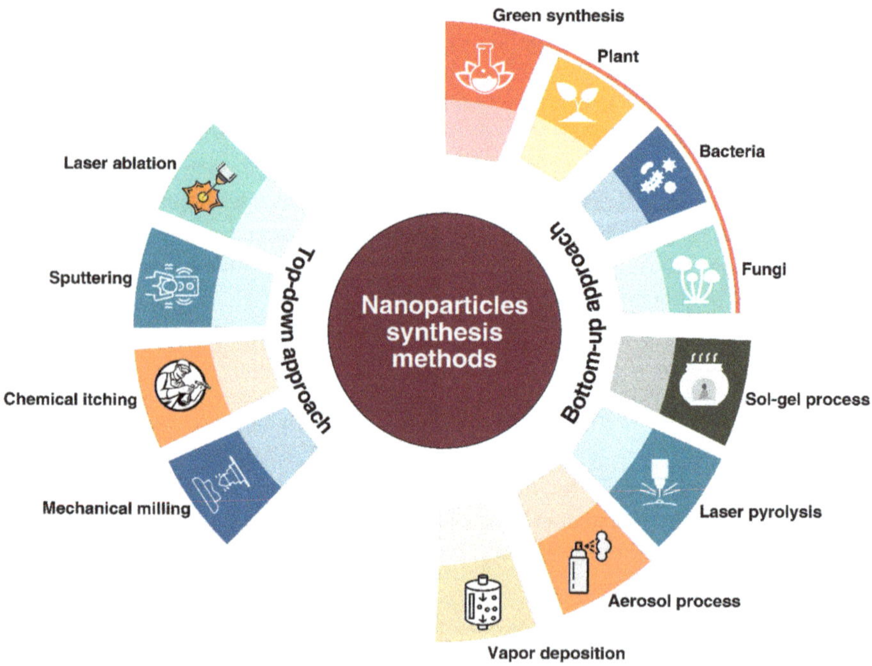


**Fig. 8.2** Different methods for synthesizing MNPs

Lee et al. 2023; Khan et al. 2023), microemulsion methods that generate nanoscale droplets (Malik et al. 2012), chemical vapor deposition (CVD) which transforms metal precursors into vapor to create NPs on substrates (Sun et al. 2021), the sol–gel technique that transitions from a liquid to a solid form to incorporate various metal ions (Emrie 2024), and co-precipitation, which involves the precipitation of metal ions through the introduction of a precipitating agent (Besenhard et al. 2020). Each technique has specific applications across various fields, including catalysis, electronics, and biomedical sciences (Table 8.1).

Physical synthesis techniques utilize physical processes to create NPs without the use of chemical reagents, making them particularly suitable for applications that require high purity and specific surface characteristics. Key techniques include laser ablation, which harnesses powerful laser pulses to vaporize target substances within a liquid or gaseous environment (Nyabadza et al. 2023); flame synthesis, which ignites metal precursors to produce metal-oxide NPs with remarkable efficiency (Serrano-Bayona et al. 2023); physical vapor deposition (PVD), which vaporizes metals in a vacuum to strategically deposit them onto substrates (Sohal et al. 2021; Abbas et al. 2024); mechanical attrition, or ball milling, which skillfully breaks down bulk materials into NPs while potentially introducing minimal contaminants (Yadav et al. 2012); and sonochemical synthesis, which employs ultrasonic waves to trigger chemical reactions through the dynamic process of acoustic cavitation (Xu et al. 2013).

Green synthesis methods utilize microorganisms or biomolecules to generate NPs in an eco-friendly manner. These techniques use the reducing and stabilizing properties of biological entities. Prominent methodologies for biological synthesis include the use of plant extracts, which utilize phytochemicals such as polyphenols and flavonoids to facilitate the conversion of metal ions into NPs (Dubey et al. 2024; Zulfiqar et al. 2024). Whereas, microbial synthesis, comprising bacteria, fungi, and algae, generate NPs through either intracellular or extracellular mechanisms. This method is particularly recognized for its biocompatibility and its promising applications in medicine, including drug delivery and cancer therapy (Singh et al. 2023a, b).

### 8.2.2 *Characterization of MNPs*

Following the synthesis of MNPs, it is essential to characterize the NPs in order to understand their distinctive characteristics, which are important for applications in environmental science, electronics, and medicine. Various techniques are employed to analyze the size, shape, structure, composition, and magnetic properties of NPs (Fig. 8.3). For instances, the shape and size of MNPs characterized by using imaging techniques and scattering techniques. The imaging techniques include transmission electron microscopy (TEM), a powerful method that provides high-resolution images, allowing for detailed analysis of size, shape, and crystalline structure. It is beneficial for studying NPs during synthesis or under operational conditions (Reimer

**Table 8.1** Various methods for the synthesis of metal nanoparticles, their principles, and applications

| Synthesis method | Principle | Applications | References |
|---|---|---|---|
| *Chemical methods* | | | |
| Chemical reduction | Reduction of metal ions using reducing agents to form NPs | Catalysis, electronics | Marinescu et al. (2022) |
| Solvothermal and hydrothermal methods | Chemical reactions in a liquid medium under high temperature and pressure for controlled morphology | Metal oxides, chalcogenides | Marinescu et al. (2022) |
| Electrodeposition | Reduction of metal ions onto a conductive substrate using an electric current | Uniform coatings, electronics, and catalysis | Dikusar et al. (2009); Lee et al. (2023); Khan et al. (2023) |
| Microemulsion | A stable mixture of oil, water, and surfactant forms nanoscale droplets as nanoreactors | Drug delivery, catalysis | Malik et al. (2012) |
| Chemical vapor deposition | Vaporization and decomposition of metal precursors on a substrate to form NPs | Semiconductor industry, electronics, photonics | Sun et al. (2021) |
| Sol–Gel | Transition from a liquid "sol" to a solid "gel" phase incorporating various metal ions | Catalysis, environmental remediation | Emrie (2024) |
| Co-precipitation | Precipitation of metal ions from solution by adding a precipitating agent | Magnetic NPs for data storage | Besenhard et al. (2020) |
| *Physical methods* | | | |
| Laser ablation | Uses high-energy laser pulses to vaporize a target material in a liquid or gas medium | Ideal for applications requiring ligand-free surfaces | Nyabadza et al. (2023) |
| Flame synthesis | Involves the combustion of metal precursors in a flame for efficient nanoparticle production | Large-scale production of metal-oxide NPs | Serrano-Bayona et al. (2023) |

(continued)

**Table 8.1** (continued)

| Synthesis method | Principle | Applications | References |
|---|---|---|---|
| Physical vapor deposition | Evaporation of metals in a vacuum followed by condensation onto a substrate to form NPs | Produces high-purity NPs with controlled size and morphology | Sohal et al. (2021); Abbas et al. (2024) |
| Mechanical attrition | Mechanical grinding of bulk materials into NPs through kinetic energy transfer | Cost-effective method for producing various NPs | Yadav et al. (2012) |
| *Biological methods* | | | |
| Plant extracts | Utilizes plant extracts as reducing and stabilizing agents due to phytochemicals like polyphenols and flavonoids | Environmentally friendly and biocompatible in medical applications like drug delivery and cancer therapy | Dubey et al. (2024); Zulfiqar et al. (2024) |
| Microbial synthesis | Involves microorganisms (bacteria, fungi, algae) that secrete enzymes and metabolites to reduce metal ions into NPs | Environmentally friendly and biocompatible in medical applications like drug delivery and cancer therapy | Singh et al. (2023a, b) |

2013; Patil and Kim 2018). Scanning electron microscopy (SEM) examines the surface morphology and topography of NPs, providing three-dimensional images that aid in understanding surface characteristics and particle distribution (Delvallée et al. 2015; Singh et al. 2024). Atomic force microscopy (AFM) is a powerful tool for measuring surface topography at the nanoscale, providing detailed analysis of particle dimensions and shapes of MNPs (Chen et al. 2022; Bellotti et al. 2022; Singh et al. 2024). Scattering techniques include dynamic light scattering (DLS), a standard method for determining size distributions in suspensions, which is particularly useful for characterizing NPs in biological and environmental samples (Singh et al. 2024). X-ray diffraction (XRD) provides information about the crystalline structure and phase composition of NPs, which is essential for understanding their catalytic and electronic properties (Holder and Schaak 2019; Singh et al. 2024).

The chemical composition and bonding structure of MNPs are characterized using various spectroscopic techniques, including X-ray photoelectron spectroscopy (XPS), which analyzes the surface chemistry of NPs, providing information about their elemental composition and chemical states (Greczynski and Hultman 2020; Singh et al. 2024). Fourier transform infrared spectroscopy (FTIR) identifies functional groups and molecular interactions on nanoparticle surfaces, which is crucial for applications in drug delivery and biosensing (Pugazhendhi et al. 2016; Altammar 2023). Energy-dispersive X-ray spectroscopy (EDX), often coupled with SEM or TEM, provides elemental analysis by detecting characteristic X-rays emitted from

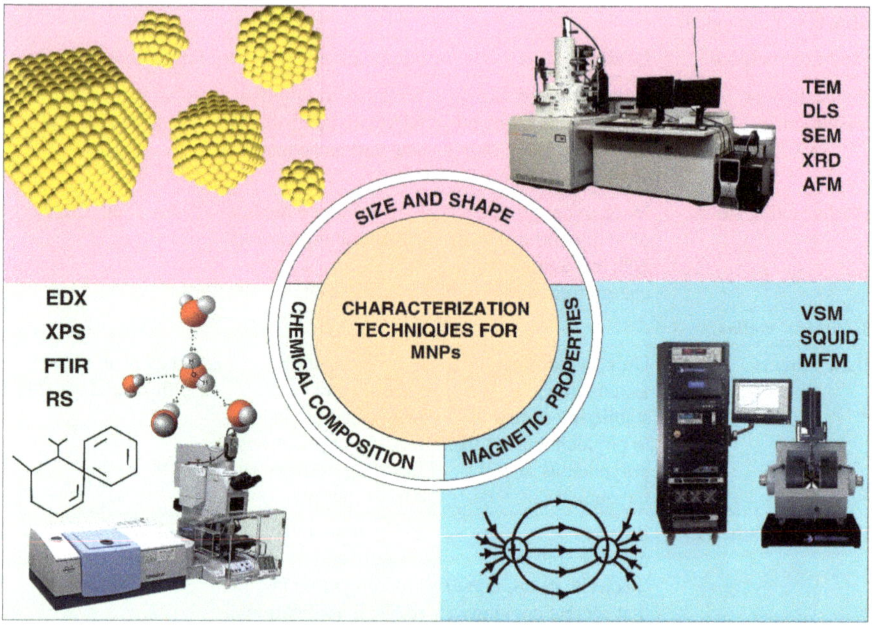


**Fig. 8.3** Characterization techniques for MNPs. EDX: energy-dispersive X-ray spectroscopy; XPS: X-ray photoelectron spectroscopy; FTIR: Fourier transform infrared spectroscopy; RS: Raman spectroscopy; TEM: transmission electron microscopy; DLS: dynamic light scattering; SEM: scanning electron microscopy; XRD: X-ray diffraction; AFM: atomic force microscopy; VSM: vibrating sample magnetometer; SQUID: superconducting quantum interference devices; MFM: magnetic force microscopy

the sample when it is bombarded with electrons. This technique helps identify the elemental composition of NPs (Tshabalala et al. 2020). Additionally, Raman spectroscopy offers insights into the vibrational modes of NPs, aiding in the analysis of their molecular structure and interactions (Singh et al. 2024).

The magnetic properties of MNPs are characterized using vibrating sample magnetometry (VSM), superconducting quantum interference device (SQUID), and magnetic force microscopy (MFM), which are pivotal in understanding these properties. Each method offers unique insights into the magnetic behavior of NPs, contributing to their effective utilization in technology and medicine. VSM is widely used for measuring the magnetic properties of NPs, such as magnetization and coercivity. It operates by vibrating the sample in a uniform magnetic field and detecting the induced voltage in pickup coils, which is proportional to the magnetization (Wu et al. 2021). This technique is beneficial for characterizing the magnetic response of FeO NPs, which are commonly used in biomedical applications, such as magnetic hyperthermia and MRI contrast agents (Wu et al. 2021). SQUID magnetometry is one of the most sensitive techniques for measuring magnetic properties, capable of detecting minimal magnetic signals. It is beneficial for studying superparamagnetic relaxation and magnetic interactions among NPs (Bennington-Gray 2019). MFM provides high-resolution imaging of magnetic domains at the nanoscale, making it an ideal tool for characterizing the local magnetic properties of individual NPs (Angeloni et al. 2020; Passeri et al. 2021). It is versatile, operating in various environments (air, vacuum, liquid) and requires minimal sample preparation. MFM can also perform mechanical or magnetic nanomanipulation of single NPs, providing insights into their magnetic moment and saturation magnetization (Angeloni et al. 2020; Passeri et al. 2021).

Although these techniques provide comprehensive insights into nanoparticle characteristics, the choice of method depends on specific properties of interest and the application context. The integration of multiple techniques often yields a more comprehensive understanding of nanoparticle behavior and functionality, which is crucial for advancing their applications in various fields (Mourdikoudis et al. 2018; Singh et al. 2024).

## 8.3 Functional Properties Enabling MNPs in Diagnostics

MNPs have distinct features that make them useful in various biomedical applications, particularly in the disease diagnosis.

### *8.3.1 Localized Surface Plasmon Resonance (LSPR)*

MNPs possess distinct optical characteristics that hold considerable importance in numerous sectors, including diagnostics. A fundamental aspect of these properties is LSPR, which arises when free electrons in noble MNPs oscillate in response to incoming light, resulting in amplified electromagnetic fields at specific wavelengths (Krasochko et al. 2020). This effect is particularly evident in noble metals, rendering them highly advantageous for use in biosensing and medical diagnostics due to their heightened sensitivity to environmental variations and biomolecule identification (Fathi et al. 2019). Moreover, the optical behavior of MNPs is intricately linked to their dimensions and morphology (Higaki and Jin 2021; Alrahili 2024), wherein varying shapes can influence plasmon resonance and absorption properties (Wang et al. 2020).

### *8.3.2 High Surface-to-Volume Ratio*

Like LSPR, the other most essential properties of MNPs are "high surface-to-volume ratio." The large surface area relative to volume allows for more active sites for chemical reactions, making MNPs highly reactive. Additionally, it facilitates the attachment of various functional groups, such as antibodies, enzymes, or fluorescent markers (Zhu and Xu 2016; Kumar et al. 2018; Narayan et al. 2019). This property enables MNPs to be engineered for effective binding with specific biomarkers, making them particularly valuable for the early detection of disease. Enhanced sensitivity arises from the increased reactivity of MNPs, allowing them to bind more effectively with target molecules, such as disease biomarkers, thereby leading to improved detection methods (Jouyandeh et al. 2022; Şimşek et al. 2023). This capability is crucial for developing biosensors that can detect low concentrations of biomarkers associated with diseases such as cancer and infections (Mughal 2022; Maity et al. 2023). Furthermore, MNPs serve as contrast agents in imaging techniques such as MRI and CT scans, where their high surface area enhances interaction with imaging agents, resulting in more explicit images (Gupta et al. 2023; kr Katare and Tripathi 2024). Their versatility enables them to be functionalized with various chemical groups, allowing them to target specific disease markers effectively across multiple diagnostic platforms, including electrochemical and optical sensors (Ahmad et al. 2022; Yashaswini et al. 2024). Additionally, MNPs contribute to cost-effective diagnostics by requiring smaller amounts of material to achieve high sensitivity (Maity et al. 2023; Şimşek et al. 2023). They also enable real-time monitoring of disease progression and treatment efficacy.

### *8.3.3 Ferromagnetism and Superparamagnetism of MNPs*

MNPs also exhibit distinct magnetic properties, primarily superparamagnetism and ferromagnetism each offering unique advantages in disease diagnosis. Superparamagnetism is a phenomenon observed in NPs typically sized between 10 and 20 nm. In this state, the NPs become magnetized in the presence of an external magnetic field but lose their magnetization once the field is removed. This property enables a rapid and strong magnetic response, making superparamagnetic iron oxide NPs (SPIONs) particularly useful in applications such as targeted drug delivery and MRI contrast enhancement (Wahajuddin and Arora 2012; Losito et al. 2024). Superparamagnetic MNPs can be utilized in highly sensitive imaging techniques, such as MRI, and can be functionalized to target specific cells or tissues, thereby enhancing the specificity of both imaging and therapy (Alcantara and Josephson 2012; Dulińska-Litewka et al. 2019; Sezer et al. 2021). On the other hand, ferromagnetism is characterized by a permanent magnetic moment, allowing ferromagnetic NPs (Fe, Co, Ni, and Au) to retain their magnetization even in the absence of an external magnetic field. This property can be advantageous for imaging techniques that require stable signals. Ferromagnetic NPs can enhance contrast in imaging modalities, such as MRI, by creating localized magnetic fields that improve signal quality, thereby aiding in the detection of lesions and abnormalities (Alcantara and Josephson 2012; Dulińska-Litewka et al. 2019; Sezer et al. 2021). Ferromagnetic NPs can provide stable signals over time, and persistent magnetization can improve the signal-to-noise ratio in various imaging techniques. Furthermore, ferromagnetic MNPs can facilitate the isolation of specific cells or biomarkers from complex biological samples, aiding in the diagnosis of diseases (Alcantara and Josephson 2012; Dulińska-Litewka et al. 2019).

## 8.4 Recent Trends in MNPs-Assisted Diagnosis

Recent trends in MNPs-assisted diagnosis underscore their increasing importance in various disease diagnostic applications, particularly in imaging modalities and POC diagnostic devices.

### *8.4.1 MNPs in Imaging Modalities*

MNPs, particularly those made from materials such as iron oxide and Au, play a crucial role in enhancing contrast in various imaging modalities, including X-ray, computed tomography (CT), ultrasound, magnetic resonance imaging (MRI), and positron emission tomography (PET) and others. Their unique physicochemical properties allow for improved visualization of tissues and pathological conditions, which leads to better diagnostic outcomes (Chouhan et al. 2024; Jagdale et al. 2024). For instance, the X-ray imaging such as, X-rays and CT scan, MNPs, particularly those made from high atomic number materials such as Au and Ag, increase X-ray attenuation. This results in better contrast compared to traditional iodine-based contrast agents (Ahmad et al. 2023). The NPs can be designed to bind specifically to certain biomarkers, thereby localizing the contrast effect to areas of interest, such as tumors or inflamed tissues. This targeted approach enhances diagnostic accuracy by increasing the visibility of specific structures within the body (Ahmad et al. 2023). In the case of targeted imaging and theranostics, complex metal nanostructures, such as nanorods and nanostars, have been engineered for targeted imaging applications. These structures can be modified with ligands and antibodies to specifically target diseased tissues, thereby enhancing the precision of imaging techniques such as near-infrared imaging and SERS (Hajfathalian et al. 2024).

The use of MNPs in ultrasound imaging benefits from the incorporation of gas-filled microbubbles or liposomes, which serve as contrast agents. These bubbles reflect ultrasound waves strongly because of their different acoustic impedance compared to the surrounding tissues. The functionalization of these bubbles enables them to target specific tissues or disease states, thereby improving resolution and facilitating the visualization of vascular structures and tumor angiogenesis (Fu and Ke 2016). Furthermore, in MRI, superparamagnetic iron oxide NPs (SPIONs) enhance contrast by affecting the relaxation times of nearby protons in tissues (Dulińska-Litewka et al. 2019; Sezer et al. 2021). The effectiveness of these NPs as contrast agents is influenced by several factors. The composition of the metal ions impacts both magnetic properties and relaxivity, with manganese-doped iron oxide NPs exhibiting higher relaxivity than pure iron oxide due to their increased magnetic moments. Surface functionalization with various ligands can also enhance biocompatibility and targeting capabilities, facilitating tumor imaging through specific antibody attachments (Alcantara and Josephson 2012; Dulińska-Litewka et al. 2019; Sezer et al. 2021). Moreover, PET imaging utilizes MNPs labeled with radioactive isotopes. These labeled NPs can be used for dual-mode PET/CT imaging, allowing for both functional and anatomical visualization of tumors (Stockhofe et al. 2014; Sarcan 2024; Uritu et al. 2024). The incorporation of high atomic number materials in nanotechnology enhances the imaging capabilities by improving signal detection and localization within the body. Overall, MNPs are revolutionizing imaging techniques by enhancing diagnostic capabilities and enabling targeted imaging.

## 8.4.2 *MNPs in Rapid and Point of Care Diagnosis (POC)*

MNPs are increasingly utilized in rapid and POC diagnostic devices due to their unique properties that enhance sensitivity and specificity in detecting various biomarkers. Some of the key devices that have recently incorporated MNPs are discussed.

### 8.4.2.1 Microfluidics Devices

Microfluidics involves the precise control and manipulation of fluids at a microscale, enabling the development of portable devices. It is foundational for creating systems, such as organ-on-chip, which mimic organ functions under controlled conditions (Syama and Mohanan 2021). The application of microfluidics is widely used in regenerative medicine, drug screening, and cancer theranostics, where it aids in the diagnosis and treatment of cancers by utilizing biomarkers and NPs (Regmi et al. 2022; Lei et al. 2024). The technology offers high precision with minimal sample volumes, making it cost-effective and efficient for various biomedical applications (Carvalho et al. 2022). MNPs integrated into microfluidic assay kits represent a significant advancement in disease diagnosis, particularly in resource-limited settings. MNPs enable the detection of low-abundance biomarkers, such as microRNAs, which are crucial for early disease diagnosis. It facilitates quick results, which are essential for making timely clinical decisions (Wen 2019). Paper-based microfluidic devices offer a cost-effective alternative to traditional diagnostics, making them accessible in developing regions (Tseng et al. 2021; Zhou et al. 2022).

### 8.4.2.2 Lab-on-Chip (LOC) Devices

LOC devices integrate microfluidics with other technologies such as nanosensors and microelectronics, to perform multiple laboratory functions on a single chip. This integration allows for automation and high-throughput screening (Jones 2006). The integration of MNPs with LOC technologies represents a significant advancement in diagnostic methodologies, particularly impactful in rapid and POC settings. For instances, MNPs are utilized in LOC devices to improve detection sensitivity and specificity. Their high surface area and unique optical properties enable the detection of low concentrations of biomarkers, which is essential for early disease diagnosis (Nor et al. 2022; Ait-Touchente et al. 2024). These NPs are often functionalized to target specific biomarkers, enhancing the precision of diagnostic assays. This functionalization allows for the simultaneous detection of multiple analytes, a process known as multiplexing, which is crucial for comprehensive disease profiling (Şimşek et al. 2023; Ait-Touchente et al. 2024). For instance, LOC technologies have been applied to diagnose a range of diseases, including infectious diseases caused by ESKAPE pathogens (Gutiérrez-Santana et al. 2024). MNPs facilitate rapid and accurate pathogen detection, which is crucial for managing antimicrobial resistance and enhancing public health outcomes (Gutiérrez-Santana et al. 2024; Moradi et al. 2024). In cancer diagnostics, MNP-based LOC systems have been used to detect specific cancer biomarkers, providing a platform for early detection and personalized treatment strategies (Firatligil-Yildirir and Yalcin-Ozuysal 2023). Overall, MNP-assisted LOC technologies represent a significant advancement in diagnostic methodologies, offering enhanced sensitivity, specificity, and rapid analysis capabilities.

#### 8.4.2.3 MNPs-Based Biosensors

Biosensors are analytical devices that convert a biological response into an electrical signal, enabling the detection of specific biomolecules. They typically consist of a biorecognition element (such as enzymes, antibodies, or nucleic acids) and a transducer that converts the biological interaction into a measurable signal. Based on the type of transducer, sensors are categorized as electrochemical, optical, piezoelectric, and calorimetric. MNPs have become integral to enhancing biosensor performance due to their unique properties.

- **Electrochemical biosensors**

The use of MNPs in electrochemical biosensors has become a trend in disease diagnosis due to their unique properties that enhance sensor performance. Au and AgNPs are particularly prominent in this field, offering various advantages in terms of sensitivity, selectivity, and cost-effectiveness. They facilitate electron transfer and provide a high surface area, which is crucial for biomolecule immobilization (Siciliano et al. 2023; Eivazzadeh-Keihan et al. 2024). The modification of electrodes with AuNPs has led to significant improvements in the sensitivity and detection limits of electrochemical biosensors, making them suitable for various biomedical applications (Siciliano et al. 2023). Furthermore, AgNPs are valued for their high electrical conductivity and ability to enhance signal amplification in biosensors. They are particularly effective in electrochemical aptamer sensors, which are used for detecting specific proteins and bioactive substances (Zhao 2023). Metal oxides such as $Fe_3O_4$, ZnO, $TiO_2$, NiO, and CuO are commonly used in biosensors. These materials are valued for their electrochemical properties and are often modified to enhance their performance in biosensors (Ansari et al. 2010; Keles et al. 2024). Metal oxides are used in biosensors for detecting various biological analytes, including glucose, cholesterol, and cancer biomarkers. Their high sensitivity and selectivity make them suitable for early diagnosis and point-of-care applications (Keles et al. 2024). The Gold-Linked Electrochemical ImmunoAssay (GLEIA) utilizes AuNPs for signal amplification, achieving a lower limit of detection and improved sensitivity in immunoassays (Osaki et al. 2022). Moreover, AgNPs, along with other metals such as Pt and Cu, are used in electrochemical sensing due to their high electrical conductivity and surface area, which enhances biomarker detection (Şimşek et al. 2023).

- **Optical biosensors**

The use of MNPs in optical biosensors has seen significant advancements, particularly with metals such as Au, Ag, and Cu, which are being increasingly utilized due to their unique properties. For instance, AuNPs are extensively used in optical biosensors due to their excellent optical properties, such as LSPR, which enhances signal detection (Nie et al. 2014; Tai et al. 2022). AuNPs have been successfully employed in the detection of cancer biomarkers, offering high sensitivity and specificity (Tai et al. 2022). Furthermore, AgNPs, though less stable than Au, offer unique optical and electrochemical properties that enhance biosensor performance. They are particularly effective in POC applications, where their ability to detect low concentrations of analytes is crucial (Beck et al. 2023). Recent strategies have focused on stabilizing AgNPs to prevent oxidation, thereby improving their reliability in diagnostic applications (Beck et al. 2023). Additionally, Cu nanoclusters are gaining attention due to their high fluorescence and cost-effectiveness. They are used in optical imaging technologies, which are crucial for early disease detection and imaging-guided treatment (Babu Busi et al. 2022). CuNPs offer advantages over Au and Ag in terms of raw abundance and cost, making them a promising alternative for widespread diagnostic use (Babu Busi et al. 2022). The interaction of metal oxides with biomolecules can be tailored through surface modifications, allowing for the development of highly specific and sensitive biosensors. These are particularly useful in applications requiring rapid and precise detection, such as in food safety and biomedical diagnostics (Maciulis et al. 2022). These NPs are integrated with digital platforms to enhance diagnostic capabilities, especially in resource-limited settings.

- **Piezoelectric biosensors**

Piezoelectric biosensors are analytical devices that utilize the piezoelectric effect to detect biological interactions by measuring changes in mass on a sensor surface (Pohanka 2018). These biosensors are highly sensitive and selective, making them suitable for various applications, including disease diagnosis. The principle of piezoelectric biosensors involves the use of piezoelectric crystals, which change their resonant frequency in response to mass changes, allowing for the detection of biomolecular interactions such as antigen–antibody binding or nucleic acid hybridization (Swain et al. 2023). The integration of MNPs, particularly AuNPs, into piezoelectric biosensors enhances their sensitivity and functionality, providing advantages in disease diagnosis. For instance, in a quartz crystal microbalance (QCM) immunosensor, AuNPs were employed to detect prostate-specific antigen (PSA) with a detection limit of 48 pg/mL, compared to 687 pg/mL without gold staining enhancement (Kwak and Lee 2019). In tuberculosis diagnosis, AuNPs-based biosensors have demonstrated superior biocompatibility and catalytic properties, rendering them effective in detecting *Mycobacterium tuberculosis* with high accuracy (Eivazzadeh-Keihan et al. 2024).

- **Calorimetric sensors**

Calorimetric biosensors are analytical devices that measure the heat change associated with biochemical reactions, providing insights into molecular interactions and reaction kinetics (Das et al. 2022). These sensors are particularly useful in biochemical, clinical, and pharmaceutical industries due to their ability to continuously monitor reactions and their reduced size compared to traditional calorimeters. MNPs, such as Au, Ag, and Pt, are known for their high surface area and excellent conductivity, which significantly enhance the sensitivity and selectivity of calorimetric sensors (Maity et al. 2023; Şimşek et al. 2023). MNPs are used in biosensors to detect infectious diseases by targeting specific DNA, proteins, or pathogens, enhancing the diagnostic capabilities of calorimetric sensors (Maity et al. 2023; Şimşek et al. 2023). MNPs exhibit unique optical properties that enable colorimetric sensing through color changes when interacting with target analytes (Geng et al. 2022; Natarajan et al. 2024). MNPs have a high surface-to-volume ratio, which enhances their interaction with biological molecules, increasing detection sensitivity (Şimşek et al. 2023).

## 8.5 Applications of MNPs in Disease Diagnosis

MNPs have garnered significant attention in the field of disease diagnosis due to their unique properties, which enhance sensitivity and specificity. Several studies have reported the diagnostic potential of MNPs against cancer and cardiovascular disease diagnosis.

### *8.5.1 Cancer Diagnosis*

The use of MNPs in cancer diagnosis has demonstrated significant promise (Table 8.2). These NPs leverage unique properties to enhance sensitivity and specificity across several diagnostic techniques. For instance, utilizing laser-induced fluorescence and diffuse reflectance techniques, AuNPs enhanced the fluorescence signal from cancer cells, achieving a sensitivity of 94% and specificity of 87%, which significantly outperformed traditional methods (Nour et al. 2022). Similarly, improvements in imaging techniques and drug delivery systems have been observed with MNPs in the diagnosis and treatment of lung cancer (Zheng et al. 2023). Xi et al. (2014) employed near-infrared (NIR) light-induced photoacoustic tomography combined with NIR dye-labeled magnetic iron oxide nanoparticles (FeO NPs) for enhanced imaging of breast tumors in murine models. This approach yielded improved specificity and resolution compared to conventional imaging methods. Zinc oxide nanoparticles (ZnO NPs) can distinguish between cancerous and normal cells, permeating cell lines such as MCF7 while showing minimal accumulation in normal cells (Ibraheem et al. 2022). ZnO-based quantum probes were developed for SERS, enabling label-free detection of cancerous cells at a single-cell level (Haldavnekar et al. 2018). Moreover, chitosan-coated MNPs conjugated with surviving antisense oligonucleotides have been developed for targeted localization in pancreatic tumors, facilitating MRI-based tumor detection (Wang et al. 2016).

FeO NPs have been employed in multiple studies for the diagnosis of pancreatic and liver cancers. For instance, FeO NPs can target the surviving proteins in pancreatic cancer cells, thereby improving MRI imaging localization (Chen et al. 2015). Similarly, Li et al. (2018) demonstrated that magnetic targeting with FeO NPs enhances imaging specificity for liver cancer cells through MRI techniques. Other applications of AuNPs include their use in improving prostate cancer detection through optical imaging and PET (Pretze et al. 2019). These NPs were effective in imaging technologies such as optical coherence tomography and reflectance confocal microscopy due to their strong surface plasmon resonance scattering properties (Jain et al. 2007). In another study, Popovtzer et al. (2008) found that concentrated assemblies of AuNPs on head and neck cancer cells lead to strong selective X-ray attenuation during CT scans. Quantum dots have also been investigated, where aluminum-based quantum probes were utilized for SERS to enhance detection sensitivity for cancer cells (Ganesan et al. 2020). Copper NPs can generate singlet oxygen for targeted imaging and therapy in HeLa cells, which are derived from cervical cancer (Joseph et al. 2022). On the other hand, Lu et al. (2016) and Chen et al. (2019) used sandwich-type electrochemical immunosensors to amplify signals from carcinoembryonic antigen tumor markers, enhancing diagnostic accuracy. These studies collectively underscore the multifunctional applications of MNPs in cancer diagnostics, enhancing imaging contrast and increasing sensitivity for biomarker detection.

### 8.5.2 Cardiovascular Diseases

Similar to cancer, MNPs are increasingly utilized in the diagnosis of cardiovascular diseases due to their ability to enhance imaging quality and sensitivity, which are crucial for managing heart diseases. Several studies have demonstrated the potential of various NPs in improving the accuracy and efficacy of heart disease diagnostics. For instance, the nanomaterials such as microbubbles, nanobubbles, and nanodroplets prepared using Au have been tested as ultrasound contrast agents for heart disease detection. These materials create a contrast due to the difference in

**Table 8.2** Diagnosis of cancer and cardiovascular disease by utilizing MNPs

| Type of MNPs | Target | Diagnosis techniques | Role of MNPs | References |
|---|---|---|---|---|
| *Cancer diagnosis* | | | | |
| FeO NPs | Pancreatic cancer cells | MRI | Targets the surviving protein for localized imaging | Chen et al. (2015) |
| FeO NPs | Liver cancer cells | MRI | Enhance contrast imaging of cancer cells | Li et al. (2018) |
| AuNPs and CuNPs | Prostate cancer cells | Optical imaging (OI) and PET | Enhance detection of prostate cancer via OI and PET | Pretze et al. (2019) |
| AuNPs | Epithelial cancer cells | Imaging technologies such as optical coherence tomography (OCT) and Reflectance confocal microscopy (RCM) | Strongly enhanced SPR scattering and absorption | Jain et al. (2007) |
| Al-based quantum probe | Cancer cells | SERS | Strongly enhanced SERS | Ganesan et al. (2020) |
| $MnFe_2O_4$@ $SiO_2$ | Cancer cells | MRI | Enhance contrast imaging of cancer cells | Kavkhani et al. (2022) |
| AuNPs | Head and neck cancer cells | X-ray and CT scan | A concentrated assembly of AuNPs that form exclusively on the targeted cancer cells yields a strong, selective x-ray attenuation | Popovtzer et al. (2008) |
| NiNPs | Carcinoembryonic antigen (CEA) tumor marker | Sandwich-type electrochemical immunobiosensor | Ni/C@$SiO_2$ has a large specific surface area and therefore has a wealth of binding sites connected to Anti-CEA and accelerates the electron transfer | Song et al. (2021) |
| AuNPs | Mi RNA tumor marker | Electrochemical assay | Amplified electrochemical assay | Lu et al. (2016) |
| AuNPs and AgNPs | Carcinoembryonic antigen (CEA) tumor marker | Electrochemical immunosensor | Amplified electrochemical assay | Chen et al. (2019) |

(continued)

**Table 8.2** (continued)

| Type of MNPs | Target | Diagnosis techniques | Role of MNPs | References |
|---|---|---|---|---|
| AuNPs, AgNPs, ZnO NPs | Biomarkers, including DNA, RNA, proteins, and lipids of cancer cells | SERS | Charge transfer resonance due to Herzberg-Teller vibrionic coupling and photo-induced charge transfer | Haldavnekar et al. (2018) |
| CuNPs | Hela cells derived from cervical cancer | Imaging | The copper nanocluster generates singlet oxygen and facilitates targeted imaging and therapy through the conjugation of a tumor-targeting agent and an anticancer drug | Joseph et al. (2022) |
| ZnO NPs | MCF7 AMJ13 and RD cancer cells | Diagnostic imaging techniques | ZnO NPs work as a fluorophore for diagnostic imaging in the early identification | Ibraheem et al. (2022) |
| AuNPs | MCF-7 and murine melanoma B16 | Electrochemical nanopipette sensor | Enhanced stability and selectivity | Timoshenko et al. (2023) |
| AuNPs | MDA-MB-231 cancer cells | MRI | Enhance contrast imaging of cancer cells | Chauhan et al. (2019) |
| Tantalum oxide (TaOx) AuNPs | Hepatocellular carcinoma cells | Gemstone spectral computer tomography (GSCT) with photoacoustic (PA) imaging | Imaging contrast agent in cancer diagnostics | Lei et al. (2022) |
| *Cardiovascular disease diagnosis* | | | | |
| FeO | Ischemic myocardium | Ultrasound, photoacoustic, MRI | Enhance contrast imaging | Chen et al. (2021a) |
| Lanthanide NPs | Cardiac Troponin I (cTnI) | Lanthanide-doped NPs-based POC diagnosis | Lanthanide-doped NPs enhance luminescent signals | Chen et al. (2022) |
| MnNPs | Acute myocardial infarction markers | MRI | Manganese carbonate NPs improve MRI contrast | Li et al. (2022) |

(continued)

**Table 8.2** (continued)

| Type of MNPs | Target | Diagnosis techniques | Role of MNPs | References |
|---|---|---|---|---|
| MnONPs | Myocardial infarction markers | MRI | MnO NPs serve as dual-function agents, providing contrast enhancement in MRI while delivering therapeutic agents | Zheng et al. (2018) |
| AuNPs | Cardiac biomarkers | Rapid diagnostic assays | AuNPs improve detection speed and sensitivity through enhanced colorimetric responses | Namdari et al. (2017) |
| AuNPs | Atherosclerotic plaques, Intravascular thrombus, or fibrotic tissue | Imaging techniques | AuNPs enhance imaging contrast | Varna et al. (2018) |
| FeO NPs | Macrophage markers | Fluorescence tomography and MRI | Iron oxide NPs enhance imaging contrast | Sosnovik et al. (2007) |
| FeNPs | Atherosclerotic plaque markers | In vitro cytocompatibility testing | Carbon-coated iron NPs improve stent biocompatibility and reduce inflammatory responses in atherosclerotic plaques | Akhmedov et al. (2021) |
| FeO NPs | Various cardiovascular markers | Gel coating techniques | $Fe_3O_4$ NPs enhance the stability and biocompatibility of gel coatings on stents, and improving performance | Hataminia et al. (2022) |
| FeONPs | Atherosclerotic plaque | MRI | FeO enhances image contrast | Makowski et al. (2011) |
| AgNPs | Cardiac troponin-I (cTnI) | SERS | Enhanced detection | Naz et al. (2024) |
| AuNPs | Diseased heart region (DHR) | Ultrasound | Creates contrast due to differences in acoustic impedance | Alphandéry (2022) |

acoustic impedance, which can be enhanced by targeting the diseased heart region (DHR) through passive, active, or magnetic methods (Alphandéry 2022). Similarly, by the use of FeO a nanoprobe (IMTP-$Fe_3O_4$-PFH NPs) has been developed for multimodal imaging, combining ultrasound, photoacoustic, and MRI to visualize ischemic myocardium in rats. These NPs show excellent targeting performance and stability, making them suitable for early detection and monitoring of ischemic heart conditions (Chen et al. 2021a). Similarly, Zheng et al. (2018), utilized MnO-based NPs for MRI and near-infrared fluorescence imaging, providing an efficient approach for detecting myocardial infarction. These NPs exhibit high relaxivity and preferentially accumulate in the infarcted myocardium, enhancing imaging specificity without cardiotoxicity.

Additionally, manganese oxide NPs serve as dual-function agents that provide contrast enhancement in MRI while delivering therapeutic agents directly to the site of myocardial infarction (Zheng et al. 2018). Li et al. (2019), used Lanthanide metal–organic framework nanoprobes for detection of cardiac disease markers, such as those associated with acute myocardial infarction (AMI). These nanoprobes offer a sensitive and specific method for early diagnosis, which is critical for reducing mortality rates. Manganese carbonate NPs also play a crucial role in MRI by improving contrast for acute myocardial infarction markers (Li et al. 2022). AuNPs enhance the speed and sensitivity of rapid diagnostic assays for various cardiac biomarkers, improving detection rates for cardiac failure (Namdari et al. 2017). They also strengthen imaging contrast, which is vital for visualizing atherosclerotic plaques and other cardiovascular structures during diagnostic procedures (Varna et al. 2018). Iron oxide NPs have been effectively used to visualize macrophage infiltration in myocardial infarction through fluorescence tomography and MRI, enhancing imaging contrast and aiding in the assessment of inflammatory responses (Sosnovik et al. 2007). Furthermore, carbon-coated iron NPs enhance stent biocompatibility and mitigate inflammatory responses in atherosclerotic plaques (Akhmedov et al. 2021). Iron oxide NPs also contribute to the stability and biocompatibility of gel coatings on stents, enhancing their performance in cardiovascular applications (Hataminia et al. 2022). Recent advancements with AgNPs have shown promise in improving sensitivity through SERS, allowing for the detection of trace amounts of cardiac troponin-I (Naz et al. 2024). Overall, MNPs significantly enhance diagnostic capabilities in cardiovascular diseases through improved imaging contrast, increased sensitivity for biomarker detection, and multifunctional applications.

### *8.5.3 Diagnosis of Infectious Diseases Using MNPs*

NPs have emerged as a novel tool for diagnosing a range of infectious diseases, including those caused by parasites, bacteria, and viruses, with enhanced sensitivity, specificity, and treatment efficacy.

- **Parasitic infections**

Several studies highlight the innovative applications of MNPs in diagnosing parasitic infectious diseases through various advanced techniques. For instance, Deris et al. (2022) demonstrated the efficacy of AuNPs-probe conjugate nanobiosensors for leishmaniasis diagnosis, utilizing both amplification and non-amplification methods to enhance signal detection. A nano-based enzyme-linked immunosorbent assay (ELISA) was employed for early detection of *Trichinella spiralis*, improving sensitivity by increasing the binding efficiency of antibodies to antigens (Gomaa 2020). Conversely, a fluorescent immunosensor using chitosan-ZnO NPs was developed for *Toxoplasmosis* detection, where MNPs amplified fluorescence signals, enabling sensitive detection at low pathogen concentrations (Medawar-Aguilar et al. 2019). Similarly, a nanogold particle-based ELISA for hydatidosis, leveraging the surface plasmon resonance properties of AuNPs to enhance colorimetric responses and improve visual quantification (Rashed et al. 2018). Rodríguez et al. (2013) focused on functionalized AuNPs conjugated with anti-gp63 antibodies for detecting the glycoprotein gp63 on *Leishmania* parasites, thereby enhancing specificity and sensitivity through targeted detection. Non-spherical AuNPs in label-free electrochemical biosensing to detect *Trichomonas vaginalis* RNA, providing a rapid method that detects target binding through changes in electrical signals (Vais et al. 2023). A piezoelectric immunoagglutination assay utilizing AuNPs to detect *Toxoplasma gondii* antibodies enhances detection sensitivity by measuring mass changes resulting from nanoparticle aggregation (Wang et al. 2004). Conversely, a smartphone-based enhanced colorimetric immunoassay for *Trichinella spiralis* infection, allowing for easy interpretation of visual color changes facilitated by MNPs (Yu et al. 2024). Additionally, Heshmati et al. (2023) explored the role of MNPs in preventing and treating parasitic diseases in poultry, highlighting their potential as effective therapeutic agents and preventive measures against infections.

- **Bacterial infections**

MNPs have proven to be transformative in enhancing disease diagnosis across various bacterial and infectious agents. For instance, AuNPs have been utilized in colorimetric detection methods, leveraging their optical properties to indicate bacterial presence through observable color changes. This technique does not require complex instrumentation, making it suitable for early diagnosis and monitoring (Buică et al. 2023). Furthermore, in detecting *Chlamydia trachomatis*, gold nanoparticle-based approaches have shown notable advancements, enabling rapid and accurate identification without the need for skilled technicians or costly equipment (Moradi et al. 2024). Additionally, palladium-platinum bimetallic NPs (Pd@Pt NPs) have been incorporated into paper-based analytical devices for detecting *Mycobacterium tuberculosis*. These nanozymes exhibit peroxidase-like activity, which enhances the oxidation rate of substrates, enabling the sensitive detection of target DNA within 15 min (Tung et al. 2024). Additionally, Chan et al. (2013) utilized gold nanoclusters in affinity mass spectrometry, which significantly improved the sensitivity and accuracy toward bacterial pathogen detection. Subsequently, microconductometric immunosensors that exploit the unique conductive properties of metals for label-free detection of Gram-negative bacteria, facilitating rapid diagnostics (El Ichi et al.

2014). In addressing antibiotic resistance, Srisrattakarn et al. (2017) introduced the GoldNano carb test, which enables the rapid phenotypic detection of carbapenemases in Enterobacteriaceae, *Pseudomonas aeruginosa*, and *Acinetobacter* spp. An ultrasensitive electrochemical detection of bacteremia using redox-active AuNPs in a nano-sieving microfluidic system, highlighting the potential of MNPs in clinical diagnostics (Lee et al. 2019). Subsequently, photonic hydrogels containing MNPs could provide synergistic visual bacterial detection and facilitate on-site photothermal disinfection (Xie et al. 2022). Furthermore, Wen et al. (2023) developed colorimetric nanosensors that couple plasmonic NPs with magnetic separation for sensitive multi-pathogen detection. Recent advancements have showcased the use of urchin-shaped Au–Ag@Pt sensors integrated into lateral flow immunoassays for the multimodal detection of bacterial infections (Huang et al. 2023). The potential of MNPs has also been illustrated in several studies, as shown in Table 8.3.

- **Viral infections**

Several studies highlight the application of MNPs in diagnosing various viral infectious diseases. For instance, AuNPs lateral flow test (GNP-LFT) for the rapid field diagnostics of foot-and-mouth disease virus (FMDV) and lumpy skin disease virus (LSDV) achieves high sensitivity and specificity, making it a valuable tool for on-site testing (Abdalhamed et al. 2022). Similarly, an AuNPs-based biosensor for the ultrasensitive diagnosis of FMDV enhances detection capabilities through improved binding and signal amplification (Hamdy et al. 2018). In the realm of advanced diagnostics, Huelck (2024) utilized SERS and LSPR for the rapid detection of SARS-CoV-2, showcasing the effectiveness of AuNPs in enhancing signal sensitivity. A low-cost plasmonic biosensor has been introduced for the label-free detection of viruses in resource-limited settings, emphasizing accessibility and ease of use (Kurul et al. 2023). Conversely, Li (2023) explored the use of encapsulated AuNPs probes for virus detection, demonstrating their potential to improve diagnostic accuracy. Liu et al. (2022a) highlighted the use of Au nanourchins to enhance virus targeting and plasmonic coupling for smartphone-based diagnostics. Another study focused on digital plasmonic nanobubble detection for rapid and ultrasensitive diagnostics (Liu et al. 2022b). A nano-assembly techniques demonstrated quick SARS-CoV-2 detection at point-of-care settings (Park et al. 2022). A rapid electronic diagnostic for the Ebola virus using synthetic nanobody-conjugated AuNPs enhanced specificity and sensitivity (Chen et al. 2021b). An ultraviolet-induced in situ AuNPs for POC testing of infectious diseases using loop-mediated isothermal amplification illustrated the versatility of MNPs in viral diagnostics (Sivakumar et al. 2021). These studies highlight the transformative role of MNPs in enhancing diagnostic mechanisms for various viral infections, thereby facilitating rapid, sensitive, and specific detection methods that are crucial for effective disease management.

**Table 8.3** MNPs in the diagnosis of infectious diseases caused by parasites, bacteria, and viruses

| Type of MNPs | Pathogens | Diagnosis techniques | Key takeaways | References |
|---|---|---|---|---|
| ZnO NPs | Parasite | LIF immunosensor | Improved the sensitivity | Medawar-Aguilar et al. (2019) |
| AuNPs | Parasite | Piezoelectric immunoagglutination Assay | Improved sensitivity and reaction rate | Wang et al. (2004) |
| AuNPs | Parasite | Electrochemical DNA biosensors | Enhance the rate of electron transfer | Vais et al. (2023) |
| AuNPs | Parasite | Nano-based ELISA | Improved sensitivity and reaction rate | Gomaa (2020) |
| AuNPs | Parasite | Colorimetric biosensor | AuNPs change color based on aggregation | Deris et al. (2022) |
| AuNPs | Parasite | Nano-based ELISA | Improved sensitivity and reaction rate | Rashed et al. (2018) |
| AuNPs | Parasite | Colorimetric immunoassay | Enhance Ag-Ab binding | Yu et al. (2024) |
| AuNPs | Parasite | Nano-based ELISA | Enhance specific interaction | Rodríguez et al. (2013) |
| AuNPs | Bacteria | Mass spectrometry | Gold nanoclusters enhance mass spectrometry sensitivity for bacterial detection | Chan et al. (2013) |
| Magnetic NPs | Bacteria | Microconductometric immunosensor | Metal surfaces enhance the conductivity change upon bacterial binding | El Ichi et al. (2014) |
| AuNPs | Bacteria | GoldNano carb test | AuNPs enhance detection through colorimetric changes | Srisrattakarn et al. (2017) |
| AuNPs | Bacteria | Electrochemical detection | Redox-active properties of AuNPs enhance electrochemical signals | Lee et al. (2019) |
| FeO | Bacteria | Photonic hydrogel assays | MNPs enhance light absorption for improved bacterial detection | Xie et al. (2022) |

(continued)

**Table 8.3** (continued)

| Type of MNPs | Pathogens | Diagnosis techniques | Key takeaways | References |
|---|---|---|---|---|
| AuNPs AgNPs | Bacteria | Colorimetric nanosensor | Plasmonic NPs enhance color changes upon target binding | Wen et al. (2023) |
| AuNPs AgNPs | Bacteria | Lateral flow immunoassay | A multimetal system enhances sensitivity and specificity in detection | Huang et al. (2023) |
| FeO | Bacteria | SERS assay | Magnetic NPs improve capture and signal amplification in detection | Li et al. (2023) |
| FeO | Bacteria | Nanosystems For Bacterial Identification | Magnetic properties enhance the isolation and identification of bacteria | Wang et al. (2018) |
| AuNPs AgNPs | Bacteria | Electrochemical Biosensor | Silver wire enhances conductivity for rapid bacterial detection | Feng et al. (2020) |
| AgNPs | Bacteria | SERS | Silver enhances Raman signals for the detection of metabolites | Žukovskaja et al. (2017) |
| AgNPs | Bacteria | Surface plasmon resonance sensor | AgNPs enhance surface plasmon resonance for detection | Zhou et al. (2018) |
| AuNPs | Virus | Calorimetric sensing | NPs enhance color changes upon target binding | Li et al. (2023) |
| AuNPs | Virus | Digital plasmonic nanobubble detection (DIAMOND) | The formation of nanobubbles enhances signal detection through light scattering | Liu et al. (2022a) |
| AuNPs | Virus | RT-PCR | Colorimetric property enhanced detection capability | Sivakumar et al. (2021) |

(continued)

**Table 8.3** (continued)

| Type of MNPs | Pathogens | Diagnosis techniques | Key takeaways | References |
|---|---|---|---|---|
| AuNPs | Virus | Lateral flow assay | Colorimetric changes indicate the presence of viral antigens | Abdalhamed et al. (2022) |
| AuNPs | Virus | RT-PCR | Plasmonic coupling increases sensitivity for real-time detection | Park et al. (2022) |
| AuNPs | Virus | Plasmonic biosensing | Enhanced detection sensitivity through plasmonic effects | Kurul et al. (2023) |
| AuNPs | Virus | Electronic diagnostics | Enhanced specificity and sensitivity via nanobody targeting | Chen et al. (2021b) |
| AuNPs | Virus | Raman spectroscopy | Enhanced Raman signals improve detection sensitivity | Huelck (2024) |
| AuNPs | Virus | Smartphone-based detection | Improved targeting and signal through plasmonic coupling | Liu et al. (2022b) |
| AuNPs | Virus | Biosensor technology | Improved sensitivity via enhanced binding and signal amplification | Hamdy et al. (2018) |

## 8.6 Challenges in Diagnosis Using Metal Nanoparticles

MNPs are increasingly utilized in biosensors and diagnostic tools due to their unique physicochemical properties; however, several factors hinder their clinical application. These include issues related to stability and reproducibility, toxicity, environmental factors, and regulatory and commercialization barriers. MNPs demand careful synthesis to ensure stability and functionality. Variations in size, shape, and surface chemistry can significantly impact their performance in diagnostic applications (Poon et al. 2024). Achieving reproducibility is essential for widespread clinical use; however, consistent results in synthesis remain elusive (Poon et al. 2024). The size and shape of MNPs are critical factors influencing their toxicity; smaller NPs tend to have a higher surface area-to-volume ratio, which can enhance reactivity and potentially lead to harm (Eker et al. 2024; Ijaz et al. 2024).

Environmental factors, such as the presence of other substances in biological samples, can interfere with the detection capabilities of MNPs, potentially leading to false positives or negatives (Nyembe et al. 2023). Moreover, the unique properties of nanomaterials necessitate specialized safety assessments. Regulatory agencies, such as the FDA and EMA, are developing guidelines to address these requirements; however, the absence of standardized protocols remains a significant hurdle (Anchordoquy et al. 2024; Souto et al. 2024). Current regulatory frameworks often struggle to accommodate the specific features of these products, resulting in delays in approval and market entry (Musazzi et al. 2023). Furthermore, commercialization is complicated by the need for cost-effective production methods that can meet the demands of large-scale manufacturing (Priya et al. 2023). To fully harness the potential of MNPs in disease diagnosis, these challenges must be systematically addressed. The development of multifunctional biosensors and theragnostic systems could offer viable solutions but will require interdisciplinary collaboration and innovation. Overcoming these barriers is essential for successfully integrating MNPs into clinical practice, which could significantly transform diagnostic and therapeutic strategies across various diseases.

## 8.7 Conclusion and Future Perspectives

Medical technology has advanced tremendously with the use of MNPs in disease identification. MNPs possess distinctive qualities that enhance the sensitivity and specificity of diagnostic tests, facilitating early illness diagnosis and enabling personalized therapeutic strategies. Recent trends of MNPs in diverse disease diagnostic tools, such as imaging modalities, microfluidics, LOC, and biosensor devices, highlight their multidimensional utility. Notwithstanding MNPs' encouraging promise, there are still obstacles that need to be overcome to properly utilize them in clinical practice, including problems with stability, toxicity, and regulatory compliance. To overcome these obstacles and maximize the application of MNPs in disease detection, further studies are necessary. Several key perspectives will shape the future of MNPs in disease diagnostics:

- **Next-generation NPs**: Future advancements will concentrate on developing upgraded MNPs with improved functionality for targeted diagnostics.
- **Integration with artificial intelligence**: The integration of artificial intelligence into the analysis of data from MNP-based diagnostics will improve accuracy and allow for customized treatment plans by finding patterns in intricate datasets.
- **Cross-disciplinary collaborations**: Innovation in nanoparticle design and application methods will be fostered by collaborations between materials scientists, biologists, chemists, and medical specialists.
- **Regulatory framework development**: Establishing strong regulatory frameworks is critical for the safe incorporation of MNPs into clinical practice.
- **Focus on safety and biocompatibility**: Ongoing research will prioritize improving MNPs' safety profiles to reduce possible adverse reactions while maintaining efficacy.

These future views highlight the revolutionary potential of MNPs in disease diagnostics, as well as the significance of ongoing research and multidisciplinary collaboration to fully realize their therapeutic uses.

## References

Abbas R, Luo J, Qi X, Naz A, Khan IA, Liu H, Yu S, Wei J (2024) Silver nanoparticles: synthesis, structure, properties and applications. Nanomaterials 14(17)

Abdalhamed AM, Naser SM, Mohamed AH, Zeedan GSG (2022) Development of gold nanoparticles-lateral flow test as a novel field diagnostic assay for detecting foot-and-mouth disease and lumpy skin disease viruses. Iran J Microbiol 14(4):574

Ahmad MY, Liu S, Tegafaw T, Saidi AKAA, Zhao D, Liu Y, Nam SW, Chang Y, Lee GH (2023) Heavy metal-based nanoparticles as high-performance x-ray computed tomography contrast agents. Pharmaceuticals 16(10):1463

Ahmad MZ, Mohammed AA, Pathak K, Gogoi U, Saikia R, Ahmad J (2022) Metallic nanomaterials for the diagnosis and treatment of infectious diseases. In: Nanotheranostics for treatment and diagnosis of infectious diseases, pp 289–317

Ait-Touchente Z, Khizar S, Zine N, Jaffrezic-Renault N, Errachid A, Elaissari A (2024) Lab-on-a-chip for in vitro diagnosis: manipulating the power of nanoparticles. In: Detection science. Royal Society of Chemistry, pp 244–271. https://doi.org/10.1039/9781837673476-00244

Akhmedov S, Afanasyev S, Trusova M, Postnikov P, Rogovskaya Y, Grakova E, Kopeva K, Carreon Paz RK, Balakin S, Wiesmann H-P (2021) Chemically modified biomimetic carbon-coated iron nanoparticles for stent coatings: in vitro cytocompatibility and in vivo structural changes in human atherosclerotic plaques. Biomedicines 9(7):802

Alcantara D, Josephson L (2012) Magnetic nanoparticles for application in biomedical sensing. In: Frontiers of nanoscience, vol 4. Elsevier, pp 269–289

Alphandéry E (2022) Nanomaterials as ultrasound theragnostic tools for heart disease treatment/diagnosis. Int J Mol Sci 23(3):1683

Alrahili MR (2024) Insights into gold nanoparticle optical behavior: size-dependent absorption and extinction peaks. J Nanoelectron Optoelectron 19(8):783–790

Altammar KA (2023) A review on nanoparticles: characteristics, synthesis, applications, and challenges. Front Microbiol 14:1155622

Anchordoquy T, Artzi N, Balyasnikova IV, Barenholz Y, La-Beck NM, Brenner JS, Chan WCW, Decuzzi P, Exner AA, Gabizon A, Godin B, Lai SK, Lammers T, Mitchell MJ, Moghimi SM, Muzykantov VR, Peer D, Nguyen J, Popovtzer R, Simberg D et al (2024) Mechanisms and barriers in nanomedicine: progress in the field and future directions. ACS Nano 18(22):13983–13999. https://doi.org/10.1021/acsnano.4c00182

Angeloni L, Passeri D, Schiavi PG, Pagnanelli F, Rossi M (2020) Magnetic force microscopy characterization of cobalt nanoparticles: a preliminary study. In: AIP conference proceedings, vol 2257(1). AIP Publishing. https://doi.org/10.1063/5.0022440

Ansari AA, Alhoshan M, Alsalhi MS, Aldwayyan AS (2010) Nanostructured metal oxides based enzymatic electrochemical biosensors. Biosensors 302:23–46

Babu Busi K, Palanivel M, Kanta Ghosh K, Basu Ball W, Gulyás B, Padmanabhan P, Chakrabortty S (2022) The multifarious applications of copper nanoclusters in biosensing and bioimaging and their translational role in early disease detection. Nanomaterials 12(3):301

Balkrishna A, Thakur N, Patial B, Sharma S, Kumar A, Arya V, Amarowicz R (2023) Synthesis, characterization and antibacterial efficacy of *Catharanthus roseus* and *Ocimum tenuiflorum*-mediated silver nanoparticles: phytonanotechnology in disease management. Processes 11(5):1479

Beck F, Loessl M, Baeumner AJ (2023) Signaling strategies of silver nanoparticles in optical and electrochemical biosensors: considering their potential for the point-of-care. Microchim Acta 190(3):91. https://doi.org/10.1007/s00604-023-05666-6

Bellotti R, Picotto GB, Ribotta L (2022) AFM measurements and tip characterization of nanoparticles with different shapes. Nanomanuf Metrol 5(2):127–138

Bennington-Gray S (2019) Magnetic characterisation of nanostructured magnetic systems using SQUID magnetometry (Doctoral dissertation). Queen's University Belfast, Faculty of Engineering and Physical Sciences

Besenhard MO, LaGrow AP, Hodzic A, Kriechbaum M, Panariello L, Bais G, Loizou K, Damilos S, Cruz MM, Thanh NTK (2020) Co-precipitation synthesis of stable iron oxide nanoparticles with NaOH: new insights and continuous production via flow chemistry. Chem Eng J 399:125740

Buică G, Ungureanu C, Baroi AM, Hosu IS, Fierascu I, Fierascu RC (2023) Colorimetric detection of bacteria using Au nanoparticles. Proceedings 90(1):12

Burlec AF, Corciova A, Boev M, Batir-Marin D, Mircea C, Cioanca O, Danila G, Danila M, Bucur AF, Hancianu M (2023) Current overview of metal nanoparticles' synthesis, characterization, and biomedical applications, with a focus on silver and gold nanoparticles. Pharmaceuticals 16(10):1410

Carvalho V, de Teixeira SFCF, Ribeiro J (2022). Micro/nanofluidic and lab-on-a-chip devices for biomedical applications. https://doi.org/10.3390/books978-3-0365-6099-1

Caspani S, Magalhães R, Araújo JP, Sousa CT (2020) Magnetic nanomaterials as contrast agents for MRI. Materials 13(11):2586

Chan P, Wong S, Lin S, Chen Y (2013) Lysozyme-encapsulated gold nanocluster-based affinity mass spectrometry for pathogenic bacteria. Rapid Commun Mass Spectrom 27(19):2143–2148

Chauhan R, El-Baz N, Keynton RS, James KT, Malik DA, Zhu M, El-Baz A, Ng CK, Bates PJ, Malik MT (2019) Targeted gold nanoparticle–oligonucleotide contrast agents in combination with a new local voxel-wise MRI analysis algorithm for in vitro imaging of triple-negative breast cancer. Nanomaterials 9(5):709

Chen C, Wu CQ, Chen TW, Tang MY, Zhang XM (2015) Molecular imaging with MRI: potential application in pancreatic cancer. Biomed Res Int 2015(1):624074

Chen P, Hua X, Liu J, Liu H, Xia F, Tian D, Zhou C (2019) A dual amplification electrochemical immunosensor based on HRP-Au@ Ag NPs for carcinoembryonic antigen detection. Anal Biochem 574:23–30

Chen X, Zhang Y, Zhang H, Zhang L, Liu L, Cao Y, Ran H, Tian J (2021a) A non-invasive nanoparticle for multimodal imaging of ischemic myocardium in rats. J Nanobiotechnol 19:1–18

Chen X, Kang S, Ikbal A, Zhao Z, Zuo J, Yao Y, Gu L, Wang C (2021b) Rapid electronic diagnostics of ebola virus with synthetic nanobody-conjugated gold nanoparticles. Biosens Bioelectron:113971

Chen L, Zhou SY, Zhu W, Liu SP, Zhang JX, Zhuang H, Zhang JL, Li YS, Gao F (2022) Highly sensitive lanthanide-doped nanoparticles-based point-of-care diagnosis of human cardiac troponin I. Int J Nanomed:635–646

Chouhan APS, Verma A, Singh V (2024) The significance of nanomaterials in medical imaging and diagnostics. In: Cutting-edge applications of nanomaterials in biomedical sciences. IGI Global, pp 301–325

Das AB, Kumari T, Sahu PP (2022) Calorimetric biosensors: core principles, techniques, fabrication and application. In: Biosensors in food safety and quality. CRC Press, pp 11–21

Delvallée A, Feltin N, Ducourtieux S, Trabelsi M, Hochepied JF (2015) Direct comparison of AFM and SEM measurements on the same set of nanoparticles. Meas Sci Technol 26(8):85601

Deris S, Osanloo M, Ghasemian A, Ataei S, Kohansal M, Samsami S, Yazdanpanah A, Ebrahimnezhad A, Ghanbariasad A (2022) The efficacy of AuNP-probe conjugate nanobiosensor in non-amplification and amplification forms for the diagnosis of leishmaniasis. BMC Infect Dis 22(1):847

Dikusar AI, Globa PG, Belevskii SS, Sidel'nikova SP (2009) On limiting rate of dimensional electrodeposition at meso-and nanomaterial manufacturing by template synthesis. Surf Eng Appl Electrochem 45:171–179
Dubey S, Virmani T, Yadav SK, Sharma A, Kumar G, Alhalmi A (2024) Breaking barriers in eco-friendly synthesis of plant-mediated metal/metal oxide/bimetallic nanoparticles: antibacterial, anticancer, mechanism elucidation, and versatile utilizations. J Nanomater. https://doi.org/10.1155/2024/9914079
Dulińska-Litewka J, Łazarczyk A, Hałubiec P, Szafrański O, Karnas K, Karewicz A (2019) Superparamagnetic iron oxide nanoparticles—current and prospective medical applications. Materials 12(4):617
Eivazzadeh-Keihan R, Saadatidizaji Z, Mahdavi M, Maleki A, Irani M, Zare I (2024) Recent advances in gold nanoparticles-based biosensors for tuberculosis determination. Talanta:126099
Eker F, Duman H, Akdaşçi E, Bolat E, Sarıtaş S, Karav S, Witkowska AM (2024) A comprehensive review of nanoparticles: from classification to application and toxicity. Molecules 29(15):3482
El Ichi S, Leon F, Vossier L, Marchandin H, Errachid A, Coste J, Jaffrezic-Renault N, Fournier-Wirth C (2014) Microconductometric immunosensor for label-free and sensitive detection of gram-negative bacteria. Biosens Bioelectron 54:378–384
Emrie DB (2024) Sol-Gel synthesis of nanostructured mesoporous silica powder and thin films. J Nanomater 2024(1):6109770
Fathi F, Rashidi M-R, Omidi Y (2019) Ultra-sensitive detection by metal nanoparticles-mediated enhanced SPR biosensors. Talanta 192:118–127
Feng Y, Zhou D, Gao L, He F (2020) Electrochemical biosensor for rapid detection of bacteria based on facile synthesis of silver wire across electrodes. Biosens Bioelectron 168:112527
Firatligil-Yildirir B, Yalcin-Ozuysal O (2023) Recent advances in lab-on-a-chip systems for breast cancer metastasis research. Nanoscale Adv 5(9):2375–2393
Fu L, Ke H-T (2016) Nanomaterials incorporated ultrasound contrast agents for cancer theranostics. Cancer Biol Med 13(3):313
Ganesan S, Venkatakrishnan K, Tan B (2020) Wrinkled metal based quantum sensor for In vitro cancer diagnosis. Biosens Bioelectron 151:111967
Geng H, Vilms Pedersen S, Ma Y, Haghighi T, Dai H, Howes PD, Stevens MM (2022) Noble metal nanoparticle biosensors: from fundamental studies toward point-of-care diagnostics. Acc Chem Res 55(5):593–604
Gomaa MM (2020) Early diagnosis of experimental *Trichinella spiralis* infection by nano-based enzyme-linked immunosorbent assay (nano-based ELISA). Exp Parasitol 212:107867
Greczynski G, Hultman L (2020) X-ray photoelectron spectroscopy: towards reliable binding energy referencing. Prog Mater Sci 107:100591
Gupta D, Roy I, Gandhi S (2023) Metallic nanoparticles for CT-guided imaging of tumors and their therapeutic applications. OpenNano 12:100146
Gutiérrez-Santana JC, Rosas-Espinosa V, Martinez E, Casiano-García E, Coria-Jiménez VR (2024) Metal nanoparticle-based biosensors for the early diagnosis of infectious diseases caused by ESKAPE pathogens in the fight against the antimicrobial-resistance crisis. Biosensors 14(7):339
Haas KH (2021) Application of metal oxide nanoparticles and their economic impact. In: Metal oxide nanoparticles: formation, functional properties, and interfaces, vol 1, pp 29–65
Hajfathalian M, Mossburg KJ, Radaic A, Woo KE, Jonnalagadda P, Kapila Y, Bollyky PL, Cormode DP (2024) A review of recent advances in the use of complex metal nanostructures for biomedical applications from diagnosis to treatment. Wiley Interdisc Rev Nanomed Nanobiotechnol 16(3):e1959
Haldavnekar R, Venkatakrishnan K, Tan B (2018) Non plasmonic semiconductor quantum SERS probe as a pathway for in vitro cancer detection. Nat Commun 9(1):3065
Hamdy ME, Del Carlo M, Hussein HA, Salah TA, El-Deeb AH, Emara MM, Pezzoni G, Compagnone D (2018) Development of gold nanoparticles biosensor for ultrasensitive diagnosis of foot and mouth disease virus. J Nanobiotechnol 16:1–12

Hataminia F, Majidi RF, Shabankareh ANT, Ghanbari H (2022) Green synthesis of oxidized starch with a novel catalyst based on $Fe_3O_4$ nanoparticles and $H_2O_2$ reagent to form thermoplastic as a stable gel coating on the cardiovascular stents. Int J Biol Macromol 219:290–303

Heshmati F, Sangar SG, Amoozadehsamakoosh A, Azadi E, Komeili N (2023) The role of metallic nanoparticles in the prevention and treatment of parasitic diseases in poultry. J World's Poult Sci 2(3):13–19

Higaki T, Jin R (2021) Optical properties of atomically precise gold nanoclusters: transition from excitons to plasmons. In: Atomically precise nanoclusters. Jenny Stanford Publishing, pp 149–163

Holder CF, Schaak RE (2019) Tutorial on powder X-ray diffraction for characterizing nanoscale materials. ACS Nano 13(7):7359–7365

Huang X, Chen L, Zhi W, Zeng R, Ji G, Cai H, Xu J, Wang J, Chen S, Tang Y (2023) Urchin-shaped Au–Ag@ Pt sensor integrated lateral flow immunoassay for multimodal detection and specific discrimination of clinical multiple bacterial infections. Anal Chem 95(35):13101–13112

Huelck S (2024) Advancing viral diagnostics: harnessing surface enhanced Raman scattering and localized surface plasmon resonance for rapid detection of SARS CoV-2. In: Biomedical spectroscopy, microscopy, and imaging III, vol 13006, pp 43–48

Ibraheem S, Kadhim AA, Kadhim KA, Kadhim IA, Jabir M (2022) Zinc oxide nanoparticles as diagnostic tool for cancer cells. Int J Biomater 2022(1):2807644

Ijaz MU, Akbar A, Ashraf A, Saqalein M, Khan HA, Hayat S, Muzammil S (2024) Toxicity of nanoparticles in biological systems. Nanopart Toxic Compat 161:245–266

Jagdale SC, Chandorkar PU, Podutwar AA, Polshettiwar SA, Chabukswar AR (2024) Nanomaterials in imaging and diagnostics. In: Cutting-edge applications of nanomaterials in biomedical sciences. IGI Global, pp 277–288

Jain PK, El-Sayed IH, El-Sayed MA (2007) Au nanoparticles target cancer. Nano Today 2(1):18–29

Jones VM (2006) Proceedings of the symposium on integrated micro nano systems: convergence of bio and nanotechnologies, Enschede, The Netherlands

Joseph M, Pathiripparambath MSR, Tharayil H, Jayasree RS, Nair LV (2022) Copper nanocluster enables simultaneous photodynamic and chemo therapy for effective cancer diagnosis and treatment in vitro. ChemMedChem 17(19):e202200201

Jouyandeh M, Sajadi SM, Seidi F, Habibzadeh S, Munir MT, Abida O, Ahmadi S, Kowalkowska-Zedler D, Rabiee N, Rabiee M (2022) Metal nanoparticles-assisted early diagnosis of diseases. OpenNano 8:100104

Kavkhani R, Hajalilou A, Abouzari-Lotf E, Ferreira LP, Cruz MM, Yusefi M, Parvini E, Ogholbeyg AB, Ismail UN (2022) CTAB assisted synthesis of $MnFe_2O_4$@ $SiO_2$ nanoparticles for magnetic hyperthermia and MRI application. Mater Today Commun 31:103412

Keles G, Sifa Ataman E, Taskin SB, Polatoglu İ, Kurbanoglu S (2024) Nanostructured metal oxide-based electrochemical biosensors in medical diagnosis. Biosensors 14(5):238

Khan MU, Ullah H, Honey S, Talib Z, Abbas M, Umar A, Ahmad T, Sohail J, Sohail A, Makgopa K, Ahmad J, Asim J (2023) Metal nanoparticles: synthesis approach, types and applications a mini review. Nano-Horizons 2:1–21

kr Katare R, Tripathi SBL (2024) Study on synthesis and characterization of metal nanoparticles for biomedical applications. Res Rev Int J Multi 9(6):123–129

Krasochko P, Korochkin R, Krasochko P, Gvozdev S, Ponaskov M (2020) Optical properties of colloidal solutions of metal nanoparticles. Sci Horiz 23(10):47–53. https://doi.org/10.48077/scihor.23(10).2020.47-53

Kumar H, Venkatesh N, Bhowmik H, Kuila A (2018) Metallic nanoparticle: a review. Biomed J Sci Tech Res 4(2):3765–3775

Kumar K, Kumar R, Jasrotia R, Kalia S, Arya V, Kumar A, Khargotra R, Singh T, Thakur N (2024) In-vitro bactericidal and anti-oxidant efficacy of biosynthesized $CuO/Cu_2O$-NiO nanocomposites against the pathogenic bacteria and DPPH free radical. Discov Appl Sci 6(3):87

Kurul F, Yazici ZA, Kocer ZA, Topkaya SN, Cetin AE (2023) Low-cost and portable plasmonic biosensor for label-free detection of viruses in resource-limited settings. In: European conference on biomedical optics, 126272O

Kwak J, Lee SS (2019) Highly sensitive piezoelectric immunosensors employing signal amplification with gold nanoparticles. Nanotechnology 30(44):445502

Lee C-W, Chang H-Y, Wu J-K, Tseng F-G (2019) Ultra-sensitive electrochemical detection of bacteremia enabled by redox-active gold nanoparticles (raGNPs) in a nano-sieving microfluidic system (NS-MFS). Biosens Bioelectron 133:215–222

Lee H, Matthews KC, Zhan X, Warner JH, Ren H (2023) Precision synthesis of bimetallic nanoparticles via nanofluidics in nanopipets. ACS Nano 17(22):22499–22507. https://doi.org/10.1021/acsnano.3c06011

Lei P, Chen H, Feng C, Yuan X, Xiong Z, Liu Y, Liao W (2022) Noninvasive visualization of sub-5 mm orthotopic hepatic tumors by a nanoprobe-mediated positive and reverse contrast-balanced imaging strategy. ACS Nano 16(1):897–909

Lei X, Ye W, Safdarin F, Baghaei S (2024) Microfluidics devices for sports: a review on technology for biomedical application used in fields such as biomedicine, drug encapsulation, preparation of nanoparticles, cell targeting, analysis, diagnosis, and cell culture. Tissue Cell 87:102339

Li B, Han L, Wang H, Zheng Y (2022) Albumin-templated manganese carbonate nanoparticles for precise magnetic resonance imaging of acute myocardial infarction. J Biomater Appl 37(3):493–501

Li J (2023) Encapsulated gold nanoparticle probes for virus detection. In: Second international conference on biological engineering and medical science (ICBioMed 2022), vol 12611, pp 311–317

Li J, Cha R, Zhang Y, Guo H, Long K, Gao P, Wang X, Zhou F, Jiang X (2018) Iron oxide nanoparticles for targeted imaging of liver tumors with ultralow hepatotoxicity. J Mater Chem B 6(40):6413–6423

Li J, Chen J, Dai Y, Liu Z, Zhao J, Liu S, Xiao R (2023) Magnetic SERS strip based on 4-mercaptophenylboronic acid-modified $Fe_3O_4$@ Au for active capture and simultaneous detection of respiratory bacteria. Biosensors 13(2):210

Li X, Zhou S, Lu S, Tu D, Zheng W, Liu Y, Li R, Chen X (2019) Lanthanide metal–organic framework nanoprobes for the in vitro detection of cardiac disease markers. ACS Appl Mater Interfaces 11(47):43989–43995

Liu Y, Ye H, Bayram A, Zhang T, Cai Q, Xie C, Huynh H, Peerzade SAMA, Kahn JS, Qin Z (2022a) Gold nanourchins improve virus targeting and plasmonic coupling for virus diagnosis on a smartphone platform. ACS Sensors 7(12):3741–3752

Liu Y, Ye H, Huynh H, Xie C, Kang P, Kahn JS, Qin Z (2022b) Digital plasmonic nanobubble detection for rapid and ultrasensitive virus diagnostics. Nat Commun 13(1):1687

Losito DW, Souza NIN, Martins TS, Britos TN, Schumacher ML, Haddad PS (2024) A review of superparamagnetic nanoparticles applications and regulatory aspects in medicine and environmental areas. J Mater Sci:1–31

Lu Z, Tang H, Wu D, Xia Y, Wu M, Yi X, Li H, Wang J (2016) Amplified voltammetric detection of miRNA from serum samples of glioma patients via combination of conducting magnetic microbeads and ferrocene-capped gold nanoparticle/streptavidin conjugates. Biosens Bioelectron 86:502–507

Maciulis V, Ramanaviciene A, Plikusiene I (2022) Recent advances in synthesis and application of metal oxide nanostructures in chemical sensors and biosensors. Nanomaterials 12(24):4413

Maity D, Murmu G, Sahoo SR, Tiwari A, Ajith S, Saha S (2023) Metal/metal oxide nanoparticles-based biosensors for detection of infectious diseases. Point-of-Care Biosens Infect Dis:147–185

Makowski MR, Varma G, Wiethoff AJ, Smith A, Mattock K, Jansen CHP, Warley A, Taupitz M, Schaeffter T, Botnar RM (2011) Noninvasive assessment of atherosclerotic plaque progression in ApoE−/− mice using susceptibility gradient mapping. Circul Cardiovasc Imaging 4(3):295–303

Malik MA, Wani MY, Hashim MA (2012) Microemulsion method: a novel route to synthesize organic and inorganic nanomaterials: 1st nano update. Arab J Chem 5(4):397–417

Marinescu L, Ficai D, Ficai A, Oprea O, Nicoara AI, Vasile BS, Boanta L, Marin A, Andronescu E, Holban AM (2022) Comparative antimicrobial activity of silver nanoparticles obtained by wet chemical reduction and solvothermal methods. Int J Mol Sci 23(11). https://doi.org/10.3390/ijms23115982

Medawar-Aguilar V, Jofre CF, Fernández-Baldo MA, Alonso A, Angel S, Raba J, Pereira SV, Messina GA (2019) Serological diagnosis of Toxoplasmosis disease using a fluorescent immunosensor with chitosan-ZnO-nanoparticles. Anal Biochem 564:116–122

Medina-Cruz D, Saleh B, Vernet-Crua A, Nieto-Argüello A, Lomelí-Marroquín D, Vélez-Escamilla LY, Cholula-Díaz JL, García-Martín JM, Webster T (2020) Bimetallic nanoparticles for biomedical applications: a review. In: Racing for the surface: antimicrobial and interface tissue engineering, pp 397–434

Moradi F, Nasoohian N, Ghorbanian N, Fooladfar Z (2024) Recent advances in laboratory detection of *Chlamydia trachomatis* using gold (Au) nanoparticle-based methods; another evolution of nanotechnology in diagnostic bacteriology. Microchem J:111373

Mourdikoudis S, Pallares RM, Thanh NTK (2018) Characterization techniques for nanoparticles: comparison and complementarity upon studying nanoparticle properties. Nanoscale 10(27):12871–12934

Mughal SS (2022) Diagnosis and treatment of diseases by using metallic nanoparticles—a review. Authorea Preprints

Musazzi UM, Franzè S, Condorelli F, Minghetti P, Caliceti P (2023) Feeding next-generation nanomedicines to Europe: regulatory and quality challenges. Adv Healthc Mater 12(30):2301956

Namdari M, Negahdari B, Cheraghi M, Aiyelabegan HT, Eatmadi A (2017) Cardiac failure detection in 30 minutes: new approach based on gold nanoparticles. J Microencapsul 34(2):132–139

Narayan N, Meiyazhagan A, Vajtai R (2019) Metal nanoparticles as green catalysts. Materials 12(21):3602

Natarajan B, Kannan P, Guo L (2024) Metallic nanoparticles-based multicolor visual detection of biomolecules sensing mechanism-based on the change in various colors. Chin J Struct Chem:100349

Naz M, Shafique H, Majeed MI, Nawaz H, Rashid N, Alshammari A, Albekairi NA, Amber A, Zohaib M, Shahid U (2024) Surface-enhanced Raman spectroscopy (SERS) for the diagnosis of acute myocardial infarction (AMI) using blood serum samples. RSC Adv 14(40):29151–29159

Nie L, Liu F, Ma P, Xiao X (2014) Applications of gold nanoparticles in optical biosensors. J Biomed Nanotechnol 10(10):2700–2721

Nor NM, Ramli NH, Zakaria ND, Razak KA (2022) Nanomaterials and lab-on-a-chip technologies. In: Bioelectronics. CRC Press, pp 73–91

Nour M, Hamdy O, Faid AH, Eltayeb EA, Zaky AA (2022) Utilization of gold nanoparticles for the detection of squamous cell carcinoma of the tongue based on laser-induced fluorescence and diffuse reflectance characteristics: an in vitro study. Lasers Med Sci 37(9):3551–3560

Nyabadza A, McCarthy É, Makhesana M, Heidarinassab S, Plouze A, Vazquez M, Brabazon D (2023) A review of physical, chemical and biological synthesis methods of bimetallic nanoparticles and applications in sensing, water treatment, biomedicine, catalysis and hydrogen storage. Adv Colloid Interface Sci:103010

Nyembe S, Mkhohlakali A, May B, Mhlanga N (2023) Application of plasmonic nanostructures in molecular diagnostics and biosensor technology: challenges and current developments. IntechOpen. https://doi.org/10.5772/intechopen.108319

Osaki S, Espulgar WV, Wakida S, Saito M, Tamiya E (2022) Optimization of electrochemical analysis for signal amplification in gold nanoparticle-probed immunoassays. Electrochim Acta 432:141180

Park Y, Ryu B, Ki S, Chen M, Liang X, Kurabayashi K (2022) Nano assembly of plasmonic probe-virus particles enabled rapid and ultrasensitive point-of-care SARS-CoV-2 detection. MedRxiv:2008–2022

Passeri D, Angeloni L, Rossi M (2021) Magnetic force microscopy and magnetic nanoparticles: perspectives and challenges. In: New trends in nanoparticle magnetism, pp 285–300
Patil MP, Kim G-D (2018) Marine microorganisms for synthesis of metallic nanoparticles and their biomedical applications. Colloids Surf, B 172:487–495
Pohanka M (2018) Overview of piezoelectric biosensors, immunosensors and DNA sensors and their applications. Materials 11(3):448
Poon K, Gupta A, Price WS, Zreiqat H, Singh G (2024) Manganese oxide nanoplatforms for disease diagnosis and treatment: progress, challenges and opportunities. Coord Chem Rev 500:215548
Popovtzer R, Agrawal A, Kotov NA, Popovtzer A, Balter J, Carey TE, Kopelman R (2008) Targeted gold nanoparticles enable molecular CT imaging of cancer. Nano Lett 8(12):4593–4596
Pretze M, van der Meulen NP, Wängler C, Schibli R, Wängler B (2019) Targeted 64Cu-labeled gold nanoparticles for dual imaging with positron emission tomography and optical imaging. J Labelled Compd Radiopharm 62(8):471–482
Priya SS, Choudhury A, Archana NK, Ananda C, Jahangir MA (2023) Efficacy of metallic nanoparticles and nanocarriers as an advanced tool for imaging and diagnosis: insight into theranostic applications. Int J Biomed Investig 6(2):1–17
Pugazhendhi S, Sathya P, Palanisamy PK, Gopalakrishnan R (2016) Synthesis of silver nanoparticles through green approach using *Dioscorea alata* and their characterization on antibacterial activities and optical limiting behavior. J Photochem Photobiol B 159:155–160
Rashed S, Naser M, Bayoumi I, Aly N, Mohamed W, ElGhannm A (2018) Evaluation of nanogold particles-based enzyme-linked immunosorbent assay for detection of hydatidosis. Benha Med J 35(2):134–138
Regmi S, Poudel C, Adhikari R, Luo KQ (2022) Applications of microfluidics and organ-on-a-chip in cancer research. Biosensors 12(7):459
Reimer L (2013) Transmission electron microscopy: physics of image formation and microanalysis, vol 36. Springer
Rodríguez P, Rojas H, Medina M, Arrivillaga J, Francisco Y, Dager F, Piscitelli V, Caetano M, Fernández A, Castillo J (2013) Study of functionalized gold nanoparticles with anti-gp63 IgG antibody for the detection of glycoprotein gp63 in membrane surface of *Leishmania* genus parasites. Am J Anal Chem 4(7A):9
Sarcan ET (2024) Nanoparticles for dual imaging: pet and fluorescence imaging. J Fac Pharm Ankara Univ 48(2):3
Serrano-Bayona R, Chu C, Liu P, Roberts WL (2023) Flame synthesis of carbon and metal-oxide nanoparticles: flame types, effects of combustion parameters on properties and measurement methods. Materials 16(3):1192
Sezer N, Arı İ, Bicer Y, Koc M (2021) Superparamagnetic nanoarchitectures: multimodal functionalities and applications. J Magn Magn Mater 538:168300
Siciliano G, Alsadig A, Chiriacò MS, Turco A, Foscarini A, Ferrara F, Gigli G, Primiceri E (2023) Beyond traditional biosensors: recent advances in gold nanoparticles modified electrodes for biosensing applications. Talanta:125280
Şimşek N, Erdoğan NÖ, Tığ GA (2023a) Metal nanoparticles-based biomarkers for clinical diagnosis. In: Handbook of nanobioelectrochemistry: application in devices and biomolecular sensing. Springer, pp 541–568
Singh A, Banerjee SL, Gantait A, Kumari K, Kundu PP (2023a) Metal-based nanoparticles: synthesis and biomedical applications. In: Nanoparticles reinforced metal nanocomposites: mechanical performance and durability. Springer, pp 365–408
Singh NA, Narang J, Garg D, Jain V, Payasi D, Suleman S, Swami RK (2023b) Nanoparticles synthesis via microorganisms and their prospective applications in agriculture. Plant Nano Biol:100047
Singh A, Chavan VA, Ariya P, Udaynadh BVSS, Adithi E (2024) Nanomaterial characterization techniques. In: Futuristic trends in chemical, material sciences & nano technology. Iterative International Publisher, pp 28–55. https://doi.org/10.58532/V3BECS13P1CH2

Sivakumar R, Dinh VP, Lee NY (2021) Ultraviolet-induced in situ gold nanoparticles for point-of-care testing of infectious diseases in loop-mediated isothermal amplification. Lab Chip 21(4):700–709
Sohal JK, Saraf A, Shukla KK (2021) Silver nanoparticles (AgNPs): methods of synthesis, mechanism of antimicrobial action and applications. Multi Res Dev:55–71
Song D, Zheng J, Myung NV, Xu J, Zhang M (2021) Sandwich-type electrochemical immunosensor for CEA detection using magnetic hollow Ni/C@ $SiO_2$ nanomatrix and boronic acid functionalized CPS@ PANI@ Au probe. Talanta 225:122006
Sosnovik DE, Nahrendorf M, Deliolanis N, Novikov M, Aikawa E, Josephson L, Rosenzweig A, Weissleder R, Ntziachristos V (2007) Fluorescence tomography and magnetic resonance imaging of myocardial macrophage infiltration in infarcted myocardium in vivo. Circulation 115(11):1384–1391
Souto EB, Blanco-Llamero C, Krambeck K, Kiran NS, Yashaswini C, Postwala H, Severino P, Priefer R, Prajapati BG, Maheshwari R (2024) Regulatory insights into nanomedicine and gene vaccine innovation: safety assessment, challenges, and regulatory perspectives. Acta Biomater 180:1–17. https://doi.org/10.1016/j.actbio.2024.04.010
Srisrattakarn A, Lulitanond A, Wilailuckana C, Charoensri N, Daduang J, Chanawong A (2017) A novel GoldNano Carb test for rapid phenotypic detection of carbapenemases, particularly OXA type, in Enterobacteriaceae, *Pseudomonas aeruginosa* and *Acinetobacter* spp. J Antimicrob Chemother 72(9):2519–2527
Stockhofe K, Postema JM, Schieferstein H, Ross TL (2014) Radiolabeling of nanoparticles and polymers for PET imaging. Pharmaceuticals 7(4):392–418
Sun L, Yuan G, Gao L, Yang J, Chhowalla M, Gharahcheshmeh MH, Gleason KK, Choi YS, Hong BH, Liu Z (2021) Chemical vapour deposition. Nat Rev Methods Primers 1(1):5
Swain J, Swain S, Singh D, Jena A, Samantaray R, Sahu R (2023) Role of piezoelectric biosensors. Point-of-Care Biosens Infect Dis:129–145
Syama S, Mohanan PV (2021) Microfluidic based human-on-a-chip: a revolutionary technology in scientific research. Trends Food Sci Technol 110:711–728
Tai J, Fan S, Ding S, Ren L (2022) Gold nanoparticles based optical biosensors for cancer biomarker proteins: a review of the current practices. Front Bioeng Biotechnol 10:877193
Thakur N, Arya V, Kumar A, Thakur N (2024) Catalytic efficiency of encapsulated coupled doped Zn-Ni $TiO_2$ nanocatalyst for the removal of organic pollutants and free radical assay. Int J Environ Anal Chem 104(20):9754–9771
Thakur N, Thakur N, Kumar K, Arya V, Kumar A, Kalia S (2022) Photocatalytic efficiency of green synthesized ZnO nanoparticles for the degradation of methyl orange dye: a review. Biomater Polym Horiz 1(4):1–11
Timoshenko RV, Gorelkin PV, Vaneev AN, Krasnovskaya OO, Akasov RA, Garanina AS, Khochenkov DA, Iakimova TM, Klyachko NL, Abakumova TO (2023) Electrochemical nanopipette sensor for in vitro/in vivo detection of $Cu^{2+}$ ions. Anal Chem 96(1):127–136
Tseng C-C, Kung C-T, Chen R-F, Tsai M-H, Chao H-R, Wang Y-N, Fu L-M (2021) Recent advances in microfluidic paper-based assay devices for diagnosis of human diseases using saliva, tears and sweat samples. Sens Actuators, B Chem 342:130078
Tshabalala ZP, Oosthuizen DN, Swart HC, Motaung DE (2020) Tools and techniques for characterization and evaluation of nanosensors. In: Nanosensors for smart cities. Elsevier, pp 85–110
Tung C-Y, Tsai T-T, Chiu P-Y, Viter R, Ramanavičius A, Yu C-J, Chen C-F (2024) Diagnosis of *Mycobacterium tuberculosis* using palladium–platinum bimetallic nanoparticles combined with paper-based analytical devices. Nanoscale 16(12):5988–5998
Uritu CM, Al-Matarneh CM, Bostiog DI, Coroaba A, Ghizdovat V, Filipiuc SI, Simionescu N, Stefanescu C, Jalloul W, Nastasa V (2024) Radiolabeled multi-layered coated gold nanoparticles as potential biocompatible PET/SPECT tracers. J Mater Chem B 12(15):3659–3675

Vais RD, Heli H, Sattarahmady N (2023) Label-free electrochemical DNA biosensing of MR TV 29 18s ribosomal RNA gene of *Trichomonas vaginalis* by signalization of non-spherical gold nanoparticles. Mater Today Commun 34:105123

Varna M, Xuan HV, Fort E (2018) Gold nanoparticles in cardiovascular imaging. Wiley Interdisc Rev Nanomed Nanobiotechnol 10(1):e1470

Vijayaram S, Razafindralambo H, Sun Y-Z, Vasantharaj S, Ghafarifarsani H, Hoseinifar SH, Raeeszadeh M (2024) Applications of green synthesized metal nanoparticles—a review. Biol Trace Elem Res 202(1):360–386

Wahajuddin N, Arora S (2012) Superparamagnetic iron oxide nanoparticles: magnetic nanoplatforms as drug carriers. Int J Nanomed:3445–3471

Wang H, Lei C, Li J, Wu Z, Shen G, Yu R (2004) A piezoelectric immunoagglutination assay for *Toxoplasma gondii* antibodies using gold nanoparticles. Biosens Bioelectron 19(7):701–709

Wang J, Wu H, Yang Y, Yan R, Zhao Y, Wang Y, Chen A, Shao S, Jiang P, Li Y-Q (2018) Bacterial species-identifiable magnetic nanosystems for early sepsis diagnosis and extracorporeal photodynamic blood disinfection. Nanoscale 10(1):132–141

Wang L, Hasanzadeh Kafshgari M, Meunier M (2020) Optical properties and applications of plasmonic-metal nanoparticles. Adv Funct Mater 30(51):2005400

Wang Z, Tong M, Chen X, Hu S, Yang Z, Zhang Y, Zhou H, Wu Y, Li X, Li D (2016) Survivin-targeted nanoparticles for pancreatic tumor imaging in mouse model. Nanomed Nanotechnol Biol Med 12(6):1651–1661

Wen C-Y, Liang X, Liu J, Zhao T-Y, Li X, Zhang Y, Guo G, Zhang Z, Zeng J (2023) An achromatic colorimetric nanosensor for sensitive multiple pathogen detection by coupling plasmonic nanoparticles with magnetic separation. Talanta 256:124271

Wen R (2019) Microfluidic-based point-of-care testing for disease diagnosis. In: E3S web of conferences, vol 131, p 1004

Wu K, Liu J, Saha R, Peng C, Su D, Wang YA, Wang J-P (2021) Investigation of commercial iron oxide nanoparticles: structural and magnetic property characterization. ACS Omega 6(9):6274–6283

Xi L, Grobmyer SR, Zhou G, Qian W, Yang L, Jiang H (2014) Molecular photoacoustic tomography of breast cancer using receptor targeted magnetic iron oxide nanoparticles as contrast agents. J Biophotonics 7(6):401–409

Xie G, Du S, Huang Q, Mo M, Gao Y, Li M, Tao J, Zhang L, Zhu J (2022) Photonic hydrogels for synergistic visual bacterial detection and on-site photothermal disinfection. ACS Appl Mater Interfaces 14(4):5856–5866

Xu H, Zeiger BW, Suslick KS (2013) Sonochemical synthesis of nanomaterials. Chem Soc Rev 42(7):2555–2567

Yadav TP, Yadav RM, Singh DP (2012) Mechanical milling: a top down approach for the synthesis of nanomaterials and nanocomposites. Nanosci Nanotechnol 2(3):22–48

Yashaswini, Swetha GA, Betageri VS, Latha MS, Vinaykumar YB (2024) Metallic nanoparticles for biomedical applications. In: Nanomaterials for biomedical and bioengineering applications. Springer, pp 459–478

Yu Y, Zhao L, Xu N, Liu X, Li L, Xu N, Bai X (2024) A smartphone-based enhanced colorimetric immunoassay for the detection of *Trichinella spiralis* infection. Vet Parasitol 110213. https://doi.org/10.1016/j.vetpar.2024.110213

Zafar N, Madni A, Khalid A, Khan T, Kousar R, Naz SS, Wahid F (2020) Pharmaceutical and biomedical applications of green synthesized metal and metal oxide nanoparticles. Curr Pharm des 26(45):5844–5865

Zahoor M, Nazir N, Iftikhar M, Naz S, Zekker I, Burlakovs J, Uddin F, Kamran AW, Kallistova A, Pimenov N (2021) A review on silver nanoparticles: classification, various methods of synthesis, and their potential roles in biomedical applications and water treatment. Water 13(16):2216

Zhao W (2023) Silver nanoparticles for advanced electrochemical aptamer sensor. In: Second international conference on biological engineering and medical science (ICBioMed 2022), vol 12611, pp 259–264

Zheng X, Wu Y, Zuo H, Chen W, Wang K (2023) Metal nanoparticles as novel agents for lung cancer diagnosis and therapy. Small 19(18):2206624

Zheng Y, Zhang H, Hu Y, Bai L, Xue J (2018) MnO nanoparticles with potential application in magnetic resonance imaging and drug delivery for myocardial infarction. Int J Nanomed:6177–6188

Zhou C, Zou H, Li M, Sun C, Ren D, Li Y (2018) Fiber optic surface plasmon resonance sensor for detection of *E. coli* O157: H7 based on antimicrobial peptides and AgNPs-rGO. Biosens Bioelectron 117:347–353

Zhou W, Tavakoli H, Ma L, Bautista C, Li X (2022) Rapid disease diagnosis using low-cost paper and paper-hybrid microfluidic devices. In: Multidisciplinary microfluidic and nanofluidic lab-on-a-chip. Elsevier, pp 325–360

Zhu Q-L, Xu Q (2016) Immobilization of ultrafine metal nanoparticles to high-surface-area materials and their catalytic applications. Chem 1(2):220–245

Žukovskaja O, Jahn IJ, Weber K, Cialla-May D, Popp J (2017) Detection of *Pseudomonas aeruginosa* metabolite pyocyanin in water and saliva by employing the SERS technique. Sensors 17(8):1704

Zulfiqar Z, Khan RRM, Summer M, Saeed Z, Pervaiz M, Rasheed S, Shehzad B, Kabir F, Ishaq S (2024) Plant-mediated green synthesis of silver nanoparticles: synthesis, characterization, biological applications, and toxicological considerations: a review. Biocatal Agric Biotechnol:103121

# Chapter 9
# Nanomaterial-Based Drug Delivery Systems

**Rosie Gomric, Daniel Phillips, Sean Kelly, Suhrud Pathak, Keyi Liu, Ashwani Kumar, James Joseph Bellarmine, and Muralikrishnan Dhanasekaran**

**Abstract** With a novel approach to targeted therapy, nanomaterial-based drug delivery systems offer unprecedented opportunities to enhance the accuracy, effectiveness, and safety of treatments for various diseases, including cancer, infectious diseases, and genetic disorders. To reduce systemic adverse effects and enhance treatment outcomes, these systems utilize the unique physicochemical properties of nanomaterials, such as liposomes, nanowires, nanoparticles, and quantum dots, to deliver therapeutic agents directly to damaged tissues or cells. Recent developments in the design and functionalization of nanomaterials for drug delivery applications are examined in this chapter, with an emphasis on their ability to target specific sites, permit controlled release, and overcome biological barriers. Basic concepts, including drug loading and release processes, nanocarrier biodistribution, and interactions with biological systems, are covered. The chapter also discusses barriers in transferring these nanocarriers from laboratory models to clinical settings, including issues with cost, manufacturing scalability, regulatory approval, and biocompatibility. Long-term impacts and patient safety are two ethical considerations of nanotechnology in medication delivery. With an emphasis on current research and potential future possibilities, the chapter concludes with a forward-looking perspective on incorporating nanomaterial-based delivery methods into conventional therapeutic medicines.

R. Gomric · D. Phillips · S. Kelly · S. Pathak · K. Liu · M. Dhanasekaran (✉)
Department of Drug Discovery and Development, Harrison College of Pharmacy, Auburn University, Auburn, AL, USA
e-mail: dhanamu@auburn.edu

A. Kumar
Patanjali Herbal Research Department, Patanjali Research Foundation, Haridwar, India

J. J. Bellarmine
Tbilisi State Medical University, Tbilisi, Georgia

M. Dhanasekaran
Graduate School, Auburn University, Auburn, AL, USA



A. Balkrishna et al. (eds.), *Nanotechnology in Environmental Science and Healthcare*,
https://doi.org/10.1007/978-981-95-6714-0_9

# 9.1 Introduction

## 9.1.1 What Are Nano-Based Drug Delivery Systems (NDDS)?

Drug bioavailability has always been of the utmost importance in the medical community, with novel ideas such as nanodrug delivery systems becoming the forefront of drug efficacy research in recent years. The concept of nanotechnology arose relatively recently. Nobel Prize winner Richard Feynman first discussed it in his lecture "There's Plenty of Room at the Bottom," which he presented at the American Physical Society's National meeting in 1959. In this presentation, he discussed the possibilities of nanotechnology, including the potential to create machines that function at the molecular level (Baig et al. 2021). Since then, nanotechnology has progressed into its field of science, with the National Nanotechnology Institute (NNI) of the United States defining it as a science, engineering, and technology conducted at the nanoscale (1 to 100 nm), where unique phenomena enable novel applications in a wide range of fields, from chemistry, physics and biology, to medicine, engineering and electronics (NNI 2023). Nanodrug delivery systems have since been researched and developed, with the FDA approving the first formulation of an NDDS in 1989 (Afzal et al. 2022).

Nano-based drug delivery systems (NDDS) refer to the utilization of nanoparticles to transport drugs within an organism to their targeted location. For a drug to be effective, it must reach its targeted location before its activation to ensure that potency and bioavailability are at their peak. Unfortunately, conventional delivery systems such as nasal, oral, mucosal, inhaled, or injectable methods often delay the movement of the drug, resulting in decreased efficacy once the target area is reached (Tibbitt et al. 2016). NDDS is a solution to this problem, removing the "middleman" for drug delivery and ensuring that a drug is most effective once it reaches its target destination.

Various other drug delivery systems have been used to increase the efficacy of drugs in the past. For instance, specific pharmaceutical agents have been administered as inactive forms, which get metabolized to their active counterparts within the body to retain potency until within an organism (Vlachopoulos et al. 2022). Although a novel idea, inactive precursors still get lost within an organism, often deployed in incorrect areas or excreted before the drug can be utilized (Builders and Arhewoh 2016). To avert this problem, pharmaceutical companies devised other drug delivery systems such as erodible materials, degradable materials, and hydrogels to protect the pharmaceutical until it reached its designated area for drug utilization (Smolensky and Peppas 2007). Though a step in the right direction, this strategy could have been more effective, resulting in high drug aggregation and poor pharmaceutical distribution throughout the body.

Then came NDDS, a way to both protect the bioavailability of a drug and ensure its distribution to its intended location. Unlike previous drug delivery systems, nanotechnology is designed to function at the molecular or atomic levels, enabling it to move freely within the body due to its minute size (Rudramurthy et al. 2016).

These nanoparticles exhibit distinctive characteristics, including structural, chemical, magnetic, electrical, and biological properties, which orchestrate their movement from one area of an organism to its desired location (Patra et al. 2018). These nanoparticles can interact with their pharmaceutical payloads in multiple ways, ranging from simple attachment to the drug to encapsulating them within the nanostructure for future release. Once the target destination is reached, the NDD dissociates with the drug, allowing interaction and pharmaceutical utilization to begin (Jahangirian et al. 2016).

Due to an increase in research in the field, multiple types of NDDS have been tested to increase the bioavailability of numerous drugs. One of the most extensively explored NDDS is liposomes, which are spherically shaped lipid vesicles formed from one or two bilayers containing an encapsulated aqueous region within their core (Jha et al. 2016). Their utilization is frequent due to factors such as ease of drug insertion, high bioavailability, biocompatibility, stability, and lack of toxicity (Nsairat et al. 2022). The liposome structure also allows numerous drugs to be associated with them due to their small diameter and ability to attach drugs to their lipophilic membrane or within their aqueous core (Fathi and Oyelere 2016). Liposomes are just one type of NDDS in an exponentially growing field, with countless possibilities to alter how pharmaceuticals are administered and delivered throughout the body.

## 9.2 NDDS for Prevention, Therapy, and Diagnosis

Advancements in nanotechnologies are currently being tested and implemented in the prevention, therapy, and diagnosis of many conditions. These nanobodies are a beneficial avenue for scientists to pursue when it comes to patient care because they provide enhanced safety features, high sensitivity, and low toxicity. Their superior properties, such as small size, high stability, strong antigen-binding affinity, water solubility, and natural origin, make them suitable for development into next-generation drugs (Jovčevska and Muyldermans 2020). These nanobody derivatives are already being used heavily in biotechnology, only 30 years after their discovery. In terms of diagnosis, these nanotechnologies can result in a cheaper, more rapid, and more accessible way to diagnose a plethora of diseases. If this technology is implemented in a diagnostic way, potential lives can be saved due to early diagnosis.

In diagnosis, nanobiotechnologies are utilized in molecular diagnostics, which is referred to as nanodiagnostics. They can be injected to target specific areas and test genetic information at the molecular level. The high surface-to-volume ratio allows them to attach to many target molecules, which improves the sensitivity of the results. This can be incorporated into a patient's personal care plan and may alter the way physicians approach diagnosis. When implemented, this modern technology would enable physicians to evaluate patients more quickly and frequently, thereby preventing certain conditions from remaining undetected for extended periods. This could save countless lives and would significantly improve the outcomes of treatment (Alharbi and Al-Sheikh 2014). However, there have been some ethical concerns.

There is some concern about the ability to obtain a patient's entire genetic makeup based on a blood sample. Some argue that individuals should consent before this information becomes easily accessible in their records. Additionally, there is concern that the accuracy of these tests is compromised to enhance other beneficial elements. This technology has not yet been perfected, but it can be greatly beneficial (Le Roux 2015).

Currently, NDDS are mainly used in the prevention, diagnosis, and treatment of various types of cancer. Cancer is an umbrella term for over 100 forms of disease. It is caused by a mutation in the patient's genetic code that turns a normal cell into a cancer cell. This cell will be able to stop dividing, forming a tumor. A benign tumor does not spread. A malignant tumor can spread into other parts of the body and disrupt important bodily processes. Cancer can be inherited, or it can be caused by exposure to certain cancer-causing substances (Cooper 2000). Cancer is among the leading causes of death worldwide. In 2022, there were almost 20 million new cases and 9.7 million cancer-related deaths worldwide (NCI 2024). One of the biggest things that affects how much someone is affected by cancer is how early or late someone is diagnosed. The earlier the cancer is caught, the more likely the treatment will be effective for the patient. The current frequently used methods for diagnosing cancer include family history, personal medical history, lab tests, imaging tests, biopsies, and genetic testing. These tests can typically take days to weeks to provide results. The implementation of nanotechnologies will allow all people to test more frequently and get their results more quickly. This will also be a less invasive option than, for example, doing a biopsy. Nanotechnology can target and accurately provide information about the exact genetic sequence occurring in a patient. This can significantly improve scientists' ability to treat a specific person's cancer.

These nanotechnologies can also be used in relation to infectious diseases. There are three primary types of infectious diseases: those caused by viruses, those caused by bacteria, and those caused by parasites. Viruses are capsules filled with genetic material that infect living hosts and replicate. Bacteria are organisms that cause disease by producing toxins and delivering them to the host. Parasites are organisms that live off another living organism, usually at the expense of said organism. Currently, three common methods for diagnosing diseases are conventionally used: culture and microscopy, immunology, and polymerase chain reaction (PCR) (Wang et al. 2017). Culture and microscopy are the most commonly used forms of testing. Collected microorganisms are cultured in a growth medium and then observed. Immunology-based testing involves testing a sample of someone's blood, mucus, or other bodily fluids to create artificial antibodies. Then, if the antibodies bind to the pathogen, that pathogen is present (IQWiG 2021). PCR testing uses Taq polymerase to amplify the DNA and RNA sequences of pathogens present. These commonly used techniques are very reliable, but they take days to even weeks to produce results. They are also not as accessible as the purpose of nanotechnologies. Nanodiagnostics allows for a quicker result. It can also enable patients to test their urine, blood, or sputum with portable devices. The low-cost, rapid, specific, and user-friendly nature of these diagnostic tools makes them very desirable in the diagnosis of infectious diseases.

This could be extremely beneficial for people in Third World countries and would help to stop the major spread of harmful infectious diseases (Wang et al. 2017).

The treatment of diseases utilizing nanotechnologies is also at the forefront of advancements in the management of many diseases. Currently, antibodies are commonly used in the treatment of disease. However, the antibodies are large and do not do a decent job of penetrating a tumor. These nanotechnologies will be a lot more effective at piercing the tumor. Accordingly, scientists have been researching and discovering many other ways that these technologies can be utilized to improve the way the world treats a wide array of diseases. Nanopharmaceuticals is the field that studies the application of nanotechnologies in the treatment of various diseases (Sheng and Huang 2011). The term for treating someone with these nanoparticles is called nanomedicine. Another example of why nanotechnologies make treatment superior to other methods is that these nanoparticles are able to penetrate the cell membrane and produce a high output of a drug intercellularly (Yan et al. 2014). The ever-growing field of nanotechnology, when it comes to drug administration and targeting certain areas, shows immense potential.

Nanotechnologies are making large strides in the field of drug-related cancer research and treatment. The commonly used drug-related treatment for cancer is chemotherapy. Chemotherapy uses alkylating agents, antimetabolites, topoisomerase inhibitors, mitotic inhibitors, antitumor antibiotics, and other drugs to treat cancer (Nygren 2001). Nanotechnologies utilize nanoparticles and are becoming so advanced that they can detect and target individual cancer cells, administering treatment to those cells. They are also being proven to enhance the treatment plan for cancer. It is typically based on pharmacogenetics, pharmacogenomics, and pharmacoproteomics, while also considering environmental factors that influence the response to therapy. Everyone reacts to treatment differently, and this technology eliminates some of the guesswork when it comes to predicting what will be best for each individual. These technologies can also improve aspects of care that are not pleasant for patients or are not conducive to the treatment working to its fullest potential. For example, nanobubbles and oxygen-releasing polymers are being used to improve hypoxia in cancer patients (Nasir et al. 2024). Hypoxia is a condition in which body cells and tissues are starved of oxygen (Bhutta et al. 2024). However, concerns and complications have arisen when administering these nanomedicines to cancer patients. The complexity of a tumor determines who can benefit from this technology, as only individuals who meet specific parameters will be eligible (Shi et al. 2017). People also worry that nanomedicines will bring residual nanoparticles that would be harmful to the digestive system. These are both aspects scientists need to consider and test for when developing these treatments. The treatment of cancer is an imperfect, ever-changing field that could benefit substantially from the incorporation of nanotechnologies.

Infectious diseases can also be treated using nanomedicines. Commonly, infectious diseases are treated with either antibiotics, antivirals, or antiparasitics. They work by killing or decreasing the growth of either a virus, bacteria, or parasite in the host. However, evolution has allowed many infections to grow to be drug-resistant. Treating these infections properly can improve the patient's health and stop the spread of infection to others. Nanotechnology will allow nanoparticles to target specific sites for treatment (Qasim et al. 2014). Additionally, nanoparticles can serve as vaccine carriers to prevent patients from contracting the infection in the first place (Huang et al. 2024). For example, tuberculosis is considered to be an infectious disease with some of the highest morbidity and mortality rates. It affects the lungs and is an intracellular infectious disease. The drugs currently being used to treat tuberculosis have their limitations, one of them being drug resistance. An alternative is nanotechnology. It offers a better ratio of the amount of active drug that reaches circulation and a reduction in the required dose. Polymeric nanoparticles show considerable promise in this treatment due to their high stability and transport capacity (Dos Santos Macedo et al. 2022). Nanotechnologies are being tested and show promise in improving the treatment of tuberculosis. That is just one of the many infectious diseases that nanotechnology is being tested on. It can offer an improved and more effective way of treating these infectious diseases.

In conclusion, nanotechnology is a growing field showing a lot of promise when diagnosing, treating, and preventing many diseases. There are some concerns that need to be addressed before implementing them for commercial use, but there are a lot of benefits that come with using them. This could be a new frontier of health care where patients can test themselves for diseases and where tests are cheaper and able to be taken more often. This will improve the rate of detection and cause better outcomes for patients. Nanotechnologies will also be able to target specific sites with drug treatment, causing more effective treatment. Also, nanoparticles can act as vaccine carriers and improve the administration of vaccines. Overall, there are many positives to making a shift in healthcare toward utilizing nanotechnologies.

## 9.3 Current FDA-Approved NDDS and Types of NDDS

Due to an increase in research, numerous NDDS have been approved by the FDA for use as a therapeutic measure when it comes to drug delivery. These NDDS include numerous types of carrier systems, one of the most prominent being liposome carriers. Depending upon the active ingredient used in the liposome carrier, the application of the NDDS can vary, ranging from treatment of fungal infections to a diverse range of cancers (Liu et al. 2022). The first FDA-approved NDDS, which also happens to be a liposomal derivative, occurred in 1995 and is known as Doxil (Barenholz 2012). Doxil is used to combat multiple types of cancer, such as ovarian and lung cancer. It works by targeting tumors and releasing its stored doxorubicin to slow the growth of cancer cells and halt tumor expansion. The most recent of these is Arikayce, an oral inhaler NDDS, used to combat lung disease caused by a group of bacteria known as Mycobacterium avium complex (Hunt 2018). This liposomal NDDS was approved in 2018; however, many more are currently in clinical trials to gain FDA approval. Some other approved liposomal derivative NDDS are Visudyne, approved in 2000, used to treat Macular degeneration, and Onivyde, approved in 2015, and used to treat pancreatic cancer (Patra et al. 2018).

Another division of FDA-approved NDDS comes in the form of polymer–protein conjugates. These conjugates combine the stability of polymers with the functional ability of biomolecules to deliver the medication safely to the allocated region of the organism. The first polymer–protein conjugate NDDS was approved in 1994 and is known as Oncaspar or pegaspargase. This NDDS targets acute lymphoblastic leukemia and utilizes a technique known as PEGylation to improve drug solubility and decrease immunogenicity, or the ability of a substance to cause a disadvantageous immune response within the organism (Veronese and Mero 2008). Some other FDA-approved polymer–protein conjugates are Krystexxa, approved in 2010 to treat chronic gout; Plegrdy, approved in 2014 to treat multiple sclerosis; and Adynovate, approved in 2015 to treat Hemophilia (Patra et al. 2018).

Hydroxyapatite is another family of FDA-approved NDDS, which all play a significant role in Bone substitution therapy by mimicking the bone structure and allowing cells to adhere to it to facilitate the growth of new tissue. Hydroxyapatite works as a bone replacement due to its cytocompatibility with bone and dental tissue, in addition to its nontoxic properties within an organism. It also has a significant loading potential, which allows it to be used as a drug transporter as well as a mechanism for drug release within an organism (Munir et al. 2021). The first hydroxyapatite to be FDA approved was Stryker in 2003 and was followed by Heraseus Kulzer in 2004, NanOss in 2005, and EquivaBone in 2009 (Patra et al. 2018).

When it comes to managing Iron deficiency pertaining to chronic kidney disease, multiple NDDSs have been approved by the FDA to relieve this issue. The first NDDS approved for this ailment was INFed, an oral supplementation that was approved in 1957 and utilizes Iron Dextran to replenish iron. Iron Dextran is composed of ferric hydroxide and dextran, which are selectively removed from the bloodstream, broken down into Iron molecules, and taken up by the reticuloendothelial system (RES), where it is distributed to the kidneys and blood cells (Parmar 2024). The NDDS DexIron also utilizes iron dextran and was similarly approved by the FDA in 1957.

Some other NDDSs that treat iron deficiencies related to chronic kidney disease are Venofer, approved by the FDA in 2000, and Ferrlecit, approved in 1999. Venofer utilizes iron sucrose to replenish iron within an organism. Venofer works intravenously, utilizing iron sucrose, which has the ability to rapidly increase transferrin, hemoglobin, and ferritin saturation levels of iron through administration (Macdougall et al. 2020). Ferrlecit works a bit differently, utilizing sodium ferric gluconate, a complex that is also administered intravenously to replenish iron within an iron-deficient individual (Baribeault 2011).

### *9.3.1 Liposomal Nanomaterial-Based Drug Delivery Systems*

Nanomaterial-based drug delivery systems utilize many different substances based on the desired effect of the pharmaceutical in question. One large family of nanomaterial-based drug delivery systems is liposomal-based delivery systems. Having been researched for over 5 decades, liposomal drug delivery systems are one of the

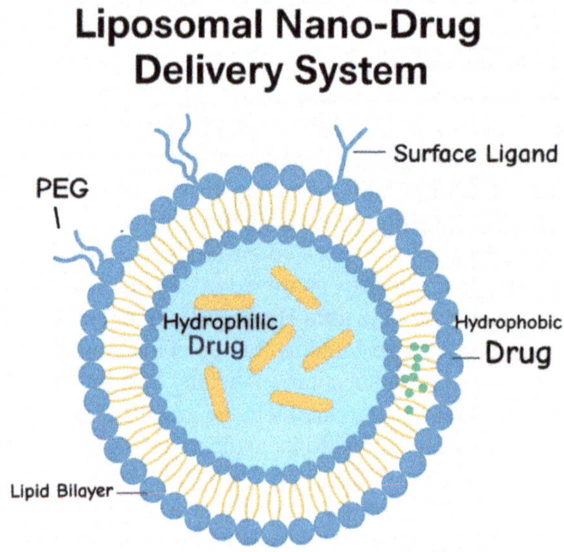


**Fig. 9.1** Liposomal-based drug delivery system

oldest nanomaterial-based drug delivery systems and are used in combination with a plethora of pharmaceuticals. Some utilizations of liposomal nanomaterial-based drug systems are drug loading, extrusion for homogeneous size, long-circulating (PEGylated) liposomes, triggered release liposomes, liposomes containing nucleic acid polymers, ligand-targeted liposomes, and liposomes containing combinations of drugs (Allen and Cullis 2013). A schematic representation of a liposomal-based drug delivery system is shown in Fig. 9.1.

From a structural standpoint, liposomal NDDS are made of multiple components. The first being the lipid Bilayer, which is composed of phospholipids and cholesterol. This creates the shell of the NDDS, protecting hydrophobic drugs between the hydrophobic tails of the phospholipids within the rings, and hydrophilic drugs within the core of the Bilayer itself (Abbasi et al. 2023). Next, surface ligands can be attached to the outer layer phospholipid layer, and act as a director for the Liposomal NDDS. These ligands can be carbohydrates, proteins, antibodies, aptamers, or other small molecules, depending on the desired end location. Lastly, polyethylene glycol (PEG) can be added to the surface of the liposome to increase repulsive forces between the liposome and serum components, increasing the circulation time of the liposomal NDDS (Riaz et al. 2018).

Liposomal Daunorubicin, also known as DuanoXome, was one of the first liposomal NDDS to be FDA-approved in 1995 and is utilized in the treatment of Kaposi's sarcoma. Daunorubicin is an anthracycline compound, a potent chemotherapeutic agent used as the foundation of many chemotherapies. However, it has numerous side effects such as nausea, cardiotoxicity, and increased immune deficiency, which can become extremely deleterious to immunocompromised cancer patients. Therefore, the utilization of Daunorubicin in conjunction with liposomal NDDS is crucial to ensure that Daunorubicin stability and specificity are increased in areas of cancer accumulation, such as tumors, limiting the adverse effects of Daunorubicin on the body (Petre and Dittmer 2007).

In 1996, Liposomal cytarabine, also known as DepoCyt, was approved by the FDA for the treatment of lymphomatous meningitis in humans. Cytarabine is an antimetabolite antineoplastic agent that inhibits the synthesis of DNA used in the treatment of multiple cancers, such as lymphomatous meningitis (Pubchem 2004). Though it is very potent, its efficacy can deteriorate due to limited drug availability and the inability to reach the site of the tumor. Therefore, Liposomal cytarabine is used to increase the bioavailability of cytarabine by encapsulating it in a liposome that enables it to target tumor areas and release the therapeutic at full efficacy once the intended area is reached (Salehi et al. 2019).

Continuing with the theme of cancer treatment, liposomal vincristine, also known as Marqibo, was approved by the FDA in 2012 and is used to treat acute lymphoblastic leukemia. Vincristine is a vinca alkaloid that inhibits cancer cell growth and proper cellular division by disrupting mitosis through interfering with microtubule polymerization (Awosika et al. 2023). However, Vincristine is not selective and can impede the cellular division of many cells within the body, not just those that have cancer. Therefore, the liposomal NDDS ensures that Vincristine reaches its intended location, the tumor cells, thereby reducing the impact that Vincristine will have on healthy cellular division.

Another cancer-associated liposomal NDDS known as Liposomal Irinotecan or Onivyde was approved by the FDA in 2015 and is used to treat pancreatic cancer. Irinotecan is a DNA topoisomerase I inhibitor that affects the S and G2 phases of the cell cycle, thereby inhibiting DNA replication. However, like the other chemotherapeutic drugs, side effects such as neutropenia, diarrhea, and fatigue can occur during its utilization (Reyhanoglu and Smith, 2025). Therefore, to limit these side effects, liposomal NDDSs are utilized to target Irinotecan to solid tumors found in the pancreas, increasing the efficacy and bioavailability of Irinotecan and minimizing the adverse effects of the topoisomerase I inhibitor on healthy cells.

Moving away from cancer, the Liposomal NDDS known as Liposomal Amphotericin B or Ambisome was approved by the FDA in 1997 to help fight fungal/protozoan infections. Amphotericin B is part of the polyene class of antifungal drugs, which has been utilized for over five decades to counteract invasive fungal infections (Noor and Preuss 2025). It works by binding ergosterol in fungal cell membranes, initiating the formation of ion channels that lead to a loss of protons and monovalent cations in the cell. The membrane potential disturbance eventually leads to concentration-dependent cell death (Rybak et al. 2019). The utilization of liposomal NDDSs in conjunction with Amphotericin B allows for improved drug distribution to reticular endothelial systems, increased daily doses, and reduced nephrotoxicity (Hamill 2013).

Liposomal NDDSs can also be used to delay the release of a drug, causing a delayed-release effect instead of instant exposure. One excellent example of this is liposomal morphine sulfate, also known as DepoDur, which was approved by the FDA in 2004. Morphine is a classic opioid analgesic used to manage moderate to severe pain in a multitude of settings (Murphy et al. 2025). However, due to its addictive nature, the utilization of Morphine is difficult because many patients become dependent on the drug, causing painful withdrawal symptoms once the drug administration halts. However, the utilization of a lipid bilayer to encapsulate the drug provides two benefits. The first being delaying the release time of the Morphine, creating a longer analgesic effect and lowering the amount of morphine administration required for a patient. In addition, it could attenuate the addictive aspects of Morphine, yet more research must be conducted to ensure this effect (Gómez-Murcia et al. 2019).

Moving into a field that hasn't been touched upon yet, liposomal NDDSs can also be used to prevent eye-related degeneration. Liposomal Verteporfin, also known as Visudyne, was approved by the FDA in 2000 and is used to decrease macular degeneration, myopia, and ocular histoplasmosis. Verteporfin is a benzoporphyrin derivative, a photosensitizing drug used in photodynamic therapy (PDT), which, when activated, improves visual acuity in patients (Scott and Goa 2000). However, Verteporfin is hydrophobic and tends to aggregate in aqueous systems, unable to reach the vascular choroid endothelial cells, causing vision impairment (Siriwardane et al. 2023). Therefore, the liposomal NDDS enables the hydrophobic therapeutic to be transported to the affected area without accumulating due to its hydrophilic surface, thereby facilitating increased therapeutic delivery and photodynamic therapy.

All things considered, liposomal NDDSs are a highly important tool for drug delivery. Being one of the oldest NDDSs, they are heavily researched and have plenty of room to keep growing. Their simple design makes them easy to produce, and the wide arrangement of surface ligands allows for many locations to be targeted by differentiated liposomal NDDSs. Being able to assist in therapeutic measures ranging from cancer therapy to fungal infections demonstrates the flexibility of liposomal NDDSs and highlights the numerous possibilities this technology holds for the future.

### *9.3.2 Polymeric Nanoparticles*

Another family of nanomaterial-based drug delivery systems is polymeric nanoparticles. A polymer is a chain of macromolecules made up of small monomers. These monomers are present in the polymer in units of repetition that make up long chains. Research has also been conducted into the efficiency of polymeric nanomaterials in nose-to-brain drug delivery. Polymeric nanoparticles have shown promise in enhancing brain bioavailability. These nanoparticles would be more beneficial when applied in patient care because their small size allows some of them to be excreted from the kidneys in the urine after therapeutic use (Awad et al. 2023).

There are three main types of polymeric nanomaterials: natural, synthetic, and dendrimers. Natural polysaccharides are ideal for drug delivery systems because they are easily modified, biodegradable, biocompatible, have low immunogenicity, are low-cost, and many more. A polymer is classified as natural if it is produced organically by organisms. Natural polymers can be utilized in the creation of nanocarriers. These nanocarriers can be applied diagnostically or therapeutically (Qiu et al. 2022).

Some examples of natural polymeric nanomaterials are chitosan, alginate, and pectin. Chitosan is a heteropolysaccharide with three six-carbon rings connected by two 1,4 glycosidic linkages. Its amino acid side chains allow it to remain soluble in acidic compounds. Consequently, chitosan is utilized as a bioadhesive that binds to cells and penetrates them, making it effective in NDDS (Fan et al. 2018). Alginate is a linear and anionic natural polysaccharide that is found in brown seaweed. It consists of two six-carbon rings with one 1,4 glycosidic linkage. It produces gels when it encounters cations that contain a two or three-plus charge. It has gotten attention as a nanoparticle used in NDDS, and it has been used especially for oral insulin administration. It is easy to manipulate the shape of alginate, which is another aspect of it that makes it attractive for use in nanodrug delivery. Finally, pectin is another example of a common natural polymeric nanoparticle. Pectin is a hydrophobic, anionic, and linear natural polysaccharide. It contains three six-carbon rings with two 1,4 glycosidic linkages. Pectin is a promising nanocarrier to be involved in the nanoencapsulation process. The nanoencapsulation process enhances inhibition of proteases, thereby protecting the administered drugs from degradation due to enzymatic proteolysis (Sharma et al. 2006).

Polymeric micelles are also an example of a natural polymeric nanomaterial studied in therapeutic use. These micelles have the typical hydrophobic core and hydrophilic shell of a micelle while being nano-sized. This size is only achievable because of their molecular geometry. Nanomicelles are typically used for drug delivery (Wen et al. 2019). Their core allows for the delivery of hydrophobic drugs, while the crust enables the particle to remain soluble. These benefits were discovered while investigating the impact of nanotechnology, specifically polymeric nanoparticles, on the treatment of diabetes.

Synthetic polymeric nanoparticles are classified as originating from human manufacturing through chemical processes. They are beneficial in use in NDDSs because scientists are able to control their degradation. They have also received FDA approval for various applications. Polylactic-co-glycolic acid (PLGA) is an example. The advantages of PLGA use in NDDS are great. They have already been approved for commercial use and are biodegradable, biocompatible, and bioreabsorbable (Prakobvaitayakit and Nimmannit 2003). Another example of a synthetic polymeric nanoparticle is polyethylene glycol (PEG). PEG is a linear polymer with active hydroxyl groups located on both ends. PEG is notably used as a laxative for children. However, nanoparticles can be treated with PEG in a process called PEGylation. This polymer can increase the half-life of a drug, and many particles conjugated with PEG exhibit desired properties. For these reasons, it is a commonly used polymer (Shi et al. 2021).

Dendrimers are another form of polymeric nanoparticles. These polymers are three-dimensional, tree-like, branched nanocarriers. Dendrimers are well-suited for nanotechnological applications due to their size, molecular structure, and cavities, which distinguish them from other polymers, as well as the availability of their functional groups. Dendrimers contain periphery charges in their branching ends and many of their terminal ends. These charges are vital in the experimentation of these dendrimers for their use in NDDS. It enables dendrimers to serve as excellent gene carriers because cationic dendrimers can bind to negatively charged DNA. Also, cationic dendrimers' positive charges cause interaction with negatively charged biological molecules. This allows them to be applicable to be utilized in intracellular drug delivery. However, this peripheral charge ability can also lead to toxicities such as cytotoxicity or hemolysis (Kesharwani et al. 2014).

### *9.3.3 Inorganic Nanoparticles*

Nanoparticles can also be grouped into the family of inorganic nanomaterials. These are nontoxic, hydrophilic, and highly stable molecules. They were designed to increase the efficiency of drug delivery while eliminating specific adverse effects of conventional drug delivery. Development in the field of nanotechnology has led to the use of inorganic nanoparticles in drug delivery. Three examples of inorganic nanoparticles are gold nanoparticles, silica nanoparticles, and magnetic nanoparticles (Paul and Sharma 2010).

Gold nanoparticles are nano-sized particles made of gold. They have electrochemical properties that are dependent on their size. At a small size, these particles can be stabilized, which makes them highly attractive to experimenters researching nanotechnology, specifically NDDS. Dendrimer molecules have been used to stabilize these particles. They are typically coated with ligands to reduce and stabilize them. However, dendrimers are also used and can completely reduce the metal ions by encapsulating them (Sardar et al. 2009). Gold nanorods and nanoshells, when conjugated with antibodies, can utilize antibody-coated surfaces to detect cancerous cells. This makes gold nanoparticles a promising tool for their use in targeted drug therapy. The use of gold nanoparticles has also been shown to be effective in introducing new substances into cells. For these reasons, gold nanoparticles are effective for use in NDDS and other nanotechnological applications, and they are being continually studied (Giljohann et al. 2010).

Silica nanoparticles are another type of inorganic nanoparticle. These are nano-sized particles that consist of silicon dioxide. There are two types of these silica nanoparticles: non-porous and mesoporous. Nonporous silica nanoparticles are solid and dense. At the same time, mesoporous silica nanoparticles contain uniform nano-sized pores. These pores range in diameter from 2 to 50 nm and can be loaded with drugs (Jeelani 2020). They can load these drugs because of the mesoporous silica nanoparticle's surface area, pore size, and chemistry. These pores increase the drug-loading capacity of these particles compared to other nanoparticles. They also allow drugs to be released gradually and in response to certain triggers. For these reasons, mesoporous silica nanoparticles are particularly useful in the field of NDDSs (Mehmood et al. 2017).

Finally, magnetic nanoparticles are inorganic nanoparticles that have magnetic properties. Iron oxide is commonly the material used to make these particles due to its biocompatibility. Many magnetic nanoparticles exhibit superparamagnetic properties, which means they are only magnetic in the presence of an external field and do not contain residual magnetism when said field is removed. Additionally, scientists can modify the surface area of these nanoparticles. They can be coated with a variety of substances such as gold, silica, and other biological molecules. These magnetic nanoparticles are beneficial because they can be loaded with drugs and introduced into the body, and guided to a specific location or region in the body via magnets. The particles can accumulate in a tumor and, when introduced to an alternative magnetic field, heat up, killing cancerous cells selectively without damage the surrounding tissue. These benefits are attractive. However, there are some safety concerns. For instance, without proper coating, they can clump together (Kodama 1999; Colombo et al. 2012).

### 9.3.4 Carbon-Based Nanomaterials

Carbon nanomaterials are widely researched due to their distinct characteristics, including a high specific surface area, notable carrier mobility, elevated electrical conductivity, flexibility, and optical transparency. These characteristics render them highly promising candidates for advanced sensing applications and drug delivery systems (Speranza 2021). Due to their favorable biocompatibility, inherent optical emission, ability to remain stably suspended in aqueous media, and ease of surface functionalization, carbon nanoparticles (CNPs) have become a leading choice for biological applications, particularly for intracellular use (Chatterjee et al. 2022). Carbon-based nanomaterials can serve as effective drug delivery systems, with examples including nanodiamonds, carbon dots, and carbon nanotubes (CNTs).

- **Nanodiamonds**

Detonational nanodiamonds (NDs) have gained attention as innovative delivery systems for small-molecule therapeutics and macromolecular biotechnical agents, owing to their ultra-small primary size (4–5 nm), chemically inert core, surface reactivity, and capacity to form hydrogel networks. While NDs have a small size, they are also characterized by a large surface area and diverse surface functionalities, which provide an excellent platform for conjugating a wide range of biochemical agents. The effectiveness of NDs as delivery vehicles is reflected not only in their high loading capacity but also in their ability to preserve and maintain the therapeutic efficacy of the compounds they carry. A high loading capacity enables the delivery of elevated payload concentrations while minimizing the required amount of carrier. Equally important is the controlled release of the payload, as the capacity to modulate cargo release enhances the potential for advanced delivery strategies, including controlled and sustained-release applications (Kaur and Badea 2013). Numerous studies have demonstrated that NDs are effective carriers for small-molecule therapeutics, enabling sustained drug release.

In 2007, the feasibility of utilizing nanodiamonds (NDs) for the delivery of doxorubicin hydrochloride (DOX) was systematically assessed. The research was based on the hypothesis that carboxylic and hydroxyl groups present on the surface of detonation NDs can effectively interact with the amine groups of DOX through ionic interactions in an aqueous environment. Adding 1% sodium chloride increased DOX loading on nanodiamonds from 0.5 to 10 wt%, and salt removal promoted DOX release. DOX-loaded nanodiamonds are thought to form loose clusters, trapping a certain amount of DOX within their cavities (Huang et al. 2007). These delivery systems, which contain loosely aggregated nanodiamonds, have the potential to address challenges related to the use of high concentrations of chemotherapeutic drugs in cancer treatments (Sinha et al. 2006). This approach may decrease the cytotoxic effects of chemotherapeutic agents, potentially improving patient outcomes.

- **Carbon Dots**

Delivering anticancer drugs to the therapeutic sites of central nervous system (CNS) tumors remains a significant challenge due to the blood–brain barrier (BBB). However, the utilization of carbon dots can facilitate the movement of a therapeutic agent across the BBB. Carbon dots (C-dots) possess tunable nanoscale dimensions, size-dependent photoluminescence, high photostability, and outstanding biocompatibility (Zheng et al. 2015). These carbon nanoallotropes are composed of symmetrical nanoparticles smaller than 10 nm, featuring a nanocrystalline graphitic core rich in $sp^2$-hybridized carbon and a surface with high oxygen content (Baker and Baker 2010). Unlike nanodiamonds, formed under extreme conditions and characterized by an $sp^3$-hybridized diamond core with an $sp^2$ carbon shell, carbon dots differ primarily from graphene quantum dots in morphology, with the former being spherical nanoparticles and the latter resembling zero-dimensional graphene disks (Nekoueian et al. 2019). C-dots commonly feature surface functional groups such as carboxyl, hydroxyl, and amino moieties. Their surface chemistry is highly influenced by the choice of precursors and synthesis conditions, resulting in substantial variations in their physicochemical properties (Calabrese et al. 2021). The inherent presence of polar functional groups on the surface of C-dots imparts excellent water solubility, making them highly suitable for applications in nanobiology and nanomedicine (Hong et al. 2015). Although the nanoparticles can readily disperse in aqueous environments without requiring additional additives, modifying the surface functional groups of the original carbon nanomaterials remains advantageous. Such modifications can improve biocompatibility or introduce new chemical functionalities. The naturally occurring reactive groups are commonly utilized in both covalent and non-covalent surface functionalization methods. These characteristics of C-dots can be applied to the delivery of drugs across the BBB.

The BBB serves as a critical barrier between the brain and circulating blood-borne compounds, preventing the entry of toxins and cells into the brain. Consequently, maintaining the integrity of the BBB is essential for preserving homeostasis and protecting the CNS (Shakeri et al. 2020). The BBB limits the transport of drugs from the bloodstream into the brain, making it crucial to develop delivery systems capable of crossing this barrier to administer therapeutic agents to the brain effectively. Recently, C-dots have garnered significant interest as potential drug delivery vehicles due to their key attributes, such as ultra-small size, low toxicity, high drug-loading efficiency, prolonged stability, and the ability to enable controlled drug release (Zhao et al. 2015). Therapeutic strategies utilizing C-dots have demonstrated enhanced capability to traverse the BBB and deliver neurological drugs to the CNS. In an in vivo study, Li et al. showed that transferrin-functionalized C-dots successfully crossed the BBB and penetrated the CNS, highlighting their potential for treating neurological disorders (Li et al. 2016). In a separate study, Seven et al. demonstrated that carbon dots synthesized from glucose and conjugated with fluorescein were able to cross the BBB in both zebrafish and rats without the use of additional targeting ligands, underscoring their potential as a CNS drug delivery platform (Seven et al. 2021). These studies highlight the promising potential of carbon dots in enabling drug transport across the BBB for the treatment of brain tumors.

- **Carbon Nanotubes**

CNTs possess distinctive characteristics, including a high surface-to-volume ratio, superior electrical conductivity and mechanical strength, biocompatibility, ease of surface modification, and favorable optical properties, making them promising candidates for novel drug delivery systems. Adding functionalities to the surface of CNTs can enhance their cellular uptake by leveraging the enhanced permeability and retention (EPR) effect, which promotes the accumulation of nanoparticles in tumor tissue due to increased blood flow and vascular permeability. Targeting can also be improved by attaching specific antigens or aptamers. For example, CNT-based drug delivery systems modified with polyethylene glycol (PEG) have been successfully utilized to transport anticancer drugs for the treatment of malignant melanoma (de Carvalho Lima 2018). Interestingly, studies have shown that even multiwalled carbon nanotubes (MWCNTs) without any anticancer drug payload can suppress tumor growth. For example, research by Garcia-Hevia et al. demonstrated that a 2-μg injection of MWCNTs into a melanoma tumor implanted on the back of a mouse resulted in a 27% reduction in tumor mass compared to control animals (García-Hevia et al. 2016). Although current studies show promising results, further investigation is necessary to fully evaluate the efficacy of CNTs as a drug delivery system.

The rapid growth of research on nanomaterials in recent years has led to substantial advancements in the field. Among various nanomaterials, CNTs, a carbon allotrope, have garnered significant attention in this area of research due to their favorable physicochemical and structural characteristics (Balkrishna et al. 2024). These characteristics have stimulated extensive research efforts aimed at exploring their potential in various biomedical applications. These investigations highlight their suitability for medical use, particularly as drug delivery systems, among other functions (Zare et al. 2021).

## 9.4 Conclusion

NDDS are changing current medicine by introducing novel approaches to focused, efficient, and precise therapy. These systems utilize the unique properties of nanoparticles to enhance the solubility, bioavailability, and site-specific delivery of medications, resulting in improved therapeutic outcomes while minimizing adverse effects. Applications such as regulated drug release, biological barrier crossing, and active targeting of affected tissues demonstrate the expanding promise of NDDS in treating complex diseases, including cancer and neurological disorders. Despite these advances, significant obstacles remain. To guarantee safety, issues such as cytotoxicity, which is commonly caused by the formation of reactive oxygen species (ROS), must be addressed.

Furthermore, negotiating complicated regulatory procedures and obtaining clinical-grade scalability present substantial challenges. The high cost of development and production might also impede accessibility and acceptance. Nevertheless, NDDS has considerable potential for individualized therapy in future. Their incorporation with cutting-edge technology, such as artificial intelligence, might improve medication administration and treatment planning for individual patients. In AI-driven systems, ethical considerations must also be taken into account, particularly those related to data privacy and long-term biocompatibility. Overcoming current obstacles and realizing the full therapeutic potential of medication delivery systems based on nanomaterials requires ongoing multidisciplinary research and collaboration.

## References

Abbasi H, Kouchak M, Mirveis Z, Hajipour F, Khodarahmi M, Rahbar N, Handali S (2022) What we need to know about liposomes as drug nanocarriers: an updated review. Adv Pharm Bull 13(1):7–23

Afzal O, Altamimi AS, Nadeem MS, Alzarea SI, Almalki WH, Tariq A et al (2022) Nanoparticles in drug delivery: from history to therapeutic applications. Nanomater 12(24):4494

Alharbi KK, Al-Sheikh YA (2014) Role and implications of nanodiagnostics in the changing trends of clinical diagnosis. Saudi J Biol Sci 21(2):109–117

Allen TM, Cullis PR (2013) Liposomal drug delivery systems: from concept to clinical applications. Adv Drug Deliv Rev 65(1):36–48

Awad R, Avital A, Sosnik A (2023) Polymeric nanocarriers for nose-to-brain drug delivery in neurodegenerative diseases and neurodevelopmental disorders. Acta Pharmaceutica Sinica B 13(5):1866–1886

Awosika AO, Below J, Das JM (2023) Vincristine. StatPearls—NCBI Bookshelf. https://www.ncbi.nlm.nih.gov/sites/books/NBK537122/

Baig N, Kammakakam I, Falath W (2021) Nanomaterials: A review of synthesis methods, properties, recent progress, and challenges. Materials Advances 2(6):1821–1871

Baker SN, Baker GA (2010) Luminescent carbon nanodots: emergent nanolights. Angew Chem Int Ed 49(38):6726–6744

Balkrishna A, Sinha S, Kumar A, Arya V, Kumar D, Dhansekhran M (2024) Nanoparticles as drug delivery systems: advances and challenges. In Nanotechnol: Agric, Environ Health (pp. 245–288). Singapore: Springer Nature Singapore

Barenholz YC (2012) Doxil®—The first FDA-approved nano-drug: Lessons learned. J Control Release 160(2):117–134

Baribeault D (2011) Short-term stability of a new generic sodium ferric gluconate in complex with sucrose. Curr Med Res Opin 27(12):2241–2243

Bhutta BS, Alghoula F, Berim I (2024) Hypoxia. In StatPearls. StatPearls Publishing

Builders PF, Arhewoh MI (2016) Pharmaceutical applications of native starch in conventional drug delivery. Starch-Stärke 68(9–10):864–873

Calabrese G, De Luca G, Nocito G, Rizzo MG, Lombardo SP, Chisari G et al (2021) Carbon dots: an innovative tool for drug delivery in brain tumors. Int J Mol Sci 22(21):11783

Chatterjee N, Kumar P, Kumar K, Misra SK (2022) What makes carbon nanoparticle a potent material for biological application? Wiley Interdiscip Rev: Nanomedicine Nanobiotechnology 14(3):e1782

Colombo M, Carregal-Romero S, Casula MF, Gutiérrez L, Morales MP, Böhm IB et al (2012) Biological applications of magnetic nanoparticles. Chem Soc Rev 41(11):4306–4334

Cooper GM (2000) The Cell: A Molecular Approach. 2nd edition. The Development and Causes of Cancer. 2000, Sunderland (MA): Sinauer Associates

de Carvalho Lima EN, Piqueira JRC, Maria DA (2018) Advances in carbon nanotubes for malignant melanoma: a chance for treatment. Mol Diagn Ther 22(6):703–715

Dos Santos Macedo DC, Cavalcanti IDL, de Souza JB, Nogueira MCDBL, Cavalcanti IMF (2022) Nanotechnology and tuberculosis: An old disease with new treatment strategies. Tuberculosis 135:102208

Fan Z, Qin Y, Liu S, Xing R, Yu H, Chen X et al (2018) Synthesis, characterization, and antifungal evaluation of diethoxyphosphoryl polyaminoethyl chitosan derivatives. Carbohyd Polym 190:1–11

Fathi S, Oyelere AK (2016) Liposomal drug delivery systems for targeted cancer therapy: is active targeting the best choice? Future Med Chem 8(17):2091–2112

García-Hevia L, Villegas JC, Fernández F, Casafont Í, González J, Valiente R, Fanarraga ML (2016) Multiwalled carbon nanotubes inhibit tumor progression in a mouse model. Adv Healthcare Mater 5(9):1080–1087

Giljohann DA, Seferos DS, Daniel WL, Massich MD, Patel PC, Mirkin CA (2010) Gold nanoparticles for biology and medicine. Angew Chem Int Ed 49(19):3280–3294

Gómez-Murcia V, Ribeiro Do Couto B, Gómez-Fernández JC, Milanés MV, Laorden ML, Almela P (2019) Liposome-encapsulated morphine affords a prolonged analgesia while facilitating extinction of reward and aversive memories. Front Pharmacol, 10, 1082

Hamill RJ (2013) Amphotericin B formulations: a comparative review of efficacy and toxicity. Drugs 73(9):919–934

Hong G, Diao S, Antaris AL, Dai H (2015) Carbon nanomaterials for biological imaging and nanomedicinal therapy. Chem Rev 115(19):10816–10906

Huang H, Pierstorff E, Osawa E, Ho D (2007) Active nanodiamond hydrogels for chemotherapeutic delivery. Nano Lett 7(11):3305–3314

Huang Y, Guo X, Wu Y, Chen X, Feng L, Xie N, Shen G (2024) Nanotechnology's frontier in combatting infectious and inflammatory diseases: prevention and treatment. Signal Transduct Target Ther 9(1):34

Hunt A (2018) FDA approves a new antibacterial drug to treat a serious lung disease using a novel pathway to spur innovation. https://www.fda.gov/news-events/press-announcements/fda-approves-new-antibacterial-drug-treat-serious-lung-disease-using-novel-pathway-spur-innovation

IQWiG (2021) In brief: What are immunological tests? https://www.ncbi.nlm.nih.gov/books/NBK373089/

Jahangirian H, Lemraski G, Webster TJ, Rafiee-Moghaddam R, Abdollahi Y (2017) A review of drug delivery systems based on nanotechnology and green chemistry: green nanomedicine. Int J Nanomedicine, 2957–2978

Jeelani PG, Mulay P, Venkat R, Ramalingam C (2020) Multifaceted application of silica nanoparticles. A Review. Silicon 12(6):1337–1354

Jha SK, Sharma P, Malviya R (2016) Liposomal drug delivery system for cancer therapy: advancement and patents. Recent PatS Drug Deliv & Formul, 10(3), 177–183

JJovčevska I, Muyldermans S (2020) The therapeutic potential of nanobodies. BioDrugs, 34(1), 11–26

Kaur R, Badea I (2013) Nanodiamonds as novel nanomaterials for biomedical applications: drug delivery and imaging systems. Int J Nanomedicine, 203–220

Kesharwani P, Jain K, Jain NK (2014) Dendrimer as nanocarrier for drug delivery. Prog Polym Sci 39(2):268–307

Kodama RH (1999) Magnetic nanoparticles. J Magn Magn Mater 200(1–3):359–372

Le Roux R (2015) A matter of accuracy. Nanobiochips in diagnostics and in research: ethical issues as value trade-offs. Sci Eng Ethics, 21(2), 343–358

Li S, Peng Z, Dallman J, Baker J, Othman AM, Blackwelder PL, Leblanc RM (2016) Crossing the blood–brain–barrier with transferrin conjugated carbon dots: A zebrafish model study. Colloids Surf, B 145:251–256

Liu P, Chen G, Zhang J (2022) A review of liposomes as a drug delivery system: current status of approved products, regulatory environments, and future perspectives. Molecules 27(4):1372

Macdougall IC, Comin-Colet J, Breymann C, Spahn DR, Koutroubakis IE (2020) Iron sucrose: a wealth of experience in treating iron deficiency. Adv Ther 37(5):1960–2002

Mehmood A, Ghafar H, Yaqoob S, Gohar UF, Ahmad B (2017) Mesoporous silica nanoparticles: a review. J Dev Drugs 6(02):100174

Munir MU, Salman S, Javed I, Bukhari SNA, Ahmad N, Shad NA, Aziz F (2021) Nano-hydroxyapatite as a delivery system: overview and advancements. Artif Cells, Nanomedicine, Biotechnol 49(1):717–727

Murphy PB, Patel P, Barrett MJ (2025) Morphine. In StatPearls. StatPearls Publishing

Nasir A, Rehman MU, Khan T, Husn M, Khan M, Khan A et al (2024) Advances in nanotechnology-assisted photodynamic therapy for neurological disorders: A comprehensive review. Artif Cells, Nanomedicine, Biotechnol 52(1):84–103

NCI (2024) Cancer Statistics. https://www.cancer.gov/about-cancer/understanding/statistics

Nekoueian K, Amiri M, Sillanpää M, Marken F, Boukherroub R, Szunerits S (2019) Carbon-based quantum particles: an electroanalytical and biomedical perspective. Chem Soc Rev 48(15):4281–4316

NNI (2023) National Nanotechnology Initiative (NNI). https://www.nano.gov/

Noor A, Preuss CV (2024) Amphotericin B. In StatPearls [Internet]. StatPearls Publishing

Nsairat H, Khater D, Sayed U, Odeh F, Al Bawab A, Alshaer W (2022) Liposomes: structure, composition, types, and clinical applications. Heliyon 8(5):e09394

Nygren P (2001) What is cancer chemotherapy? Acta Oncol 40(2–3):166–174

Parmar SFSNM (2024) Iron Dextran. Treasure Island (FL): StatPearls Publishing

Patra JK, Das G, Fraceto LF, Campos EVR, Rodriguez-Torres MDP, Acosta-Torres LS, Shin HS (2018) Nano based drug delivery systems: recent developments and future prospects. J Nanobiotechnology 16(1):71

Paul W, Sharma C (2010) Inorganic nanoparticles for targeted drug delivery. In Elsevier eBooks (pp. 204–235)

Petre CE, Dittmer DP (2007) Liposomal daunorubicin as treatment for Kaposi's sarcoma. Int J Nanomed 2(3):277–288

Prakobvaitayakit M, Nimmannit U (2003) Optimization of polylactic-co-glycolic acid nanoparticles containing itraconazole using 23 factorial design. AAPS PharmSciTech 4(4):71

PubChem (2004) Cytarabine. https://pubchem.ncbi.nlm.nih.gov/compound/Cytarabine

Qasim M, Lim DJ, Park H, Na D (2014) Nanotechnology for diagnosis and treatment of infectious diseases. J Nanosci Nanotechnol 14(10):7374–7387

Qiu A, Wang Y, Zhang G, Wang H (2022) Natural polysaccharide-based nanodrug delivery systems for treatment of diabetes. Polymers 14(15):3217

Reyhanoglu G, Smith T (2023) Irinotecan. StatPearls—NCBI Bookshelf. http://www.ncbi.nlm.nih.gov/books/NBK554441/

Riaz MK, Riaz MA, Zhang X, Lin C, Wong KH, Chen X et al (2018) Surface functionalization and targeting strategies of liposomes in solid tumor therapy: A review. Int J Mol Sci 19(1):195

Rudramurthy GR, Swamy MK, Sinniah UR, Ghasemzadeh A (2016) Nanoparticles: alternatives against drug-resistant pathogenic microbes. Molecules 21(7):836

Rybak JM, Fortwendel JR, Rogers PD (2019) Emerging threat of triazole-resistant *Aspergillus fumigatus*. J Antimicrob Chemother 74(4):835–842

Salehi B, Selamoglu ZS, Mileski K, Pezzani R, Redaelli M, Cho CW et al. (2019) Liposomal cytarabine as cancer therapy: from chemistry to medicine. Biomolecules, 9(12), 773

Sardar R, Funston AM, Mulvaney P, Murray RW (2009) Gold nanoparticles: past, present, and future. Langmuir 25(24):13840–13851

Scott LJ, Goa KL (2000) Verteporfin. Drugs Aging 16(2):139–146

Seven ES, Seven YB, Zhou Y, Poudel-Sharma S, Diaz-Rucco JJ, Cilingir EK et al (2021) Crossing the blood–brain barrier with carbon dots: uptake mechanism and in vivo cargo delivery. Nanoscale Advances 3(13):3942–3953

Shakeri S, Ashrafizadeh M, Zarrabi A, Roghanian R, Afshar EG, Pardakhty A et al (2020) Multi-functional polymeric nanoplatforms for brain diseases diagnosis, therapy and theranostics. Biomedicines 8(1):13

Sharma BR, Naresh L, Dhuldhoya NC, Merchant SU, Merchant UC (2006) An overview on pectins. Times Food Process J 23(2):44–51

Sheng WY, Huang L (2011) Cancer immunotherapy and nanomedicine. Pharm Res 28(2):200–214

Shi J, Kantoff PW, Wooster R, Farokhzad OC (2017) Cancer nanomedicine: progress, challenges and opportunities. Nat Rev Cancer 17(1):20–37

Shi L, Zhang J, Zhao M, Tang S, Cheng X, Zhang W et al (2021) Effects of polyethylene glycol on the surface of nanoparticles for targeted drug delivery. Nanoscale 13(24):10748–10764

Sinha R, Kim GJ, Nie S, Shin DM (2006) Nanotechnology in cancer therapeutics: bioconjugated nanoparticles for drug delivery. Mol Cancer Ther 5(8):1909–1917

Siriwardane DA, Jiang W, Mudalige T (2023) Profiling in-vitro release of verteporfin from VISUDYNE® liposomal formulation and investigating verteporfin binding to human serum albumin. Int J Pharm 646:123449

Smolensky MH, Peppas NA (2007) Chronobiology, drug delivery, and chronotherapeutics. Adv Drug Deliv Rev 59(9–10):828–851

Speranza G (2021) Carbon nanomaterials: Synthesis, functionalization and sensing applications. Nanomaterials 11(4):967

Tibbitt MW, Dahlman JE, Langer R (2016) Emerging frontiers in drug delivery. J Am Chem Soc 138(3):704–717

Veronese FM, Mero A (2008) The impact of PEGylation on biological therapies. BioDrugs 22(5):315–329

Vlachopoulos A, Karlioti G, Balla E, Daniilidis V, Kalamas T, Stefanidou M et al (2022) Poly (lactic acid)-based microparticles for drug delivery applications: an overview of recent advances. Pharmaceutics 14(2):359

Wang Y, Yu L, Kong X, Sun L (2017) Application of nanodiagnostics in point-of-care tests for infectious diseases. Int J Nanomedicine, 4789–4803

Wen N, Lü S, Xu X, Ning P, Wang Z, Zhang Z et al (2019) A polysaccharide-based micelle-hydrogel synergistic therapy system for diabetes and vascular diabetes complications treatment. Mater Sci Eng, C 100:94–103

Yan L, Zhang J, Lee CS, Chen X (2014) Micro-and Nanotechnologies for Intracellular Delivery. Small 10(22):4487–4504

Zare H, Ahmadi S, Ghasemi A, Ghanbari M, Rabiee N, Bagherzadeh M et al (2021). Carbon nanotubes: Smart drug/gene delivery carriers. Int J Nanomedicine, 1681–1706.

Zhao A, Chen Z, Zhao C, Gao N, Ren J, Qu X (2015) Recent advances in bioapplications of C-dots. Carbon 85:309–327

Zheng XT, Ananthanarayanan A, Luo KQ, Chen P (2015) Glowing graphene quantum dots and carbon dots: properties, syntheses, and biological applications. Small 11(14):1620–1636

# Chapter 10
# Nano-Drug Delivery Approaches Against Neurodegenerative Diseases with a Focus on Alzheimer's Disease

**Ishika Patel, Suhrud Pathak, Keyi Liu, Ashwani Kumar, James Joseph Bellarmine, and Muralikrishnan Dhanasekaran**

**Abstract** The intricacy of neurodegenerative diseases' pathogenesis and the limitations of available treatments make them a major issue in contemporary medicine, especially Alzheimer's disease (AD). A potential method for overcoming the blood–brain barrier (BBB) and enhancing the effectiveness of medicinal drugs is the use of nano-drug delivery systems. By providing improved medication stability, controlled release, and targeted delivery to specific brain areas, nanosystems comprising liposomes, dendrimers, and nanoparticles help patients by reducing adverse effects and enhancing outcomes. These methods have the potential to deliver a range of therapeutic medicines, including anti-inflammatory, anti-amyloid, and neuroprotective chemicals, directly to the brain regions affected by Alzheimer's disease. Nanotechnology-based solutions can enhance medication bioavailability, mitigate issues related to poor solubility and degradation, and increase patient compliance by reducing the frequency of doses, thereby enabling precise targeting of brain tissue. Despite these advancements, several challenges remain, including surface modifications, particle size optimization, and addressing potential toxicity issues. The mechanics of drug release, targeting tactics, and therapeutic promise of the most recent developments in nano-drug delivery systems specifically designed for Alzheimer's disease are examined in this article. To transform these technologies into effective medicines, we highlight significant advancements and provide a summary of current clinical studies. The treatment of Alzheimer's disease and other neurodegenerative diseases might be entirely transformed by nano-drug delivery in future.

I. Patel · S. Pathak · K. Liu · M. Dhanasekaran (✉)
Department of Drug Discovery and Development, Harrison College of Pharmacy, Auburn University, Auburn, AL, USA
e-mail: dhanamu@auburn.edu

A. Kumar
Patanjali Herbal Research Department, Patanjali Research Foundation, Haridwar, India

J. J. Bellarmine
Tbilisi State Medical University, Tbilisi, Georgia



A. Balkrishna et al. (eds.), *Nanotechnology in Environmental Science and Healthcare*,
https://doi.org/10.1007/978-981-95-6714-0_10

# 10.1 Introduction

## *10.1.1 Neurodegenerative Diseases*

Neurodegenerative diseases (ND) are a group of chronic disorders that negatively affect the brain and nervous system through the gradual degeneration or death of nerve cells (neurons). Neuronal degeneration results in a range of symptoms, depending on the specific condition and the affected region of the brain or spinal cord. For example, Alzheimer's disease shows memory loss, confusion, and cognitive decline; Parkinson's disease shows tremors in the hands, difficulty balancing the body and coordination; frontotemporal dementia (FTD) affects the frontal and temporal lobes of the brain, posing obstacles in personality, language, and behavior. These symptoms also cause impairment of basic bodily functions, such as movement and memory. Typical examples of such diseases include amyloidosis, tauopathies, α-synucleinopathies, and TDP-43 proteinopathies, which differ from the static neuronal loss seen in toxic or metabolic disorders. Proteins with abnormal conformational properties aggregate in specific brain regions, leading to the development of ND. Multiple autopsies on affected individuals have shown how abnormal proteins extend through anatomically connected pathways. Primary clinical features of these diseases include Parkinson's disease, dementia, motor neuron disease, neurodegeneration, and anatomic distribution (such as frontotemporal degeneration, extrapyramidal disorders, or spinocerebellar degeneration), as well as the primary molecular abnormality (Dugger and Dickson 2017). While ND diseases are incurable, current treatment plans within the medical field are geared toward slowing down the progression of their symptoms.

## *10.1.2 Epidemiology of ND*

Neurodegenerative diseases are increasingly prevalent in various countries in Sub-Saharan Africa, the Middle East, South and North America, and Asia. A few examples of affected countries include Ethiopia, Italy, Norway, and the United States of America. Approximately one hundred million individuals experience chronic pain, 70 million experience insomnia, and 69 million have coordination or balance diseases. Altogether, these numbers account for 54% of ND. High-risk familial genes, toxicity, and level of sanitation in residential areas are genetic and environmental factors that contribute to the development of these diseases. Typically, neurodegenerative diseases that exhibit sex differences show a higher percentage of males affected by Parkinson's. In comparison, a higher rate of females are affected by Alzheimer's and multiple Sclerosis, and Huntington's affects both genders equally. Diseases such as Alzheimer's, Parkinson's, and motor neuron disease drastically increase with age, and the prevalence and incidence of these cases will increase shortly.

### 10.1.3 Different Neurodegenerative Diseases

- **Alzheimer's Disease**

Alzheimer's disease (AD) is a type of progressive brain disorder that causes issues in memorization, thought processes, and behavioral conduct. AD results in synapse loss and neuronal atrophy dominantly throughout the hippocampus and cerebral cortex (Sheppard and Coleman 2020). It can be caused by atrophy (shrinkage of brain tissue), head injuries, advanced age, family genetics, neurofibrillary tau tangles, and amyloid plaques. Increasing confusion, obsessive behavior, aphasia, depression, memory loss, and forgetfulness are associated with Alzheimer's.

- **Parkinson's Disease**

Parkinson's disease (PD) is a progressive, chronic disorder that affects the nervous system and causes tremors in one hand, stiffness, or slowing of movement. The exact cause of this disease is still unknown, but it could be due to Lewy bodies and loss or apoptosis (cell death) of dopaminergic cells in the substantia nigra (Sumien et al. 2021). Painful muscle contractions, difficulty speaking, and coordination problems are associated with Parkinson's.

- **Amyotrophic Lateral Sclerosis (ALS)**

Amyotrophic lateral sclerosis (ALS) is a progressive neurological disorder that causes weakened muscles and deformity. It is also known as motor neuron disease and is characterized by the degeneration of both upper and lower motor neurons, resulting in weakness and eventual paralysis (Hardiman et al. 2017). Gene mutations, autoimmune reactions, heredity, gender, age, and protein mishandling are potential causes. Difficulty swallowing and breathing, weight loss, muscle cramps, clumsiness, and slurred speech are associated with ALS.

- **Huntington's Disease**

Huntington's disease (HD) is an inherited disease characterized by neuropsychiatric symptoms, a movement disorder (like chorea), and progressive cognitive impairment (Stoker et al. 2022). It causes apoptosis of neurons and attacks regions of the brain that help mitigate voluntary movement. A gene defect inherited by both parents leads to HD. Problems with balance and movement, thought processes, emotions, mood, and behavioral and physical changes are associated with HD.

- **Multiple Sclerosis**

Multiple sclerosis (MS) is a chronic, nontraumatic, disabling, and central nervous system inflammatory disease that affects the immune system. The immune system mistakenly attacks and damages the myelin sheaths covering neurons, causing slow signal transmission between neurons. Low serum levels of vitamin D, smoking, childhood obesity, and infection with the Epstein–Barr virus are likely causes (Dobson and Giovannoni 2019). Bladder and bowel problems, muscle spasms, paralysis, numbness, and vision changes are associated with MS.

- **Lewy Body Dementia**

Lewy body dementia (LBD) is a progressive disorder characterized by abnormal deposits of protein in areas of the nervous system and brain, commonly mistaken with PD. It stems from the formation of Lewy bodies, which contain aggregates of the misfolded protein α-synuclein (Lee et al. 1987). These Lewy bodies accumulate, causing neuronal apoptosis. Apathy, bradykinesia (slowed movement), depression, visual hallucinations, confusion, changes in memory and mood, and rigid muscles are associated with LBD.

- **List of Other ND Diseases**

Other ND diseases include Adrenoleukodystrophy, Agenesis of the Corpus Callosum, Alexander Disease, Alpers Disease, Ataxia Telangiectasia, Batten Disease, Charcot-Marie-Tooth Disease, Creutzfeldt-Jakob Disease, Frontotemporal Dementia, Gerstmann Straussler Scheinker Disease, Giant Axonal Neuropathy, Infantile Neuroaxonal Dystrophy, Kennedy's Disease, Krabbe Disease, Mechado-Joseph Disease, Menkes Disease, Multiple System Atrophy, Rett Syndrome, Troyer Syndrome, and Zellweger Syndrome (Lee et al. 2020).

### *10.1.4 Different Types of Dementia*

- **Alzheimer's Disease**

Alzheimer's disease is the most common type of dementia that gets worse over time and is defined by a progressive decline in memory, reasoning, and thinking. Aggregated amyloid-beta proteins and neurofibrillary tangles of accumulated tau proteins cause AD. This tau protein aggregation and changes in amyloid precursor protein (APP) cleavage and production of the APP fragment beta-amyloid (Aβ) lead to AD (Lopez et al. 2019). Symptoms include trouble sleeping, confusion about time or place, language difficulties, weight loss, seizures, infections, changes in behavior and personality, and trouble reasoning or handling complex tasks.

- **Vascular Dementia**

Vascular dementia (VaD), also known as multi-infarct dementia, is a dementia syndrome likely to be the consequence of lesions of the brain, vascular in origin, irrespective of their ischemic, hemorrhagic, or hypoxic nature (Parnetti 1999). The size, location, and number of vascular changes affect cognition and brain functioning, including memory, behavior, and thinking. Due to decreased blood flow to the brain, abnormalities occur in the blood vessels and white matter between brain regions, which several factors, including infarcts, hemorrhages, cerebral hyperperfusion, strokes, and white matter hyperintensities, can cause. The brain does not receive adequate oxygen or nutrients. Symptoms include delusions and hallucinations, inability to control the bowel or bladder, difficulty following instructions, trouble managing finances, getting lost in surroundings, reduced planning and concentration, and short-term memory loss.

- **Lewy Body dementia**

Lewy body dementia (LBD) is an age-associated ND producing progressive cognitive decline that interferes with everyday life and daily activities. It is characterized by an accumulation of aggregated α-synuclein protein in Lewy bodies and Lewy neurites (Outeiro et al. 2019). This accumulation of protein is deposited in the brain, causing a loss of neuron functioning that, in turn, affects the production of the acetylcholine and dopamine neurotransmitters. Symptoms include difficulty sleeping, impaired balance, cognitive problems, poor blood pressure regulation, excessive sweating, digestive issues, rigid muscles, visual hallucinations, shaky legs, tremors, and bradykinesia.

- **Parkinson's Disease**

Parkinson's disease (PD) is the second most common progressive ND that initially causes tremors in one hand, slowing movement. The exact cause of PD is still unknown. It results from a pathophysiologic loss or degeneration of dopaminergic neurons in the substantia nigra of the midbrain, leading to the development of neuronal Lewy bodies (Beitz 2014). This degeneration disrupts the normal balance between dopamine/acetylcholine levels in the basal ganglia, causing various symptoms. Symptoms include constipation, mood disorder, rigidity, bradykinesia, resting tremor, rapid eye movement, sleep behavior disorder, urinary frequency, hyposomnia, postural instability, or poor balance.

- **Frontotemporal Dementia**

Frontotemporal dementia (FTD) is an umbrella term that encompasses clinical disorders that include changes in behavior, language, executive control, and often motor symptoms. Disorders such as semantic variant PPA, behavioral variant FTD, and nonfluent/agrammatic variant primary progressive aphasia are some examples (Olney et al. 2017). These disorders can occur when the frontal lobes of the brain undergo shrinkage (atrophy), resulting in changes in basic activities of daily living. Symptoms include difficulty walking, difficulty performing work, unusual behaviors, trouble communicating, and emotional problems.

- **Creutzfeldt–Jakob Disease**

Creutzfeldt–Jakob disease (CJD), a rare, transmissible, and rapidly progressive prion disease, is an ultimately fatal central nervous system infection caused by the accumulation of abnormally shaped prion proteins in neurons (Vacca 2016). At high levels of accumulation, these prions cause irreversible damage to neurons within the brain, resulting in quick death, and there is no known cure. Symptoms include slurred speech, abnormal jerking movements, gradual loss of brain functioning and mobility, loss of coordination and balance, personality changes, vision problems and blindness, and loss of memory and intellect.

- **Wernicke–Korsakoff Syndrome**

Wernicke–Korsakoff syndrome is a chronic amnesia resulting from unrecognized or undertreated Wernicke encephalopathy, caused by thiamine (B1) deficiency (Wijnia 2022). A lack of thiamine affects the peripheral and central nervous systems, and there is atrophy of the diencephalon (thalamus, hypothalamus, and mammillary bodies) structures of the brain. It is typically caused by alcohol consumption or poor nutrition, and another factor rarely causes it. Symptoms include unsteady gait, fatigue, amnesia, behavioral changes such as anger, lack of attention, hallucinations, disorientation, and confabulation.

- **Mixed Dementia**

Mixed dementia (MD) is a combination of more than 1 type of dementia, such as Alzheimer's and Vascular dementia (the most common) or Alzheimer's with Lewy body dementia. For example, in AD mixed with VaD, blood vessel problems coexist with abnormal protein deposits in the brain, causing simultaneous cerebral changes. Its rate of progression depends on the overall health and genetic background of every individual. Symptoms include changes in speech, behavioral changes, memory problems, alterations in personality or mood, difficulties with self-care, and trouble with daily activities or tasks.

- **Normal Pressure Hydrocephalus**

Normal pressure hydrocephalus (NPH) is a steadily progressive disease characterized by an expansion of the ventricles (cavities) of the brain, despite normal intracranial pressure (ICP) values (Gavrilov et al. 2019). The abnormal accumulation of cerebrospinal fluid (CSF) in the ventricles causes enlargement of the ventricles, which puts pressure on the brain. This pressure only occurs if CSF is disrupted or blocked in its flow to the brain and spinal cord. Symptoms include bladder control problems, changes in urination frequency, memory and cognitive issues, and alterations in balance or gait.

- **Huntington's Disease**

Huntington's disease (HD) is the most common monogenic neurodegenerative disease and the most common genetic dementia in the world. The causative genetic mutation is an expanded CAG trinucleotide repeat in the gene encoding the Huntingtin protein, which leads to a prolonged polyglutamine stretch at the N-terminus of the protein (Ghosh and Tabrizi 2018). This mutation leads to the gradual decay of neurons over time, resulting in several symptoms. Symptoms include trouble moving, speaking, breathing, swallowing, mood swings and personality changes, clumsiness, depression, difficulty concentrating, and involuntary jerking or fidgety movements.

## 10.2 Most Common Type of Dementia: Alzheimer's Disease

Alzheimer's disease, the most common type of dementia, has mechanisms of amyloid plaques, neurofibrillary tangles, and inflammation. The amyloid precursor protein (APP) creates a breakdown, which is known as the beta-amyloid protein. Specifically, β-amyloid 42 develops in abnormal levels, leading to the aggregation of this toxic protein. Microglia, a type of glial cell, also contribute to the formation of these amyloid plaques when they fail to properly clear waste and toxins from the brain. The protein tau also detaches from microtubules and sticks to other tau molecules, forming threads or neurofibrillary tangles inside neurons. The neuron's transport system becomes blocked due to these tangles, damaging neuronal communication (US Department of Health and Human Services 2017). Vascular problems also exist, and necessary transporter molecules such as glucose are prevented from crossing the blood–brain barrier (BBB), leading to inflammation. These mechanisms result in several different symptoms in individuals with AD. Symptoms include memory loss, cognitive difficulties, wandering or getting lost, trouble handling finances like paying bills or handling money, repeating questions, taking longer to carry out daily life activities or tasks, and personality and behavioral changes (US Department of Health and Human Services 2017). No cure exists for AD, but there are treatments available for affected individuals. Three cholinesterase inhibitor drugs (galantamine, rivastigmine, and donepezil) are used to increase acetylcholine production in the brain. Kisunla injections target and remove amyloid plaques, slowing cognitive decline.

Preventative approaches, such as lifestyle changes in nutrition, daily exercise, avoidance of toxic substances and drugs, an active social life, meditation, stress and cortisol control, and healthy aging, also contribute to the slowing of AD symptoms (Del Din et al. 2021). For example, changes in nutrition might include consuming a diet rich in polyunsaturated fatty acids, antioxidants, and B vitamins. These approaches are beneficial because they promote physical activity, support maintaining a healthy weight, and enhance sleep quality.

Diagnosis of AD involves several diagnostic, cognitive, and blood tests, such as MRI (magnetic resonance imaging), CT (computerized tomography), and PET (positron emission tomography). It is also diagnosed through the DSM-5-TR (Diagnostic Statistical Manual of Mental Disorders); it is required to have a decline in memory or learning and at least one other cognitive domain out of six domains, and also interference with activities of daily living (ADL) for an individual to be diagnosed by a healthcare professional (Yokoi 2023).

## 10.3 Nano-Drug Delivery Systems

Nanomedicine, nanotechnology, and nano-drug delivery systems (NDDS) are relatively new but rapidly developing scientific materials that serve as diagnostic tools or deliver therapeutic agents to specifically targeted sites in a controlled manner

(Patra et al. 2018). It is a system of specifying the drug moiety directly into its target body area (organ, cellular, and subcellular level of specific tissue) to overcome the particular toxic effect of conventional drug delivery and reduce the required drugs for therapeutic efficiency (Tewabe et al. 2021). Drugs were initially delivered via shots, nasal, oral, and inhalation methods, but they harmed unaffected areas and were time-consuming for the drug effects to work. Therefore, NDDS is more widely used to treat chronic illnesses because the medication is delivered to site-specific and target-oriented regions within the body or brain. Nanoparticles consist of drugs and are characterized by a size of approximately 1–100 nm, typically having affinity ligands, such as antibodies or DNA/RNA, attached to their surfaces. These nanoparticle-carrying drugs are then inserted into the target area, where the nanoparticle recognizes and binds to the target cells through their surface receptors. After the nanoparticle binds to the target cells, the drug is released inside the cells, where it attacks foreign objects and diseases as necessary. Thus, NDDS is considered a practical approach to creating treatment plans and options for individuals affected by ND diseases, as it improves drug bioavailability and absorption time, reduces release time, eliminates drug aggregation, and enhances drug solubility in the blood (Afzal et al. 2022). NDDS additionally allows for improved pharmacokinetics, as it lengthens drug circulation within the body, which stabilizes the drug and increases the chances of effective treatment.

The current FDA-approved NDDS are as follows: Zelsuvmi, Exblifep, Letybo, Tevimbra, Rezdiffra, Tryvio, Duvyzat, Winrevair, Vafseo, Voydeya, Zevtera, Lumisight, Anktiva, Ojemda, Xolremdi, Imdelltra, Rytelo, Iqirvo, Sofdra, Piasky, Ohtuvayre, Kisunla, Leqselvi (U.S. Food & Drug Administration 2025).

- **Alzheimer's Disease**

A few disease-reversing/modifying molecules used to treat AD symptoms are donepezil, rivastigmine, galantamine, carbidopa, and levodopa. FDA-approved tracers include flortaucipir (Tauvid) for tau fibril imaging, flutemetamol (Vizamyl), florbetaben (Neuraceq), and florbetapir (Amyvid) for amyloid imaging (Suri et al. 2024).

- **Parkinson's Disease**

A few disease-targeting drugs used to treat PD include levodopa/carbidopa, PRX002 (prasinezumab), PD01A, NPT200-11 (UCB0599), Ambroxol, LTI-291, PR001, Exanatide, Sustained release Exanatide (PT320), Semaglutide, Liraglutide, Lixisenatide, NLY01, K0706, Radotinib, Ceftriazone, Blarcamesine (ANAVEX 2–73), Deferiprone, Idebenone, Azathioprine, Simvastatin, Lovastatin, Ursodeoxycholic acid (Murakami et al. 2023).

- **Amyotrophic Lateral Sclerosis (ALS)**

A few drugs used to treat ALS are Qulsody, Radicava, Rilutek, Tiglutik, Exservan, and Nuedexta (ALS Association 2025). FDA-approved Riluzole, Edaravone, Masitinib, Tofersen, Ravulizumab-cwvz, tofersen, Mesenchymal stem cell (MSC)-neurotrophic factor (NTF) cells (Chen 2020).

- **Huntington's Disease**

Recent and ongoing studies include Amaryllis and APACHE (phosphodiesterase 10A inhibitor), PRIDE-HD (dopamine stabilizer, sigma 1 receptor agonist, vestibular monoamine transporter 2 inhibitor), TRIHEP3 (anaplerotic therapy), Kinect-HD (VMAT2 inhibitor), HD-DBS (neuromodulation), STAIR (vasopressin 1a receptor antagonist), Action-HD (norepinephrine/dopamine reuptake inhibitor), ETON (polyphenol), and VCAS-HD (nicotine agonist) (Dash and Mestre 2020). A few drugs used to treat HD include Tetrabenazine, Deutetrabenazine, Olanazapine, Risperidone, Citalopram, Fluoxetine, Sertraline, Lamotrigine, and Carbamazepine (Ferguson et al. 2022).

- **Multiple Sclerosis**

Clinical studies for MS include ongoing Phase III trials, Phase I trials, and Phase II trials (Bourque and Hawiger 2021). A few drugs used to treat MS are Copazone, Tecfidera, Ocrevus, Rebif, Ampyra, Gilenya, Tysabri, Avonex, Betaseron, Kesimpta, Vumerity, Aubagio, Mavenclad, Zeposia, Briumvi, Mayzent, Extavia, Glatopa, Plegridy, and Zinbryta.

- **Lewy Body Dementia**

There are currently no FDA-approved NDDS for LBD.

## 10.4 NDDS in Alzheimer's Disease

Several different NDDS mechanisms are involved in AD, including neurotoxicity, apoptosis, inflammation, oxidative stress, mitochondrial dysfunction, and cholinergic and glutamatergic transmission. Apoptosis, a preprogrammed form of cell death, occurs when there is an imbalance in autophagy regulated by the PI3K/Akt/mTOR pathway. The aggregation or accumulation of hyperphosphorylated tau or amyloid-β (Aβ) protein leads to synaptic loss and neuronal death. Consequently, two types of apoptosis can occur: physiological and pathological apoptosis. In physiological apoptosis, mediators such as p21, p38, nitrogen-activated protein kinase (MAPK), c-Jun-NH2-terminal protein kinase, p53, caspases 2, 3, 8, and 9, BCL-XS, Bax, apoptosis-inducing factor, and Par-4 result in dysfunctions in neural connections, leading to cell death (Obulesu and Lakshmi 2014). These mediators induce morphological changes in cells through the condensation and fragmentation of chromatin and nucleosomal DNA, cell volume contraction, increased mitochondrial activity, blebbing of the plasma membrane, phagocytosis of neighboring cells, and overall shrinkage. The JNK pathway is upregulated, leading to a decrease in the expression of anti-apoptotic proteins, such as Bcl-XL and B-cell lymphoma/leukemia-2 protein (Bcl-2). Additionally, the GSK-3B pathway activates proapoptotic factors (Kumari et al., 2023). This self-destruction of the cells is triggered by two activation mechanisms: the intrinsic pathway (mitochondrial pathway) and the extrinsic pathway.

Both pathways are said to be interlinked; however, they have separate functions. The intrinsic pathway responds to damage to the vascular endothelium, a specialized cell layer that acts as a barrier between the blood and tissues. It constructs the internal vessel wall of capillaries, veins, and arteries, making up the endothelial cells (EC). Damage to the VE results in the exposure of clotting factors to negatively charged subendothelial surfaces, activating the intrinsic pathway because of a disruption or induced stress to the interaction of proapoptotic and anti-apoptotic cells within the mitochondrial organelle. The extrinsic pathway responds to external trauma, such as loss of circulating blood, causing apoptosis. In other words, the intrinsic pathway involves caspase-9 activation when cytochrome c is released from the mitochondria, whereas the extrinsic pathway involves caspase-8 or -10 activation through surface death receptors.

Inflammation is also a mechanism in AD, characterized by the immune system's recognition and elimination of harmful species, which aids the healing process after acting as a defense mechanism. However, when inflammation persists even after the foreign stimulus or threat (i.e., infection or injury) has passed, it can damage DNA material and lead to various diseases, including cardiac diseases, cancers, atherosclerosis, rheumatoid arthritis, and type 2 diabetes. In AD, NPs activate the inflammatory response, which is regulated by increased microglia, highly dynamic cells responsible for phagocytosis and eliminating dead cells. When these microglia remain in activated states for long periods, inflammasomes (i.e., NOD-, LRR-, and pyrin domain-containing 3 (NLRP3), which are multiprotein complexes that control cytokine production, cause proinflammatory cytokine secretion, where cytokines are secreted into muscle cells, fibroblasts, macrophages, neuritis, and other damaged muscle areas (Twarowski and Herbet 2023). For example, since fibroblasts secrete this additional cytokine, mast cells on the epithelial surfaces over-initiate inflammatory responses. Some examples of cytokines and neurotoxic agents that are released include IL-1β, TNF, IL-6, nitric oxide (NO), and reactive oxygen species (ROS), which directly or indirectly cause neuronal cell death (Xie et al. 2022). The release of such neurotoxins into the immune system harms other neurons, resulting in neuroinflammation.

Additionally, insoluble amyloid β peptide deposits and neurofibrillary tau tangles act as inflammatory stimulants, directly causing neuronal injury (Akiyama et al. 2000). These neurofibrillary tangles or Aβ tangles need to be cleared and are only possible when the body is at homeostasis (acute response). Acute microglial activation leads to a decrease in the aggregation of Aβ tangles, resulting in successful Aβ clearance and neuroprotective effects. However, in AD (chronic response), inflammasomes, such as NLRP3, contribute to the increased production of cytokines, leading to impaired Aβ clearance and, ultimately, neurodegeneration. When neurodegeneration begins to occur, there is synaptic dysfunction, neuronal stress, ongoing Aβ production, and proinflammatory cascades, serving as neurotoxic effects to the AD brain.

Furthermore, oxidative stress is another mechanism present in AD and is linked to several diseases, such as diabetes, cardiac disease, inflammatory conditions, hypertension, atherosclerosis, and neurodegenerative diseases like AD. It has several molecules and free radicals with one unpaired electron in the outer shells, sharing the same spin, which is the direction in which electrons move along the atom's orbital. These free radicals, also known as highly reactive oxygen species (ROS), chemically consist of diatomic oxygen ($O_2$), water, and hydrogen peroxide. Some examples of ROS include free nitrogen radicals, oxygen, hydroperoxyl radicals, hydroxyl radicals, and superoxide anion radicals. In AD, B amyloid (AB) accumulation of neurofibrillary tau tangles (NFTs) not only results in increased production of ROS but also induces oxidative stress, leading to a dysfunctional antioxidant system where there is an imbalance between the redox states of ROS and antioxidants within the body under normal physiological conditions. Since there are excess levels of ROS, it is difficult for antioxidants to neutralize them, causing them to harm organelles, such as the mitochondria. The mitochondria are responsible for synthesizing RNA and proteins; however, persistent damage to these organelles leads to changes in mitochondrial function. It leads to chronic ROS exposure, DNA mutations in the genome, and overall damage to surrounding proteins, lipids, and nucleic acids. There are adverse effects on the electron-transport chain (ETC), specifically its complexes I, II, and III, as well as the tricarboxylic acid cycle enzyme aconitase. It also affects the iron-sulfur (Fe-S) centers by inactivating them in the ETC, causing a decrease or immediate stop to mitochondrial energy production and increasing the production of neurotoxic levels of AB (Guo et al. 2013). ROS are capable of reacting very easily with other molecules, such as lipids (lipid oxidation), DNA (DNA oxidation), and sugars (glycoxidation), eventually reacting with proteins to form protein-bound carbonyls (Fig. 10.1). These protein carbonylation levels are often utilized to measure the amount of ROS damage done to proteins (Gella and Durany 2009).

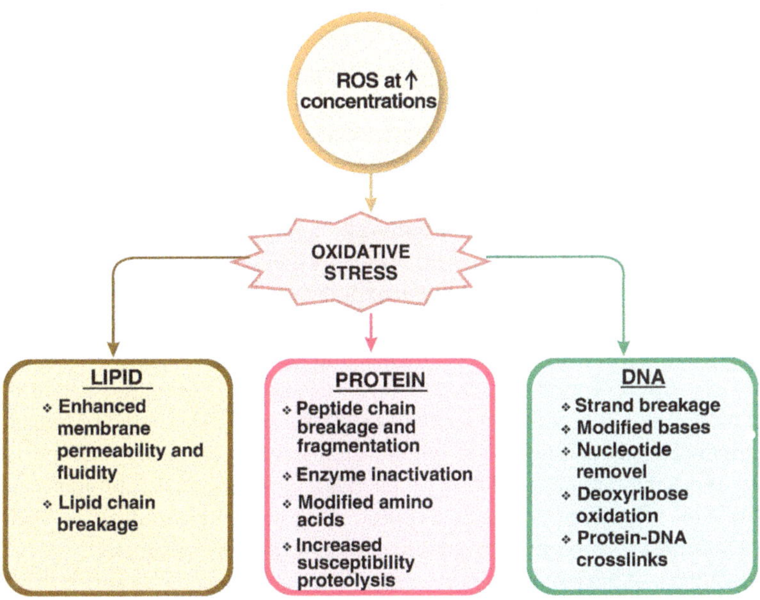


**Fig. 10.1** ROS-induced damage to DNA, proteins, and lipids

Lastly, cholinergic neurotransmission is another mechanism found in AD, involved in the synthesis and release of acetylcholine (ACh), a vital excitatory neurotransmitter responsible for transmitting chemical messages from the brain to the rest of the body through nerve impulses. ACh is a primary neurotransmitter in the parasympathetic nervous system (PNS) and is synthesized and released to the cerebral cortex, basal ganglia, and basal forebrain. It is crucial for higher neural functions like attention span, learning ability, and memory, and is secreted by cholinergic neurons to help aid cell homeostasis and plasticity. There is a series of rapid and multidirected transmissive steps in cholinergic neurotransmission. First, choline acetyltransferase (ChAT) uses acetyl-CoA in the cytoplasm to synthesize ACh. ACh is then transferred to synaptic vestibular follicles through the vestibular acetylcholine transporter (VACht). Depolarization of presynaptic neurons facilitates the extracellular secretion of acetylcholine (ACh), enabling ACh to bind to nicotinic or muscarinic receptors, two types of acetylcholine receptors (AChRs) that are integral cholinergic membrane proteins activated by ACh binding (Chen et al. 2022). As a result, excess ACh at the synaptic cleft becomes hydrolyzed by the enzyme acetylcholinesterase (AChE), producing choline and acetate, which is then reuptake or recycled into the presynaptic cholinergic neurons by the high-affinity choline transporter (CHT1) (Ferreira-Vieira et al. 2016). Recycling back into the presynaptic nerve terminal leads to a decrease in acetylcholine production because there is a delay in postsynaptic neurotransmission signals. Due to this discrepancy in cholinergic transmission, there is a gradual loss of the neocortical and limbic areas of the brain, contributing to cognitive decline, a symptom seen in AD.

Thus, apoptosis, inflammation, oxidative stress/mitochondrial dysfunction, and cholinergic neurotransmission are several mechanisms present in AD, and dysfunctioning systems within them contribute to symptoms such as cognitive decline, memory loss, confusion, difficulty with financial management, difficulty with daily life activities like cleaning, and inability to comprehend factual information. There is also the risk of developing amyloid-related imaging abnormalities (ARIA), such as vision changes and blurriness, headaches, confusion, nausea and vomiting, and dizziness. ARIA symptoms typically arise when small bleeding spots occur in or on the surface of the brain due to temporary swelling or inflammation. These symptoms could be explained by genetics and the ApoE ε4 gene risk factor, which doubles or triples the risk of AD. It is important to note, however, that having the ApoE ε4 allele does not necessarily mean an individual will develop the disease. Although there is no cure for AD, the FDA has approved numerous NDDS (nano-drug delivery systems) via different methods, such as oral, nasal, and injection, to help provide relief of progressive symptoms. Current FDA-approved NDDS include Aducanumab, Donanemab (Kisunla), Lecanemab, cholinesterase inhibitors (i.e., Aricept, Exelon, and Razadyne), glutamate regulators (Namenda), cholinesterase inhibitor and glutamate regulator (Namzaric), Orexin receptor antagonist (Belosomra) for behavioral and psychological symptoms, and atypical antipsychotics (Rexulti).

For instance, Aducanumab is an anti-Aβ monoclonal antibody that binds to fibrillar and oligomeric forms of amyloid-beta, preventing Aβ aggregation and plaque deposition within the brain through microglial-mediated removal. It is typically administered via IV treatment at a dose of 10 mg/kg over 1 h, once a month. Donanemab (Kisunla) is a type of monoclonal antibody, meaning it binds to destructive proteins, such as the amyloid-beta protein, to eliminate plaque. Once the antibodies of the drug build up, it helps slow down the progression of symptoms. It is an anti-amyloid antibody intravenous (IV) infusion therapy administered to patients every four weeks for 12 months. Another drug is Lecanemab, a humanized monoclonal antibody with high affinity to soluble amyloid-beta (Aβ) protofibrils, metastable peptides that can pose neurotoxic effects. Antibiotic binding to these soluble protofibrils helps destroy pre-existing Aβ aggregates. Patients typically receive an IV treatment of 10 mg per kilogram of body weight every two weeks for 18 months or a placebo (van Dyck et al. 2023). Cholinesterase inhibitors are a group of medicines that inhibit the cholinesterase enzyme from hydrolyzing (breaking down) acetylcholine (ACh) into choline and acetate. These drugs increase the availability of acetylcholine (ACh) in the brain, allowing for enhanced brain function and memory.

Aricept (donepezil hydrochloride), for example, is administered orally at 5 mg once daily, with or without food. Tablets need to be dissolved on the tongue for usage, and dosages (mg) may increase according to the stage (mild, moderate, severe) of AD. Exelon (rivastigmine) is available as oral capsules, oral liquid, and transdermal patches. For oral capsules and liquid, 3 to 6 mg daily by mouth twice a day is administered, and dosage increases by 1.5 mg (initial start is 1.5 mg) per dose (maximum is 12 mg) every two weeks. For transdermal patches, a 4-to 6-mg patch is applied daily for a month, and then the dosage increases to a 9.5-mg patch and eventually to a 13.3-mg patch (Patel and Gupta 2023). Razadyne (galantamine hydrobromide) has two dosage forms in either a tablet or solution (4 mg/ml): IR (immediate-release) and ER (extended-release). IR medications require frequent intake and act quickly as they enter the system to relieve pain immediately; however, symptomatic relief does not last more than a few hours. ER medications require less intake and have a predesignated amount of drug within the pill to release at a slower rate, providing symptomatic relief for more extended periods. IR Razadyne is administered in tablets of 4 mg (off-white), 8 mg (pink), and 12 mg (orange-brown), whereas ER is in 8 mg (white), 16 mg (pink), and 24 mg (caramel) (National Library of Medicine 2010). Namenda (memantine) is a glutamate regulator that acts as an uncompetitive, low-affinity antagonist, meaning it blocks the NMDAR (N-methyl-D-aspartate receptor) from entering the receptor-associated ion channel during its open state, preventing the overactivation of glutamate receptors on neuronal and non-neuronal cells and damage of normal synaptic neurotransmission (Kuns et al. 2024). It is available in IR (5 mg once daily, 5 mg twice daily, 10 mg once daily, and 10 mg twice daily) and ER (7 mg daily, 14 mg daily, 21 mg daily, and 28 mg daily). Namzaric (donepezil and memantine) is a cholinesterase inhibitor and glutamate regulator that combines the actions of blocking glutamate effects and the breakdown of ACh in the brain. It is administered orally as 28 mg/10 mg twice daily or 14 mg/10 mg once daily. Belsomra (suvorexant) is a highly selective antagonist drug used for psychological and behavioral issues that targets and inhibits orexin, a neurotransmitter responsible for wakefulness. It prevents orexin from transmitting wake signals to other regions of the brain. It is administered in tablet form and should be taken once daily, 30 min before bedtime, with or without food. Rexulti (brexpiprazole) is an atypical antipsychotic that rebalances dopaminergic and serotonin levels within the brain by acting as a partial agonist at 5-HT1A and dopamine D2 receptors, and antagonist at 5-HT2A receptors, to which the drug binds with subnanomolar affinities (the binding affinity of the drug for its target in relation to Gibbs binding energy). It is available as an oral tablet in 0.25 mg, 0.5 mg, 1 mg, 2 mg, 3 mg, and 4 mg strengths, taken once daily with or without food. Dosages can be increased after each week of intake if tolerated well.

## 10.5 Conclusion

In conclusion, by overcoming the obstacles presented by the blood–brain barrier and improving therapeutic effectiveness, nano-drug delivery systems have great potential to advance the treatment of Alzheimer's disease. These technologies have the potential to improve patient outcomes while reducing adverse reactions through safer, more effective, and more tailored therapeutic approaches. Notwithstanding the fascinating developments, problems, including surface modification, particle size optimization, and toxicity concerns, require further investigation. The ongoing development and clinical validation of these nano-based strategies may transform the treatment of Alzheimer's disease and other neurodegenerative diseases as the field advances, offering promise for future, more personalized and effective therapies.

## References

Afzal O, Altamimi AS, Nadeem MS, Alzarea SI, Almalki WH, Tariq A, Mubeen B, Murtaza BN, Iftikhar S, Riaz N, Kazmi I (2022) Nanoparticles in drug delivery: From history to therapeutic applications. Nanomaterials 12(24):1–27

Akiyama H, Barger S, Barnum S, Bradt B, Bauer J, Cole GM, Cooper NR, Eikelenboom P, Emmerling M, Fiebich BL, Finch CE, Frautschy S, Griffin WST, Hampel H, Hull M, Landreth G, Lue L-F, Mrak R, Mackenzie IR, McGeer PL, O'Banion MK, Pachter J, Pasinetti G, Plata-Salaman C, Rogers J, Rydel R, Shen Y, Streit W, Strohmeyer R, Tooyoma I, Van Muiswinkel FL, Veerhuis R, Walker D, Webster S, Wegrzyniak B, Wenk G, Wyss-Coray T (2000) Inflammation and Alzheimer's disease. Neurobiol Aging 21(3):383–421

ALS Association (2025) Medications for treating ALS. https://www.als.org/navigating-als/living-with-als/medications

Beitz JM (2014) Parkinson's Disease: a Review. Front Biosci (Schol Ed) 6(1):65–74

Bourque J, Hawiger D (2021) Current and future immunotherapies for multiple sclerosis. Mo Med 118(4):334–339

Chen JJ (2020) Overview of current and emerging therapies for amytrophic lateral sclerosis. Am J Manag Care 26(9):S191–S197

Chen ZR, Huang JB, Yang SL, Hong FF (2022) Role of cholinergic signaling in Alzheimer's disease. Molecules 27(6):1816

Dash D, Mestre TA (2020) Therapeutic update on Huntington's disease: symptomatic treatments and emerging disease-modifying therapies. Neurotherapeutics 17(4):1645–1659
Del Din S, Kirk C, Yarnall AJ, Rochester L, Hausdorff JM (2021) Body-worn sensors for remote monitoring of Parkinson's disease motor symptoms: vision, state of the art, and challenges ahead. J Parkinson's Dis 11(s1):S35–S47
Dobson R, Giovannoni G (2019) Multiple sclerosis–a review. Eur J Neurol 26(1):27–40
Dugger BN, Dickson DW (2017) Pathology of neurodegenerative diseases. Cold Spring Harb Perspect Biol 9(7):1–22
Ferguson MW, Kennedy CJ, Palpagama TH, Waldvogel HJ, Faull RL, Kwakowsky A (2022) Current and possible future therapeutic options for Huntington's disease. J CentL Nerv Syst Dis 14:1–27
Ferreira-Vieira TH, Guimaraes IM, Silva FR, Ribeiro FM (2016) Alzheimer's disease: targeting the cholinergic system. Curr Neuropharmacol 14(1):101–115
Gavrilov VG, Gaydar VB, Svistov VD, Korovin EA, Samarcev NI, Churilov PL, Tovpeko VD (2019) Idiopathic normal pressure hydrocephalus (Hakim-Adams syndrome): clinical symptoms, diagnosis and treatment. Psychiatria Danubina, 31(suppl 5), 737–744
Gella A, Durany N (2009) Oxidative stress in Alzheimer disease. Cell Adh Migr 3(1):88–93
Ghosh R, Tabrizi SJ (2018) Clinical features of Huntington's disease. Polyglutamine Disord, 1–28
Guo C, Sun L, Chen X, Zhang D (2013) Oxidative stress, mitochondrial damage and neurodegenerative diseases. Neural Regen Res 8(21):2003–2014
Hardiman O, Al-Chalabi A, Chio A, Corr EM, Logroscino G, Robberecht W, Shaw PJ, Simmons Z, Van Den Berg LH (2017) Amyotrophic lateral sclerosis. Nat Rev Dis Primers 3(1):1–19
Kumari S, Dhapola R, Reddy DH (2023) Apoptosis in Alzheimer's disease: insight into the signaling pathways and therapeutic avenues. Apoptosis 28(7):943–957
Kuns B, Rosani A, Patel P, Varghese D (2024) Memantine. In StatPearls [Internet]. StatPearls Publishing
Lee EH, Tsai MJ, Tang YP, Chai CY (1987) Differential biochemical mechanisms mediate locomotor stimulation effects by caffeine and nicotine in rats. Pharmacol Biochem Behav 26(2):427–430
Lee D, Jo MG, Kim SY, Chung CG, Lee SB (2020) Dietary antioxidants and the mitochondrial quality control: Their potential roles in Parkinson's disease treatment. Antioxidants 9(11):1056
Lopez JAS, González HM, Léger GC (2019) Alzheimer's disease. Handb Clin Neurol 167:231–255
Murakami H, Shiraishi T, Umehara T, Omoto S, Iguchi Y (2023) Recent advances in drug therapy for parkinson's disease. Intern Med 62(1):33–42
National Library of Medicine. (2010). RAZADYNE-galantamine hydrobromide tablet, film coated
Obulesu M, Lakshmi MJ (2014) Apoptosis in Alzheimer's disease: an understanding of the physiology, pathology and therapeutic avenues. Neurochem Res 39:2301–2312
Olney NT, Spina S, Miller BL (2017) Frontotemporal dementia. Neurol Clin 35(2):339–374
Outeiro TF, Koss DJ, Erskine D, Walker L, Kurzawa-Akanbi M, Burn D, Donaghy P, Morris C, Taylor J-P, Thomas A, Attems J, McKeith I (2019) Dementia with Lewy bodies: an update and outlook. Mol Neurodegener 14:1–18
Parnetti L (1999) Pathophysiology of vascular dementia and white matter changes. Revue Neurologique 155(9):754–758
Patel PH, Gupta V (2023) Rivastigmine. StatPearls—NCBI Bookshelf. https://www.ncbi.nlm.nih.gov/books/NBK557438/
Patra JK, Das G, Fraceto LF, Campos EVR, Rodriguez-Torres MDP, Acosta-Torres LS, Diaz-Torres LA, Grillo R, Swamy MK, Sharma S, Habtemariam S, Shin HS (2018) Nano based drug delivery systems: recent developments and future prospects. J Nanobiotechnology 16:1–33
Sheppard O, Coleman M (2020) Alzheimer's disease: etiology, neuropathology and pathogenesis. Exon Publications, 1–21
Stoker TB, Mason SL, Greenland JC, Holden ST, Santini H, Barker RA (2022) Huntington's disease: Diagnosis and management. Pract Neurol 22(1):32–41

Sumien N, Cunningham JT, Davis DL, Engelland R, Fadeyibi O, Farmer GE, Mabry S, Mensah-Kane P, Trinh OTP, Vann PH, Wilson EN, Cunningham RL (2021) Neurodegenerative disease: roles for sex, hormones, and oxidative stress. Endocrinology 162(11):1–14

Suri K, Ramesh M, Bhandari M, Gupta V, Kumar V, Govindaraju T, Murugan NA (2024) Role of amyloidogenic and non-amyloidogenic protein spaces in neurodegenerative diseases and their mitigation using theranostic agents. Chem Bio Chem 25(13):e202400224

Tewabe A, Abate A, Tamrie M, Seyfu A, Abdela Siraj E (2021) Targeted drug delivery—from magic bullet to nanomedicine: principles, challenges, and future perspectives. J Multidiscip Healthc, 1711–1724

Twarowski B, Herbet M (2023) Inflammatory processes in alzheimer's disease—pathomechanism, diagnosis and treatment: a review. Int J Mol Sci 24(7):1–28

U.S. Food & Drug Administration (2025) Novel drug approvals for 2024. https://www.fda.gov/drugs/novel-drug-approvals-fda/novel-drug-approvals-2025

US Department of Health and Human Services (2017) What happens to the brain in Alzheimer's disease. National institute on aging. https://www.nia.nih.gov/health/alzheimers-causes-and-risk-factors/what-happens-brain-alzheimers-disease

Vacca Jr VM (2016) CJD: understanding Creutzfeldt-Jakob disease. Nursing 46(3), 36–42

Van Dyck CH, Swanson CJ, Aisen P, Bateman RJ, Chen C, Gee M, Kanekiyo M, Iwatsubo T (2023) Lecanemab in early Alzheimer's disease. N Engl J Med 388(1):9–21

Wijnia JW (2022) A clinician's view of Wernicke-Korsakoff syndrome. J Clin Med 11(22):1–13

Xie J, Van Hoecke L, Vandenbroucke RE (2022) The impact of systemic inflammation on Alzheimer's disease pathology. Front Immunol 12:1–18

Yokoi T (2023) Alzheimer's Disease is a Disorder of Consciousness. Gerontol Geriatr Med 9:1–3

# Chapter 11
# Cancer Nanomedicine: Trends and Future Perspectives

**Acharya Balkrishna, Amita Kumari, Amar Jeet, Deepika Srivastava, Hemant Sharma, Vedpriya Arya, Ashwani Kumar, Lacy Loveleen, Riya Kumari, Surendra Nimesh, and Muralikrishnan Dhanasekaran**

**Abstract** Conventional cancer treatments are frequently constrained by systemic toxicity, inadequate targeting, and drug resistance, making it a significant worldwide health concern. By facilitating precise, regulated, and targeted medication administration, nanomedicine offers a potential alternative that enhances therapeutic efficacy and reduces adverse effects. This chapter reviews recent advances in cancer nanomedicine, highlighting various nanocarriers such as liposomes, dendrimers, micelles, polymeric nanoparticles, hydrogels, carbon nanotubes, virus-like particles, gold nanoparticles, and graphene oxide that facilitate both passive and active targeting of tumors. The integration of nanotechnology with combination therapies and immunotherapy has expanded treatment potential, enabling co-delivery of drugs, genes, and immune modulators. Nano-immunotherapy, in particular, shows potential in reshaping the tumor microenvironment and enhancing immune responses. Despite these advancements, challenges in toxicity, scalability, and clinical translation persist. Future directions emphasize the development of smart, biodegradable, and patient-specific nanoplatforms, supported by AI and precision medicine. This chapter highlights the importance of interdisciplinary collaboration in accelerating the integration of nanomedicine into mainstream clinical oncology, offering promise for more effective, targeted, and safer cancer therapies.

A. Balkrishna · A. Kumari · A. Jeet · D. Srivastava (✉) · V. Arya · A. Kumar
Patanjali Herbal Research Department, Patanjali Research Foundation, Haridwar, India
e-mail: deepika.srivastava@patanjali.res.in

H. Sharma · L. Loveleen · R. Kumari · S. Nimesh
Department of Biotechnology, School of Life Sciences, Central University of Rajasthan, Ajmer, Rajasthan, India

M. Dhanasekaran
Department of Drug Discovery and Development, Harrison College of Pharmacy, Auburn University, Auburn, AL, USA



A. Balkrishna et al. (eds.), *Nanotechnology in Environmental Science and Healthcare*,
https://doi.org/10.1007/978-981-95-6714-0_11

## 11.1 Introduction

According to the WHO, throughout the globe, there are about 20 million new cases and nearly 9.7 million deaths in 2022 (including non-melanoma skin cancers), driving the need for innovative and effective therapeutic strategies (Bray et al. 2024). On a regular basis, over 52,900 individuals are diagnosed with various types of cancer, and more than 27,000 people lose their lives (Sung et al. 2021). WHO reports that cancer is the second leading cause of death worldwide. According to the National Cancer Institute (NCI), it is projected that by 2040, approximately 30 million new cases and 16 million deaths are expected worldwide (NCI 2024). The treatments, including surgery, radiotherapy, chemotherapy, and immunotherapy, have undergone significant evolution. While surgery and radiotherapy are primarily used for localized, early-stage cancers, chemotherapy and immunotherapy target metastatic or treatment-resistant tumors. However, these therapies often affect healthy, rapidly dividing cells, resulting in severe side effects. Additionally, drug delivery challenges, such as the limitations of the blood–brain barrier and cellular resistance, hinder treatment efficacy, necessitating novel, targeted approaches (Hersh et al. 2016).

Nanomedicine has emerged as an effective alternative, addressing the lacunae of conventional therapies by enhancing drug delivery, minimizing side effects, and improving tumor targeting (Kurul et al. 2025). Nanomedicine typically refers to cargos or drug molecules at the nanoscale, ranging from 1 to 100 nm. These may include small-molecule drugs, peptides, proteins, nucleic acids, and other bioactive molecules (Balkrishna et al. 2023; Thakur et al. 2023, 2024). Nanomedication has emerged as a promising approach to enhance the efficacy and specificity of oncological interventions. The vivid applications of nanomedicine (NM) in cancer therapy encompass a wide range of nanocarriers and technologies, including lipid-based, inorganic, and polymeric, as well as viral vectors and antibody–drug conjugates (Yetisgin et al. 2020).

The origins of nanomedicine date back to ancient times when colloidal gold was used for medicinal purposes (Bhagyaraj et al. 2018). Traditional practices involved nanoscale drug formulations, such as pulverization (Klemm et al. 2018). The concept of nanotechnology was formally introduced by Richard Feynman in 1959, and the term "nanomedicine" was popularized by Robert A. Freitas in 1999 (Ranganathan et al. 2012; Bodunde et al. 2023). The field emerged to overcome limitations in conventional cancer treatments, leading to the development of nanocarriers for targeted drug delivery and controlled release (Chamundeeswari et al. 2019). Early advancements included doxorubicin-loaded liposomes, polymers, dendrimers, and siRNA-based nanoparticles, followed by innovations such as gold nanoparticles and quantum dots for bioimaging (Quader and Kataoka 2017; Senapati et al. 2018; Alshehri et al. 2020).

Nanomedicine has revolutionized drug delivery, improving the drug penetration across the blood–brain barrier and enabling innovations like painless injections (Singh et al. 2024; Shen et al. 2024). Liposomal formulations, such as Doxil®, have demonstrated success in cancer treatment. Immunotherapy has also benefited from the utilization of extracellular vesicles and virus-modified exosomes (Fulton and Najahi-Missaoui 2023; Gong et al. 2025). The European Medicines Agency (EMA) and the Food and Drug Administration (FDA) have approved nanomedicines, primarily liposomal-based, that enhance the accumulation of drugs at tumor sites through the enhanced permeability and retention (EPR) effect, thereby reducing toxicity and improving efficacy. While early nanomedicines were passively targeted, recent efforts focus on improving active delivery mechanisms (Rodríguez et al. 2022; Jia et al. 2023; Chen et al. 2024; Liu et al. 2024a, b).

The future of nanomedicine involves theranostics, integrating therapy and diagnostics into a single nanomaterial platform. Companies such as Cristal Therapeutics and Nanobiotix conduct clinical trials that utilize nanoparticle-based imaging and radiation-enhanced therapies (Muthu et al. 2014; Silva et al. 2019). However, challenges remain, particularly in optimizing nano–bio interactions for routine clinical use. Emerging nanomedicines are incorporating pathway inhibitors (Rapamycin) and gene therapy agents (siRNA, mRNA, gene editing) for enhanced cancer treatment (Liu et al. 2024a, b). With advancements in smart nanoparticles, the field aims to revolutionize therapy by increasing drug precision, reducing side effects, and improving patient outcomes. Continued innovation promises to expand the versatility and effectiveness of nanomedicine in oncology and beyond.

Currently, 15 cancer nanomedicines are approved globally, with over 80 in clinical trials (Fan et al. 2023). None of these approved nanomedicines utilize active targeting strategies. Passive targeting through the EPR effect has been the foundation of nanomedicine development, but its variability among patients and tumor types limits effectiveness (Wu 2021). Advanced strategies, such as tumor vasculature targeting, cell-mediated delivery, and biomimetic approaches, are being explored to enhance tumor specificity. To improve cellular uptake, second-generation nanomedicines incorporate tumor-specific targeting ligands, including antibodies, peptides, and small molecules. Third-generation nanomedicines focus on organelle-specific delivery, maximizing therapeutic outcomes while overcoming multidrug resistance (MDR). Unlike small-molecule drugs, nanomedicines offer controlled release, prolonged circulation, and enhanced bioavailability, thereby bolstering anti-tumor efficacy (Fan et al. 2023). However, concerns regarding nanoparticle toxicity require rigorous regulatory assessment to understand the interaction of nanomaterials with biological tissues and organs. Specific nanomaterials, such as carbon nanotubes and dendrimers, face approval challenges due to unresolved safety concerns. Prevailing over these barriers is essential for advancing tumor-targeted drug delivery. Cancer nanomedicines aim to address the limitations of traditional cancer diagnostics and treatments. Despite significant interest and promising features, clinical translation remains challenging. In this viewpoint, we examined the current trends and future perspectives of cancer nanomedicine.

## 11.2 Challenges of Conventional Cancer Therapies

Conventional cancer therapies, such as surgery, chemotherapy, and radiotherapy, remain fundamental to treatment strategies, determined by cancer type and stage. Despite their success, they face significant challenges, including a lack of specificity, poor drug penetration in solid tumors, and issues with solubility and resistance. Surgery is most effective in the early stages, while chemotherapy and radiation often damage the healthy cells, leading to severe side effects like immunosuppression and organ toxicity (Amjad et al. 2023). Chemotherapy's inability to fully reach tumor cores and rapid clearance from circulation further reduce its efficacy (Davodabadi et al. 2023). Drug resistance mechanisms, such as P-glycoprotein overexpression, also allow cancer cells to evade treatment, complicating long-term outcomes (Karthika et al. 2022).

A significant challenge in cancer treatment is targeting cancer stem cells (CSCs), which drive tumor growth and recurrence. CSCs possess unique properties, including self-renewal and resistance to chemotherapy, making them challenging to eliminate. They express ABC transporter proteins that expel toxic drugs, reducing treatment effectiveness. Their quiescent nature also limits their susceptibility to cytotoxic agents, which primarily target rapidly dividing cells. Epigenetic modifications, such as DNA methylation and histone alterations, further complicate treatment, as current epigenetic therapies lack specificity and could activate harmful genes (Phi et al. 2018). Early cancer diagnosis remains a challenge, particularly for cancers like esophageal, pancreatic, and prostate, which often present no symptoms in the initial stages. Late-stage detection makes treatment more difficult, while unreliable biomarkers limit early prognosis. Although advances in clinical proteomics hold promise, effective biomarkers for early detection and monitoring remain elusive. Metastasis presents another significant hurdle as cancer spreads beyond the primary site. Micrometastasis, which is undetectable with current diagnostic tools, complicates treatment efforts. Given these limitations, research into precision medicine and targeted therapies is crucial to overcoming conventional treatment barriers and improving patient outcomes (Chakraborty and Rahman 2012).

## 11.3 Nano Drug Delivery Vehicles: Mechanistic Insights

Drug delivery systems (DDSs) have undergone significant advancements since their initiation in the 1960s. Initially focused on constant-rate drug delivery devices, DDS technology evolved into microscopic systems utilizing poly(hydroxy acids) (Hoffman 2008). A key milestone in DDS development occurred with the commercialization of Lupron Depot® for contraception (Okada et al. 1994). Over time, innovations such as liposomes and polymer–drug conjugates led to the development of nano-engineered drug delivery systems (nDDS), which offer enhanced specificity, reduced side effects, and prolonged drug release. The approval of AmBisome® in the

1990s marked the first commercial success of an nDDS formulation, and liposomes and micelles have since been used in the treatment of diseases such as cancer and infections (Boisseau and Loubaton 2011). Depending on the release time of the drug to the targeted site, DDS is categorized into nano-, micro-, and macroscale systems. Nanoscale DDS typically provides drug delivery over several hours, while micro- and macroscale DDS offer sustained drug release over extended periods (Degertekin et al. 2002; Liistro et al. 2002; LaVan et al. 2003; Huang et al. 2016; Chang et al. 2018).

Recently, drug delivery vehicles have gained significant attention due to their ability to efficiently deliver a range of therapeutic agents, including proteins, nucleic acids, small molecule chemotherapeutics, and imaging agents, directly to the targeted site in a controlled manner (Shao et al. 2015; Li et al. 2016; Fan et al. 2018; Balkrishna et al. 2024). These carriers, typically made from polymers and lipids, integrate nanomedicines using techniques such as covalent bonding, physical encapsulation, electrostatic interactions, or coordination complexation (Rianasari et al. 2013; Zhang et al. 2018a; Bauer et al. 2021; Siemer et al. 2021; Kappel et al. 2021; Huang et al. 2021b). This engineered system enables precise delivery of therapeutic agents, enhancing drug concentration in target cells while minimizing toxicity toward healthy tissues through either passive or active targeting mechanisms. These targeted DDSs address critical limitations of traditional drug therapies, including low solubility, instability, drug resistance, and high toxicity (Merisko-Liversidge and Liversidge 2008; Muthu and Feng 2009; Siemer et al. 2021).

In cancer treatment, drug delivery vehicles improve pharmacokinetics, facilitate cellular uptake, and enhance tumor targeting (Barua and Mitragotri 2014; Jin et al. 2020b). Furthermore, they enable the co-delivery of diagnostic and therapeutic agents, providing synergistic effects that can significantly improve treatment outcomes (Ma et al. 2013). The targeted approach enhances tumor immunogenicity and stimulates T-cell-mediated immune responses, thereby boosting overall antitumor efficacy (Huang et al. 2021a). Over the past decade, substantial progress has been made in developing NPs as effective drug delivery carriers. Various types of biodegradable polymeric nanoparticles (NPs), including nanospheres, nanocapsules, ceramic NPs, polymeric micelles, dendrimers, and liposomes, have been extensively studied for their role in drug delivery systems (Yih and Al-Fandi 2006). In addition, carriers such as carbon nanotubes, carbon nanohorns, or silica NPs with drug molecules bound to their surface have also been explored (Chen et al. 2004; Kam et al. 2005). Magnetic NPs, which can be directed to specific areas of the body under a magnetic field, are another promising DDS platform (Asmatulu et al. 2005). Furthermore, implantable nanoscale devices, filled with drug molecules encapsulated by a nanoporous membrane, can function as effective DDS, with the potential to couple specific biological molecules, such as antibodies, enzymes, hormones, and pharmaceutical drugs, to enhance their therapeutic potential (Brown et al. 2010; Duncan et al. 2010).

## 11.4 Types and Traits of Cancer Nanotherapeutics

Nanomedicine-based drug delivery systems (NDDS) offer promising strategies for cancer chemoimmunotherapy by facilitating the efficient uptake of therapeutic agents by immune cells and potentially reprogramming the tumor microenvironment. Due to their unique physical and chemical properties, NDDS enhance the immune response, improving therapeutic efficacy (Zang et al. 2017). These systems can improve the solubility and circulation time of therapeutic agents, allowing for enhanced accumulation within tumor tissues. As a result, NDDS led to improved therapeutic outcomes, better pharmacokinetics, and reduced side effects (Langer et al. 2003; Hu et al. 2016; Xie et al. 2017). Furthermore, NPs-mediated drug delivery addresses the limitations associated with conventional free drug administration, such as low bioavailability and the emergence of drug resistance in cancer cells. NPs enter target cells through receptor-mediated endocytosis, releasing drugs in the lysosomal compartments, triggering apoptotic pathways, and reducing drug efflux in the cytoplasm, thereby minimizing resistance. Additionally, NPs can be functionalized with specific ligands in order to enhance targeted delivery, improve internalization, and enable organelle-specific drug targeting (Patel et al. 2013; Gosh and Biswas 2021).

A variety of polymer-based (organic) drug carriers, including polymeric nanoparticles (NPs) (for example, chitosan and dendrimers), micelles, hydrogels, carbon nanotubes, and poly(lactic-co-glycolic acid) (PLGA)-based NPs, have been investigated for cancer treatment (Yih and Al-Fandi 2006). Metal-based (inorganic) NPs, such as gold NPs, metal oxide NPs, quantum dots, and mesoporous silica NPs (MSN), have also been explored as potential drug delivery systems (Susewind et al. 2015; Wang et al. 2021). Some of these systems have received approval from regulatory agencies, including the FDA and EMA, for clinical use in cancer therapy, while others, such as polymeric and micelle-based systems, remain under clinical investigation. For instance, liposomal doxorubicin has been approved for the treatment of Kaposi's sarcoma, ovarian cancer, and multiple myeloma (Allen and Martin 2004; Bobo et al. 2016). In contrast, iron oxide NPs have been approved in Europe for glioblastoma treatment and serve as contrast agents for magnetic resonance imaging (Bobo et al. 2016). A brief description of different types of nanocarriers is provided below in Fig. 11.1:

### *11.4.1 Polymeric NPs*

- **Chitosan**

Chitosan, a cationic polysaccharide from chitin found in crustaceans and insects, has gained notable attention in cancer drug delivery due to its biocompatibility, biodegradability, and mucoadhesive properties. Composed of D-glucosamine and N-acetyl-D-glucosamine units, chitosan contains reactive functional groups, such as amino ($NH_2$) and hydroxyl (OH) groups, which enhance its potential as a drug

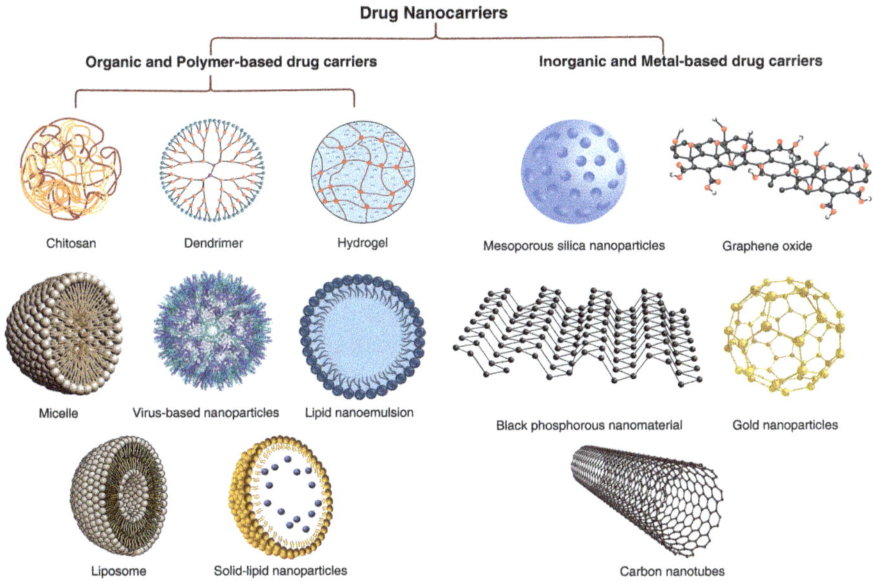


**Fig. 11.1** Different types of drug nanocarriers

delivery system (Alemu et al. 2023). Chitosan NPs offer advantages, including improved stability, enhanced cellular uptake, increased solubility of hydrophobic anticancer drugs, and controlled release kinetics. Chitosan NPs can be modified by targeting ligands or stimuli-responsive groups, enabling selective delivery to cancer cells or tissues (Grewal and Salar 2024). For instance, trimethyl chitosan conjugated with the ligand CSKSSDYQC has been utilized for oral delivery in breast cancer treatment, enhancing bioavailability through targeted uptake by intestinal goblet cells (Chen et al. 2018). Chitosan derivatives can also be conjugated with nanocomposites or chemical agents to further enhance their anticancer activities. For instance, conjugating 3-amino-2-phenyl-4(3H)-quinazolinone with PPC-silver chloride nanocomposites demonstrated increased bioavailability and sustained drug release to tumor cells (Salahuddin et al. 2017). Moreover, chitosan-based formulations, including chitosan–thymine conjugates and carboxymethyl chitosan, have shown promising anticancer effects by targeting tumor cells and inducing an antiproliferative response through mechanisms such as antiangiogenesis, apoptosis, and immune modulation (Adhikari et al. 2018; Sharma et al. 2022). However, chitosan's inherent hydrophilicity, limited solubility, and poor thermal stability pose challenges for efficient drug transport, potentially limiting its clinical efficacy. To address these limitations, modifications such as acylation, carboxymethylation, and quaternization have been developed to improve the practical application of chitosan in drug delivery systems (Shanmuganathan et al. 2019; Jafernik et al. 2023).

- **Dendrimers**

Dendrimers are uniform, hyperbranched synthetic polymers or nanomaterials with a tree-like architecture, characterized by branches extending radially from a central core (Jang et al. 2009; Chen et al. 2013). Their terminal functional groups can be anionic, neutral, or cationic, and these groups can be further modified to achieve specific applications (Wiener et al. 1996; Purohit et al. 2001; Kobayashi et al. 2003). The number of monomeric branches added to the core determines the generation of a dendrimer. As the generation increases, the dendrimer expands and adopts a globular shape, forming large cavities suitable for encapsulating therapeutic agents (Yiyun and Tongwen 2005). Dendrimers are classified based on structure, branching, solubility, terminal groups, and generation (Devarakonda et al. 2004), and the most commonly studied types for anticancer drug delivery include polyamidoamine (PAMAM), poly(propyleneimine) (PII), and poly(L-lysine) (PLL). PAMAM dendrimers are particularly favored due to their straightforward synthesis and superior properties, while PPI dendrimers are also widely explored for their potential as drug delivery agents' capabilities (Yiyun and Tongwen 2005; Escalante et al. 2021). PLL dendrimers are biocompatible, biodegradable, and highly soluble in water, making them promising candidates for cancer therapy.

The effectiveness of dendrimers in drug delivery vehicles is significantly enhanced by functionalization, which allows the incorporation of multiple active sites to improve their multifunctional capabilities (Fischer and Vögtle 1999). In drug delivery, dendrimers offer several advantages: They enhance the solubility and bioavailability of hydrophobic anticancer drugs, improve the stability of sensitive compounds by preventing degradation, and increase drug permeability, thereby facilitating better cellular uptake (Kitchens et al. 2005; Kulhari et al. 2013; Teow et al. 2013). Their biocompatibility and extended circulation time contribute to more effective treatment, while their capacity for controlled release and selective targeting of cancer sites helps minimize side effects and improve therapeutic outcomes (Parekh 2007; Van Der Poll et al. 2010; Chaudhari et al. 2016; Kulhari et al. 2016). Moreover, dendrimers can be functionalized for a variety of applications, including drug delivery, gene therapy, and bioimaging (Pooja et al. 2020). However, despite these advantages, dendrimers face several challenges in cancer therapy, including toxicity, poor degradability, and the potential for tumor cells to develop resistance (Sherje et al. 2018).

- **Hydrogels**

Cross-linked polymer networks with 3D structures, known as hydrogels, have the capacity to absorb large volumes of fluid (Ho et al. 2022). Hydrogels are perfect for a variety of biomedical applications because of their high water content, soft texture, and porous nature, which closely resemble the properties of real tissues (Zhang et al. 2017). Hydrogels can be supplied by a variety of methods, including intravenous injection, in situ injection, transdermal, oral, pulmonary administration, and transarterial chemoembolization. They come in many shapes, such as macrogels, microgels (0.5–10 $\mu$m), and nanogels (<200 nm). [98, 99]. This flexibility allows for targeted delivery to specific cancer sites, enhancing drug targeting, reducing required dosages, and improving treatment efficiency (Ho et al. 2022).

Hydrogels are appealing options for drug delivery in the treatment of cancer because of their exceptional biocompatibility and biodegradability. In cancer treatment, hydrogel offers three main benefits. First, they enable precise, controlled drug release, providing sustained delivery of chemotherapeutics agents, radionuclides, immunosuppressants, and other therapeutic agents, making them useful in various cancer treatments, such as chemotherapy, radiotherapy, and immunotherapy (Tan et al. 2021). Second, hydrogels can be tailored in size and delivery modality to target specific cancer types, improving drug localization, reducing drug dosage, and enhancing treatment outcomes (Kass and Nguyen 2022). Finally, hydrogels can be designed to react to environmental cues, allowing anticancer drugs to be released when the tumor microenvironment changes (Jiang et al. 2020). These properties position hydrogels as a promising innovation in cancer treatment, offering the potential for improved patient outcomes.

However, despite their significant potential, several challenges exist to the widespread clinical adoption of hydrogels. The development of more economical and ecologically friendly manufacturing techniques is required because the current synthesis and gelation processes are frequently ineffective, expensive, and unsuitable for large-scale production. Additionally, most hydrogel studies have been conducted in vitro and in subcutaneous ectopic cancer models, which do not fully replicate the complexity of the in vivo tumor microenvironment. Therefore, further in vivo testing is crucial to evaluate the physiological stability, biocompatibility, biodegradability, toxicology, and therapeutic efficacy of hydrogels before their clinical application (Ho et al. 2022).

- **Micelles**

Polymeric micelles are self-assembled NPs with an amphiphilic structure, typically smaller than 100 nm, which enables them to penetrate both tumors and the human body efficiently. The hydrophobic core of these micelles can encapsulate and solubilize hydrophobic drugs, while the hydrophilic outer shell facilitates their delivery and stability (Guo 2023). In aqueous environments, water-insoluble drugs are encapsulated during the self-assembly process (Kedar et al. 2010), while in non-aqueous solutions, water-soluble drugs such as proteins, peptides, and nucleic acids can be incorporated (Allen 1998; Momekova et al. 2007). Polymeric micelles are particularly effective for delivering hydrophobic chemotherapeutic agents, enhancing drug stability, prolonging circulation time, and reducing the systemic side effects commonly associated with traditional therapies (Fatfat et al. 2022).

Poly(lactide) (PLA), poly(caprolactone) (PCL), poly(lactide-co-glycolide) (PLGA), polyesters, poly(amino acids), lipids, and other polymers are frequently utilized to construct the micelle core. To aid with drug transport, hydrophobic polymers such chitosan, dextran, hyaluronic acids, poly(ethylene glycol) (PEG), and poly(oxazolines) are usually used in the outer shell. While various polymers can be used to create micelles, drug delivery applications necessitate biocompatible and biodegradability, which makes PEG a preferred choice due to its neutral, non-toxic, and water-soluble properties (Alexander-Bryant et al. 2013). By conjugating drugs to the distal ends of these polymeric chains, the resulting systems can enhance solubility and stability and enable the co-delivery of multiple therapeutic agents.

The nanoscale size of micelles facilitates their accumulation in tumors via the EPR effect. Additionally, stimuli-sensitive breakdown, triggered by factors such as pH, hypoxia, and upregulated enzymes, ensures the targeted release of effective agents at the tumor site (Ghosh and Biswas 2021). Micelles' low toxicity and possibility for renal clearance, provided the molecular weight of the polymer chains stays below the renal filtration threshold, are important advantages (Yokoyama 2011). Nevertheless, micelles encounter a number of difficulties, including as low drug-loading effectiveness, restricted in vivo physical stability, and insufficient cellular uptake by cancerous cells, especially when they are neutral (Kim et al. 2010). Additionally, compared to liposomes, micelles generally have a smaller drug-loading capacity (Alexander-Bryant et al. 2013).

### *11.4.2 Virus-Based NPs*

Virus-like nanoparticles (VLPs) are naturally occurring, polymer-based nanomaterials that mimic viral structures by self-assembling viral coat protein. However, they lack viral genomes, rendering them non-infectious. These NPs, typically ranging in size from 20 to 200 nm, have garnered significant attention in cancer therapy due to their biocompatibility, biodegradability, and low toxicity (Chen et al. 2001; Trifonova et al. 2017; Timmermans et al. 2022). VLPs have demonstrated outstanding potential in drug delivery platforms (Kim et al., 2023) due to their ability to control size, shape, and surface properties precisely. This control is achieved through modification to the capsid proteins via genetic engineering (manipulating terminal amino acids of the capsid proteins) and chemical engineering (changing the pH of the buffer and thermal conditions) (Chen et al. 2001; Timmermans et al. 2022). Such customization enables the incorporation of targeting ligands, such as transferrin or folic acid, which facilitate the selective delivery of therapeutic agents to cancer cells (Hofheinz et al. 2005). Additionally, VLPs can be loaded with therapeutic compounds or imaging agents like quantum dots, doxorubicin (DOX), and small interfering RNA (siRNA), providing a flexible platform for both diagnostic and therapeutic uses (Min et al. 2013; Bruchman et al. 2014).

Noteworthy VLPs, such as those derived from bacteriophages, tobacco mosaic virus (TMV), and human papillomavirus (HPV), have been employed for targeted cancer treatment, with promising results in preclinical models (Biabanikhankahdani et al. 2016; Wang et al. 2016a; Pang et al. 2019a). For instance, TMV has shown efficacy in drug delivery and imaging in breast cancer models (Bruckman et al. 2018), while VLPs derived from the Johnson grass chlorotic stripe mosaic virus have successfully delivered DOX in murine models, highlighting their potential as an effective drug carrier platform for cancer therapy (Alemzadeh et al. 2017).

### 11.4.3 Carbon Nanotubes

Carbon nanotubes (CNTs), characterized by a cylindrical structure composed of hexagonal-arranged carbon atoms (Kushwaha et al. 2013), have garnered significant interest in drug delivery due to their unique chemical, physical, and biological properties. These consist of their hollow monolithic structure, nanoneedle-like shape, and capacity to functionalize their exterior surfaces with different chemical groups (Sahoo et al. 2011). Single-walled CNTs (SWCNTs), which consist of a single carbon layer, have diameters ranging from 0.4 to 2 nm, while multiwalled CNTs (MWCNTs) consist of multiple concentric carbon layers with a diameter between 2 and 100 nm (Bekyarova et al. 2005). Their high surface area, aspect ratio, and ease of functionalization with diverse chemical groups make CNT excellent drug carriers capable of high drug-loading capacity and stability (Sahoo et al. 2011). Functionalization can also enhance water solubility and serum stability, thereby facilitating the precise targeting of specific cells or tissues (Sahoo et al. 2011; Beg et al. 2011). SWCNTs, due to their superior flexibility and structural stability, are particularly advantageous for carrying large drug loads and enhancing bioavailability. MWCNTs are particularly effective in thermal cancer treatment, where they absorb near-infrared light and convert it into vibrational energy, inducing localized heating that can effectively destroy cancer cells (Hirsch et al. 2003). This technique reduces the required drug quantity, minimizes the need for solvents, and ensures targeted drug delivery, positioning CNTs as a promising tool for safer and more efficient drug therapies. Additionally, the unique properties of CNTs, including their ability to penetrate cells through passive diffusion or endocytosis, further establish them as powerful candidates for both intracellular and extracellular drug delivery, particularly in the context of cancer treatment (Lamprecht et al. 2009; Sahoo et al. 2011).

### 11.4.4 Lipid-Based Drug Carriers

- **Liposomes**

Liposomes are lipid-based NPs that have demonstrated significant potential in enhancing cancer immunotherapies by effectively delivering a wide range of cancer therapeutics, including peptides, proteins, antibodies, and chemotherapeutics (Vahed et al. 2017; Gao et al. 2019). These self-assembled bilayer vesicles feature an aqueous core encased by phospholipid membranes, which confer versatility for controlled drug release, protection from rapid metabolism, and reduction of drug-related side effects (Forssen and Tökès 1981; Sercombe et al. 2015). Liposomes can encapsulate drugs, thereby increasing their plasma half-life, improving targeted delivery to cancer cells, enhancing therapeutic efficacy, and minimizing toxicity (Papahadjopoulos et al. 1972; Drummond et al. 1999). They typically range in size from 10 nm to 2.5 $\mu$m, with most formulations falling between 50 and 450 nm, making them well suited for cellular uptake (Malam et al. 2009).

Despite their advantages, liposomes face challenges, including potential carrier toxicity and the complexity associated with large-scale production with consistent properties (Mozafari 2005; Missaoui et al. 2018). However, specialized liposomal formulations, such as immunoliposomes and pH-sensitive liposomes, offer additional targeted drug delivery capabilities. Immunoliposomes can selectively target specific cancer cells, while pH-sensitive liposomes release their contents in response to the acidic microenvironment of tumor tissues (Park et al. 1997; Ferreira et al. 2013). While liposomes are considered safe and biocompatible, they face challenges such as limited stability and high production costs (Kshirsagar et al. 2005; Chen et al. 2010). However, a number of formulations of liposomes are presently on the market or undergoing clinical trials, making them one of the most effective drug delivery systems and showing great promise for enhancing cancer treatments (Bozzuto and Molinari 2015).

- **Solid Lipid NPs**

SLNs are a class of stable colloidal carriers composed of solid lipids that maintain their dorm at both room and body temperature, making them particularly suitable for drug delivery applications, especially in cancer treatment. The therapeutic agent is either dissolved or distributed within the solid lipid core of these NPs, which is stabilized by a surfactant (Weber et al. 2014; Loira-Pastoriza et al. 2014). SLNs offer several advantages, including enhanced stability compared to other formulations, the ability to encapsulate both hydrophobic and hydrophilic drugs, and the potential for sustained, controlled drug release (Karunanidhi et al. 2012; Mishra et al. 2018). Additionally, SLNs provide opportunities for active targeting and multifunctional therapy, making them ideal candidates for cancer drug delivery (German-Cortes et al. 2023).

SLNs have demonstrated improved therapeutic efficacy in various studies. For instance, glyceryl tripalmitate SLNs loaded with paclitaxel have been shown to inhibit cell proliferation in breast and lung cancer models (Leiva et al. 2017). Similarly, SLNs encapsulating temozolomide (TMZ) demonstrated superior effectiveness compared to free TMZ in reducing tumor growth and enhancing survival in melanoma models (Clemente et al. 2018). Despite these promising results, SLNs face limitations, such as issues with polymorphic transitions, particle growth, and potential expulsion of drugs due to lipid crystallization (Ghasemiyeh and Mohammadi-Samani 2018; Satapathy et al. 2021). Furthermore, the manufacturing process for SLNs can be costlier than that of conventional formulations (German-Cortes et al. 2023).

Despite these challenges, SLNs are considered biocompatible and exhibit lower toxicity than polymeric NPs, making them a promising option for pharmaceutical drug delivery. SLNs are also suitable for administration through various routes, including oral, intravenous, dermal, and ophthalmic delivery.

- **Lipid Nanoemulsions**

Lipid nanoemulsions represent a promising class of drug carriers for cancer treatment, particularly for molecules with low water solubility, by utilizing safe-grade excipients. These colloidal dispersions consist of oil, surfactant, and an aqueous phase, forming stable nanometer-sized droplets that enhance drug solubility, stability, and bioavailability (Gi et al. 1992; Mason et al. 2006). The properties of the nanoemulsion's core are critical in determining the physicochemical characteristics, particle size, and stability, which directly influence the therapeutic payload. Encapsulation within the nanoemulsions core protects the drug from degradation, prolongs its plasma half-life, and reduces toxicity, while simultaneously improving targeted delivery to cancer cells (Maeda et al. 2000). This enhanced targeting increases drug efficiency and minimizes systemic side effects. Despite these advantages, the clinical translation of lipid nanoemulsions remains hindered by challenges in large-scale production and formulation stability, which limits their widespread adoption in cancer therapies (Wilhelm et al. 2016). Nevertheless, their demonstrated safety and potential for improving drug delivery continue to position lipid nanoemulsions as a promising area of research for future cancer treatment strategies.

### *11.4.5 Mesoporous Silica Nanoparticles (MSNs)*

MSNs are tiny, porous particles that provide special benefits for medication delivery systems (Pande et al. 2023). MSNs have emerged as a promising candidate for drug delivery in cancer therapy following the introduction of MCM-41-type mesocarpous silica as a drug delivery system in 2001 (Chen and Fu 2012). The advantageous properties of MSNs including scalable and straightforward fabrication, non-toxicity, large surface area, high pore volume, and a silanol-containing surface, make them ideal for drug-loading applications (Yang et al. 2012). These textural features enable the encapsulation of a higher amount of anticancer drugs within the nanopores. At the same time, the silanol groups on the surface can be modified with various molecules to enhance the pharmacokinetics and pharmacodynamics of the therapeutic agents (Yang et al. 2012). MSNs exhibit biocompatibility and biodegradability, with their degradation occurring within a month in simulated body fluids, within cells, and in vivo, as confirmed by Rastegari et al. (2021). Furthermore, MSNs demonstrated resistance to various stresses, including mechanical, thermal, and pH-induced variations (Rastegari et al. 2021). The surface of MSNs can be functionalized with gatekeeper molecules, such as folic acid, to develop stimuli-responsive systems for controlled drug release (Huang et al. 2021c). MSNs are very useful in resolving the solubility issues of hydrophobic medications with low in vivo bioavailability, such paclitaxel (PTX) and camptothecin (CTP) (Lu et al. 2007; He et al. 2017). By encapsulating these drugs within the nanopores, MSNs significantly enhance their solubility and cytotoxic effects, as demonstrated by increased efficacy against human pancreatic cancer Capan-I cells and liver cancer HepG2 cells (Lu et al. 2007; He et al. 2017). Additionally, PEGylation of MSNs enhances their circulation time and facilitates preferential accumulation at the tumor site, thereby improving the therapeutic index of the loaded drug (Zhao et al. 2018). Multidrug resistance (MDR) inhibitors, including TPGS copolymers, can be added to MSNs to circumvent MDR processes in cancer cells (Zhao et al. 2018). The stability of MSNs protects the encapsulated drugs from premature release, thereby contributing to enhanced therapeutic efficacy (Mitran et al. 2015; Sabio et al. 2019). Additionally, to create single- and multi-modal therapeutic systems, MSNs can be functionalized with co-therapeutic agents such as chemotherapeutics, photosensitizers, and photothermal reagents (Han et al. 2020; Rastegari et al. 2021). However, despite their promising preclinical and in vivo results, the industrial-scale synthesis of MSNs faces significant challenges, particularly concerning nanoparticle size distribution and batch-to-batch variability, due to the complex, multistep synthesis process (Esfahani et al. 2022). Recent clinical experiments on MSNs with diameters below 10 nm have demonstrated potential for decreasing drug doses and eliminating treatment costs, offering a roadmap toward widespread clinical translation (Esfahani et al. 2022).

### *11.4.6 Graphene Oxide*

Graphene oxide (GO) and its derivatives have gained significant attention as promising nanocarriers for drug delivery, particularly in cancer therapy, due to their distinctive physicochemical properties. GO is a single-layered material derived from oxidized graphene, enriched with oxygen functional groups. Unlike graphene, which is hydrophobic, graphene oxide is hydrophilic, enabling it to disperse in aqueous solutions. GO is a great option for loading and delivering a range of therapeutic compounds because of its large surface area, remarkable mechanical and chemical stability, two-dimensional planar structure, superb cytocompatibility, and outstanding conductivity (Malekimusavi et al. 2019; Wei et al. 2021). The planer structure of GO facilitates the efficient loading of a wide range of substances, including biomolecules, metal ions, and small drug molecules. At the same time, the presence of free $\pi$-electrons contributes to its efficacy as a drug carrier (Bagheri et al. 2022). Additionally, the surface properties of GO can be modified to exhibit both hydrophilic and hydrophobic characteristics, providing flexibility for functionalization with various groups that enhance drug loading and stability (Maiti et al. 2019; Bagheri et al. 2022). Functionalization with oxygenated groups, such as carboxyl groups, significantly enhances the solubility and biocompatibility of GO, making it suitable for medical applications (Huang et al. 2011). Studies have shown that GO functionalized with carboxyl groups (GO-COOH) exhibits superior biocompatibility, safety, and drug delivery efficacy compared to unmodified GO (Karki et al. 2018; Gholami et al. 2020). For instance, a targeted drug delivery system using GO conjugated with methotrexate (MTX) and folic acid (FA) as a targeting ligand demonstrated 96% drug release and significantly enhanced cytotoxicity against the MDA-MB-231 breast cancer cell line compared to free MTX, effectively weakening the cancer cells (Yanikoglu et al. 2024). Furthermore, a pH-responsive GO-based drug delivery system loaded with 5-fluorouracil (5-FU) exhibited controlled drug release at the acidic pH of 5.8, which is characteristic of the tumor microenvironment, while the release was considerably reduced at physiological pH (7.4), highlighting GO's potential for targeted, controlled drug delivery in cancer therapy (Aliabadi et al. 2018).

### 11.4.7 Black Phosphorus Nanomaterials

Black phosphorus (BP), an allotrope of phosphorus, has garnered significant attention as a promising nanomaterial for drug delivery systems (DDSs) in cancer therapy due to its unique properties, including biocompatibility, biodegradability, and high surface area (Ling et al. 2015). White or red phosphorus is converted under high pressure and high temperature to create BP (Sun et al. 2012). Because of its exceptional surface properties, BP has a high drug-loading efficiency, which makes it a perfect option for delivering a range of therapeutic agents, such as small interfering RNA (siRNA), clinical anticancer medications, and inorganic components like Au, $Fe_3O_4$, Pt, and upconversion NPs (UCNP) (Yang et al. 2017; Liu et al. 2021). The negative charge and corrugated surface structure of BP nanomaterials facilitate efficient electrostatic interactions with small, positively charged drugs, thereby enhancing their drug-loading capacity. For instance, BP has shown a remarkable drug-loading capacity (950 wt%) for doxorubicin (DOX), leading to synergistic therapeutic effects (Li et al. 2022). Through electrostatic interactions, the powerful cationic medication methotrexate (MTX) can also be absorbed onto BP nanoparticles (BPNs), greatly reducing the growth of tumors (Guo 2023). Additionally, BP's puckered crystalline structure provides a large surface area-to-volume ratio, thereby enhancing the loading efficiency of therapeutic agents (Lakshmi et al. 2017; Xie et al. 2018). Notably, BP nanosheets (BPNSs) have demonstrated high drug-loading capacities for genetic materials, such as 98.7% loading efficiency for CRISPR-associated protein 9 (Cas9) ribonucleoproteins (Zhou et al. 2018). Despite these advantages, challenges persist in controlling the size, thickness, and reproducibility of BP nanosheets during exfoliation, as well as the high cost and low yield of synthesis (Li et al. 2018; Ji et al. 2019]. These factors currently limit the clinical applications of BP-based DDSs. Future research should focus on optimizing the preparation of BP nanosheets with precise control over size, thickness, and surface coating to minimize batch-to-batch variability and improve scalability (Wang and Cheng 2019; Liu et al. 2021).

### 11.4.8 Gold Nanoparticles (AuNPs)

Because of their distinct physicochemical characteristics, minimal toxicity, and simplicity in synthesis and functionalization, AuNPs are well known for their promise in cancer therapeutic delivery (Davis et al. 2008; Astruc et al. 2010). AuNPs are employed in cancer immunotherapy primarily in two ways: as carriers for targeted drug delivery and as agents for photothermal therapy. As drug delivery systems, AuNPs leverage the enhanced permeability and retention (EPR) effect, which facilitates the accumulation of NPs in tumor tissues, thereby enhancing drug efficacy (Fang 2022). The drugs can be loaded onto AuNPs through covalent or non-covalent interactions, improving their stability, solubility, and circulation time, which also ensures controlled drug release within cancer cells upon exposure to stimuli such as changes in pH or a reducing environment (Karve et al. 2012; Wang et al. 2013; Khan et al. 2014; Lee et al. 2015; Cui et al. 2017). Surface modifications using targeting ligands such as antibodies, aptamers or carbohydrates enable the selective delivery of drugs to tumor sites, further enhancing the specificity and effectiveness of the therapy (Goddard et al. 2020). However, AuNPs' toxicity profiles are greatly influenced by their physicochemical properties, which include their size, charge, shape, and surface chemistry. Studies have shown that smaller AuNPs tend to exhibit higher cytotoxicity compared to larger particles, necessitating careful optimization of GNP properties to minimize potential toxicity while maximizing therapeutic efficacy (Pan et al. 2007). The best properties of AuNPs for targeted drug delivery in cancer treatment require more investigation, with an emphasis on reducing toxicity and improving therapeutic results.

## 11.5 Nanoformulations for Cancer Therapy

To circumvent the limits of conventional cancer therapies, researchers are creating new nanotechnology-based drug delivery systems that boost oncotherapy by conveying anticancer drugs in higher concentrations directly to specific areas (Jabir et al. 2012; Gharpure et al. 2015). Because nanocarriers can overcome issues including systemic toxicity, low bioavailability, and non-specific medication distribution, they have become a viable tool in the treatment of cancer. (Patra et al. 2018). These nanocarrier systems offer several advantages, including extended plasma half-life, improved bio-distribution, and precise drug delivery to the tumor microenvironment through endothelial barriers. By enabling sustained, targeted, and controlled drug release, they enhance drug accumulation at the tumor site and improve cellular uptake, thereby minimizing toxic side effects and avoiding interactions with healthy cells. Engineered NPs can also be designed for site-specific targeting, gene delivery, and as contrast agents in imaging techniques like magnetic resonance imaging, contributing to both therapeutic and diagnostic advancements. Their unique properties such as quantum effects, high surface-to-volume ratios, and selective targeting capabilities make them distinct therapeutic options in oncology (Edis et al. 2021). Although various nanocarrier formulations have shown success in preclinical and some clinical studies, their widespread integration into clinical oncology is still evolving. Nonetheless, these nanoformulations represent a crucial step toward bridging the gap between traditional cancer treatments and the demand for safer, more effective, and personalized therapeutic strategies.

### 11.5.1 Mechanisms of Cellular Targeting

Effective cancer treatment requires nanoparticle (NP) drug or gene delivery systems that precisely target tumor cells while sparing healthy cells, thereby enhancing therapeutic efficacy and reducing cytotoxicity. NPs must navigate physiological and biological barriers, including epithelium, endothelium, and mechanical, physicochemical, and enzymatic obstacles, to reach the tumor microenvironment (TME). To prevent non-specific targeting and guarantee stability in the bloodstream, avoidance of reticuloendothelial system (RES) and mononuclear phagocyte system (MPS) clearance, accumulation in the tumor microenvironment (TME) via tumor vasculature, penetration into tumor fluid under high pressure, and specific interaction with tumor cells, nanoparticles (NPs), which range in diameter from 10 to 100 nm, require specific size, biocompatibility, and surface chemistry. Optimized engineering is crucial for effective delivery to subcellular targets, such as the cytosol or nucleus. Targeting mechanisms, categorized as passive or active, depend on surface functionalization, physicochemical properties, and tumor pathophysiology (Lassalle and Ferreira 2007; Gavas et al. 2021). Figure 11.2 illustrates the passive and active drug targeting mechanisms.

- **Passive Targeting**

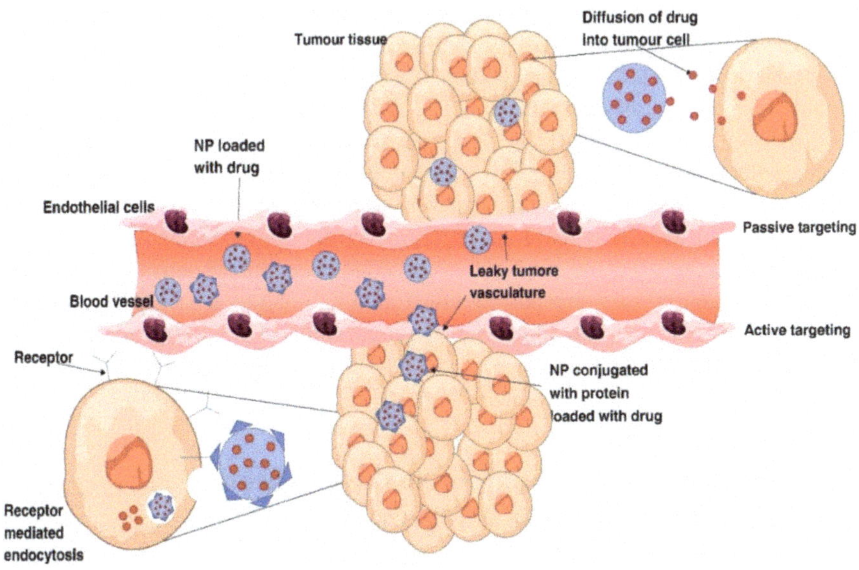


**Fig. 11.2** Comparative overview of passive and active drug targeting. Passive targeting uses the EPR effect for NPs accumulation and drug diffusion in tumors, while active targeting employs ligands for receptor-mediated uptake and drug release (Holder et al. 2023)

In the late 1980s, it was discovered that specific macromolecules preferentially accumulate in cancer cells. Poly(styrene-co-maleic acid)-neocarzinostatin (SMANCS) was the first to be reported by Matsuura and Maeda. This accumulation was linked to fenestrations in damaged tumor blood vessels and poor lymphatic drainage, known as the "enhanced permeation and retention (EPR) effect." In situations like inflammation or hypoxia, the blood vessel's endothelium layer becomes more porous (Matsumura and Maeda 1986). Rapidly growing tumors can induce the formation of new, leaky blood vessels through a process called neovascularization, featuring large pores (200–2000 nm) that allow nanoparticles (NPs) to escape and gather in tumor cells. In normal tissues, extracellular fluid drains into lymphatic vessels at 0.1–2 μm/s, but tumors disrupt this, leading to minimal fluid uptake and increased NP retention in the tumor interstitium (Hobbs et al. 1998). Small compounds are frequently enclosed in nano-sized drug carriers to increase circulation, improve pharmacokinetics, and lessen negative effects. Passive targeting of tumors also involves the tumor microenvironment (TME) and relies on tumor characteristics like vascularity and leakiness, rather than specific ligands. Key factors affecting the EPR effect include angiogenesis, lymphangiogenesis, perivascular tumor invasion, and intratumor pressure, alongside the physicochemical properties of the NPs (Pelicano et al. 2006; Attia et al. 2019).

- **Active Targeting**

Active targeting, also known as ligand-mediated targeting, utilizes specific ligands, such as transferrin or folate, to bind to overexpressed receptors on target cells, including cancer cells, thereby enhancing the affinity of NPs and improving drug penetration (Mukwaya et al. 1998). First observed in 1980 with antibody-grafted liposomes, this approach utilizes ligands such as proteins, peptides, or vitamins to trigger receptor-mediated endocytosis (Padera et al. 2004). Common receptors include transferrin, folate, glycoproteins, and the epidermal growth factor receptor (EGFR). For example, transferrin-modified NPs target overexpressed transferrin receptors in A2780 ovarian carcinoma cells. Alternatively, targeting angiogenic endothelial cells near tumors can induce hypoxia and necrosis by disrupting blood supply. Tumor tissues, being more acidic due to the Warburg effect, support pH-sensitive liposomal drug delivery. NP efficacy depends on multivalent ligand interactions, NPs architecture, ligand density, size, and administration route (Saha et al. 2010; Amreddy et al. 2015; Santi et al. 2017).

## 11.6 Nanomedicine: A Solution to the Challenges of Conventional Cancer Therapies

Nanomedicine offers an innovative solution to the challenges posed by conventional cancer therapies, enhancing drug delivery, reducing systemic toxicity, and increasing treatment specificity. Engineered nanoparticles facilitate the targeted delivery of chemotherapeutic agents directly to tumor cells, minimizing damage to healthy tissues and reducing adverse side effects. Real-time monitoring of treatment responses is made possible by multifunctional nanoparticles, which further combine therapy and diagnostics (Al-Thani et al. 2024). Furthermore, stimuli-responsive nanoparticles increase accuracy and efficacy by releasing medications in response to certain tumor microenvironmental cues (Li et al. 2019). Nanoparticles like gold, iron oxide, and quantum dots aid early detection by binding to cancer biomarkers, improving imaging techniques such as magnetic resonance imaging (MRI) and positron emission tomography (PET) scans (Al-Thani et al. 2024). Nanotheranostics, which combines diagnostic and therapeutic tools, offers promise in detecting cancer at early stages. For example, Resovist, an MRI contrast agent composed of superparamagnetic iron oxide nanoparticles, improves liver imaging (Dadfar et al. 2019). Similarly, quantum dots enhance imaging precision for colorectal, liver, and pancreatic cancers due to their fluorescence properties (Pareek et al. 2025).

Abraxane®, an albumin-bound nanoparticle formulation of paclitaxel, has proven highly effective in treating breast cancer by delaying tumor growth and increasing intracellular drug uptake (Miele et al. 2009; Yuan et al. 2020). The ability of nanocarriers to improve bioavailability, overcome drug resistance, and enable combination therapies underscores their superiority over conventional treatments. Notably, nano-enabled drugs have a higher approval rate (6%) compared to conventional oncology drugs (3.4%), as they enhance drug permeability and prolong half-life, countering resistance mechanisms. Dual-drug loading and physical modalities enable more effective combinatorial treatments for cancer (Germain et al. 2020; Zhao et al. 2023).

Nanomedicine has also made significant strides in cancer immunotherapy, particularly in vaccine development. Unlike traditional therapies targeting cancer cells, immunotherapy stimulates the immune system to recognize and attack tumors. Nanoparticle-based vaccines incorporating tumor-specific antigens (TSAs) and tumor-associated antigens (TAAs) enhance stability and prevent antigen degradation (Kemp and Kwon 2021). APCEDEN®, an autologous dendritic cell vaccine, approved by the Central Drugs Standard Control Organization in India has shown success in treating prostate, colorectal, ovarian, and lung cancers, offering significant tumor remission and extended survival (Kumar et al. 2017, 2019; Mastelic-Gavillet et al. 2019). Viral-based vaccines, such as NG0348, which utilize oncolytic adenoviral viruses to modify T-cells, are currently in human clinical trials (Coey 2024).

FDA-approved liposomal formulations, such as Doxil (doxorubicin) and Abraxane (paclitaxel), have already demonstrated efficacy in treating breast and lung cancer. Emerging platforms, including solid lipid nanoparticles (SLNs) and exosomal nanomedicines, are gaining attention for their ability to carry natural biomolecules, making them ideal candidates for targeted drug delivery (Satapathy et al. 2021; Sharma and Mukhopadhyay 2024). Inorganic nanomedicines, which include carbon-based materials like graphene and carbon nanotubes, as well as nanoparticles derived from iron, silver, and gold oxides, have great promise for use in diagnosis and treatment (Mekuye and Abera 2023). Gold nanoparticles facilitate photothermal therapy, where localized heating eradicates tumor cells (Huang and El-Sayed 2010). Iron oxide NPs enhance MRI by improving tumor visualization (Aboushoushah 2025).

Furthermore, materials such as titanium dioxide and zinc oxide generate reactive oxygen species (ROS) upon exposure to light, inducing cancer cell death (Bogdan et al. 2017). Meanwhile, silica NPs show promise in activating immune responses for tumor clearance (Bielecki et al. 2021). These inorganic nanomaterials are being actively explored to improve targeted cancer therapies and diagnostic accuracy. Polymeric nanomedicines, particularly those composed of poly(lactic-co-glycolic acid) (PLGA), have demonstrated effectiveness in cancer treatment. PLGA NPs enable controlled drug release and improved cancer cell targeting, demonstrating potential in treating triple-negative breast cancer (TNBC) and ovarian cancer (Rezvantalab et al. 2018; Zhang et al. 2022). Dendrimers, another class of polymeric NPs, enhance targeted drug delivery, ensuring precise tumor site localization and improved therapeutic outcomes (Crintea et al. 2023; Arif et al. 2024). An overview of FDA- and EMA-approved nanodrug formulations for cancer treatment is illustrated in Fig. 11.3.

Recent studies further highlight nanomedicine's potential in overcoming treatment challenges in specific cancers. For example, hybrid lipid–polymer NPs delivering doxorubicin and sorafenib have improved therapeutic effects in hepatocellular carcinoma by increasing bioavailability and reducing drug resistance (Far et al. 2023). Similarly, PEG-PLA NPs have enhanced the oral administration of tyrosine kinase inhibitors like gefitinib, enabling better stability and localization in treating non-small cell lung cancer (Alshetaili 2021). The benefits of nanomedicines include enhanced drug solubility, stability, and controlled release, resulting in reduced side effects and improved therapeutic efficacy. Nanomedicines transcend biological barriers like the blood–brain barrier and optimize pharmacokinetics by selectively targeting cancerous tissues while preserving healthy ones. However, significant challenges remain in their clinical translation. Issues related to biological toxicity, stability in physiological environments, and the complexity of distribution, metabolism, and clearance mechanisms must be addressed. High production costs and regulatory hurdles further limit the widespread clinical adoption of these technologies. Moreover, concerns about the long-term health and environmental impact of inorganic nanomaterials pose additional challenges.

Despite challenges such as immune clearance, biological barriers, and regulatory constraints, collaborative research can accelerate the clinical translation of nanomedicine. A patient-centered approach, improved understanding of the tumor microenvironment, and enhanced transcytosis strategies can enhance the effectiveness of nanomedicine. Expanding its applications beyond breast and lung cancers will ensure broader accessibility. Moreover, refining regulatory frameworks and establishing internationally accepted standards will facilitate faster approvals and mitigate manufacturing concerns. The future of cancer nanomedicine is promising, with advances in personalized medicine, artificial intelligence-driven drug optimization, and biodegradable nanomaterials shaping the development of next-generation therapies. Combining nanotechnology with conventional treatments is expected to enhance therapeutic outcomes while reducing adverse effects. Although high production costs and regulatory complexities remain, strategic collaborations among researchers, policymakers, and industry stakeholders will be essential for integrating nanomedicine into clinical practice. Nanomedicine is poised to revolutionize cancer

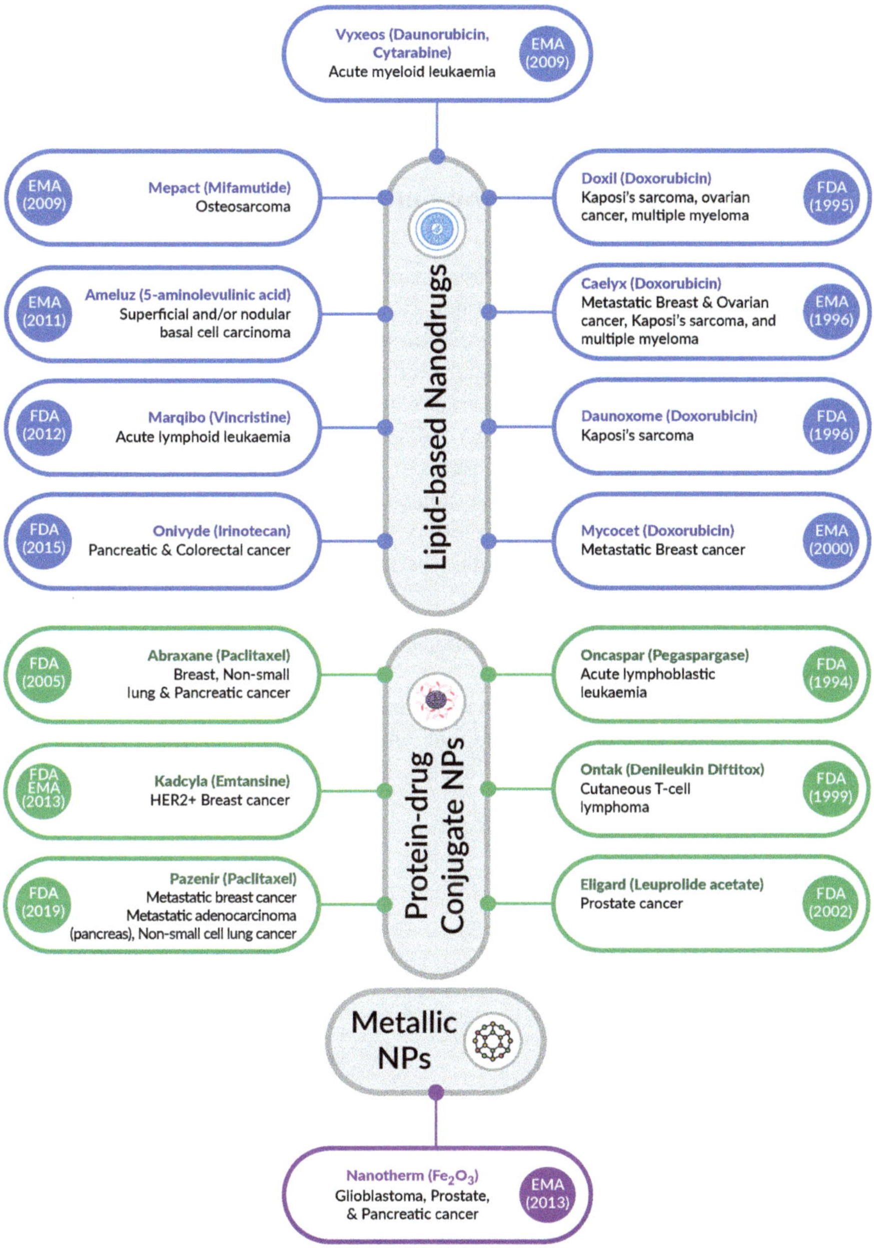


**Fig. 11.3** Approved cancer drug therapies based on nanotechnology. *Data source*: Liu et al. 2025

diagnosis and treatment by bridging the gap between laboratory research and practical applications, as evidenced by the increasing approval rates and early-stage trial success of nano-enabled medications.

## 11.7 Nanoparticle-Based Combination Therapy for Cancer

Over recent decades, significant progress in cancer biology and innovative treatment methods has improved patient survival rates. However, cancer prevalence continues to rise. One of the treatment strategy, combinational therapy, has faced challenges like varying drug pharmacokinetics, optimal combinations, and non-specific delivery. Nanotechnology has addressed these issues by enhancing drug kinetics, enabling controlled release, selective targeting of diseased tissues, and responsiveness to stimuli (Gurunathan et al. 2018; Shrestha et al. 2019), as shown in Fig. 11.4.

Nanoparticle-based combination therapies have emerged as a potent strategy in cancer treatment, leveraging nanotechnology to enhance therapeutic outcomes and minimize side effects. Three key strategies for enhancing chemotherapy effectiveness are as follows:

- Combining multiple chemotherapeutic agents to target cancer cells synergistically.
- Integrating chemotherapeutic agents with nucleic acids or therapeutic proteins to improve targeted delivery and treatment outcomes.
- Combination of chemotherapeutic agents and energy-based therapies (Shrestha et al. 2019).

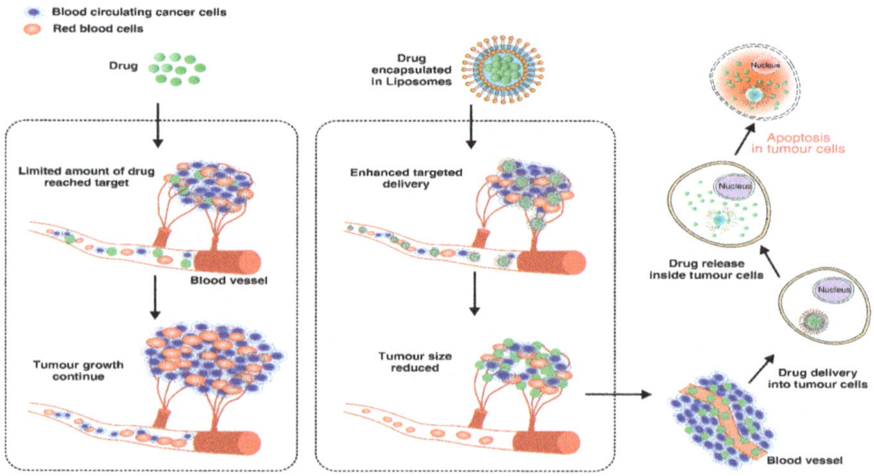


**Fig. 11.4** Nanoparticle-based drug delivery enhances anti-tumor activity through targeted and sustained drug release (Gurunathan et al. 2018)

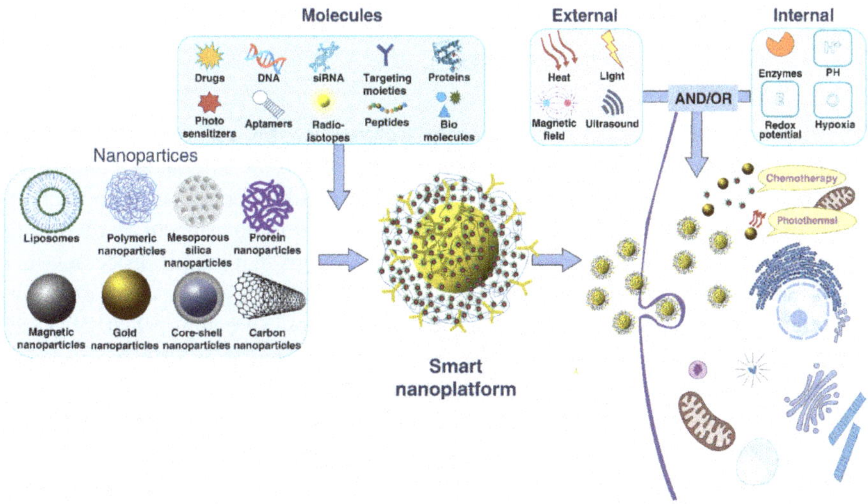


**Fig. 11.5** Schematic representation of smart nanoplatform design (Shrestha et al. 2021)

- Smart nanoparticles integrate therapeutic agents, targeting moieties, and stimuli-responsive mechanisms to enable targeted and controlled therapy. For instance, gold nanoparticles functionalized with pH-sensitive drug linkers and targeting molecules can release drugs in acidic tumor environments and act as photothermal agents for cancer ablation (Shrestha et al. 2021), as shown in Fig. 11.5.

## *11.7.1 Multiple Chemotherapeutic Agents*

Combination chemotherapy, or drug cocktail therapy, involves using multiple chemotherapeutic agents with different mechanisms to reduce dose-limiting toxicity and multidrug resistance while achieving synergistic or additive effects. This strategy is especially common in advanced cancer treatment, where maintaining an optimal drug molar ratio is crucial, as deviations can cause antagonistic or reduced therapeutic effects (Hare et al. 2013). Nanoparticle-based delivery systems, such as liposomes (e.g., CPX-351 delivering cytarabine and daunorubicin at a fixed 5:1 ratio) and polymer nanoparticles, enable precise ratio control and targeted delivery, improving efficacy and minimizing side effects (Lancet et al. 2010). However, polymer systems face challenges such as batch variability and inconsistent release. In contrast, hybrid systems (e.g., lipid–polymer or mesoporous silica nanoparticles) offer enhanced delivery capabilities but require further optimization in terms of size, shape, and surface properties *for clinical translation* (Zhang et al. 2010; Iram et al. 2017).

### 11.7.2 Chemotherapeutic Agents with Nucleic Acids or Therapeutic Proteins

For immunotherapies and targeted cancer treatments, this combination approach combines chemotherapeutic medications with genes or therapeutic proteins, including growth factors, tumor antigens, peptides, and monoclonal antibodies. Its goal is to enhance the sensitivity of cancer cells or the immune system by disrupting key signaling pathways, triggering intracellular reactions that inhibit processes such as angiogenesis or proteasome activity, thereby improving efficacy and reducing multidrug resistance for synergistic therapeutic outcomes (Shrestha et al. 2019). Gene therapy shows promise in cancer treatment by addressing mutation-driven causes, but its potential remains underutilized. Chemo–gene therapy enhances drug sensitivity and achieves synergistic effects by targeting distinct pathways, yet multidrug resistance (MDR) remains a significant challenge (Wang et al. 2011; Adair et al. 2014). MDR in cancer includes pump resistance, caused by drug-efflux transport proteins like P-glycoprotein, and non-pump resistance, driven by antiapoptotic proteins such as Bcl-2 and inhibitor of apoptosis proteins (IAPs). Overcoming MDR involves silencing or replacing these resistance-related proteins (Gottesman et al. 2002; Holohan et al. 2013).

### 11.7.3 Chemotherapeutic Agents and Energy-Based Therapies

Recent nanotechnology advances show that metallic nanoparticles (gold or magnetic NPs) can dissipate energy as heat under external stimuli like light or magnetic fields, enabling targeted cancer treatment. These NPs can be integrated into nanosystems for energy-based combination therapies, which, when combined with chemotherapeutics, enhance outcomes through approaches like photodynamic therapy, hyperthermia, or radiotherapy. Clinically used since the 1970s, hyperthermia therapy targets cancer cells by exploiting their higher heat sensitivity. At 41–45 °C, it enhances chemotherapy and radiotherapy by altering the tumor microenvironment and promoting apoptosis. Above 46 °C, it causes irreversible damage and necrosis (thermal ablation). Precise temperature control is crucial to avoid harming healthy tissues (Shrestha et al. 2019).

Poor cell-type specificity in current cancer treatments results in limited therapeutic efficacy and inefficient medication distribution at tumor locations. Existing anticancer medications' physicochemical characteristics, such as their toxicity and hydrophobicity, limit dosage levels to prevent damage to healthy tissues. Direct drug administration is also less successful due to tumor microenvironmental characteristics such multidrug resistance and excessive drug efflux. According to recent studies, combining several medicines may have synergistic effects that improve the results of cancer treatment. Because they enhance the effectiveness of drugs, nanoparticles are essential in these combination treatments. They enhance drug stability, biocompatibility, retention, and targeted accumulation, thereby reducing side effects and boosting therapeutic efficacy. Additionally, nanoparticles allow for the integration of non-pharmacological treatments or the co-delivery of several therapeutic agents. For example, nanoparticle-mediated co-delivery of multiple therapeutics can simultaneously target different cancer pathways, overcoming drug resistance and inhibiting tumor growth, progression, and metastasis. Similarly, combining nanoparticle-based drug delivery with energy-based therapies, such as those using metallic nanoparticles, enhances tumor oxygenation, drug permeability, and tumor suppression by delivering therapeutic energy locally. Moreover, nanoparticles can incorporate imaging capabilities or be engineered as theranostic systems. These nanoparticle-mediated combination therapies have shown synergistic improvements in both in vitro and in vivo studies, addressing key challenges in conventional cancer treatments. However, clinical translation faces hurdles, including batch-to-batch variability, high production costs, and the absence of standardized testing protocols and models. Advancing these therapies to clinical applications requires collaborative efforts among academia, clinicians, the pharmaceutical industry, and regulatory authorities. Despite these challenges, nanoparticle-mediated combinatorial therapies hold immense potential to revolutionize cancer treatment.

## 11.8 Nanoparticle-Based Immunotherapy for Cancer

Advancements in understanding cancer pathogenesis have led to new therapeutic approaches. Tumor cells interact with the immune system and surrounding stroma, creating an environment that promotes tumor growth, invasion, metastasis, drug resistance, and immune evasion. As tumor cells often disrupt or suppress immunity, immunotherapy has emerged as a promising alternative, particularly through immune checkpoint inhibitors, which have enhanced survival and achieved long-term remission in solid tumors (Zhang et al. 2022). Immunotherapy has shown significant advantages over conventional treatments, but it still faces limitations, including suboptimal efficacy, safety concerns, and low response rates (Urruticoechea et al. 2010; Pucci et al. 2019; Cao et al. 2020; Jiang et al. 2020). Certain immunotherapy medicines also have drawbacks, such as short half-lives, low solubility, and poor stability. Moreover, effectively delivering these agents through the tumor-permissive microenvironment remains a significant challenge (Lee and Margolin 2011; Chen et al. 2019). To overcome these hurdles, NPs-based strategies have been developed. The integration of immunotherapy with nanomedicine, termed nano-immunotherapy, offers an efficient delivery system for immune agents, reducing toxicity while improving solubility, stability, and half-lives.

### *11.8.1 Nano-Immunotherapy for Cancer Treatment: A Brief Overview*

Combining immunotherapy and nanotechnology is known as nano-immunotherapy. An increasing number of nanoformulations have been developed to enhance the effectiveness of cancer immunotherapy while reducing its side effects (Moon et al. 2012; Riley et al. 2019). As a result, one of the main areas of interest in cancer treatment is the incorporation of nanotechnology into immunotherapy. Three primary targeting strategies can be used to broadly classify the application of nanoparticles (NPs) in cancer immunotherapy. Lipid-based nanoparticles, for instance, have been utilized to increase T-cell responses against melanoma by delivering mRNA encoding neoantigens to dendritic cells (Kranz et al. 2016). Gold nanoparticles functionalized with tumor-associated antigens have served as carriers to activate dendritic cells and promote immune responses in breast cancer models (Javadi et al. 2025). Poly(lactic-co-glycolic acid) (PLGA) nanoparticles have been utilized to co-deliver checkpoint inhibitors and tumor antigens for synergistic immunomodulation in lung cancer (Zhang et al. 2022). In another approach, iron oxide nanoparticles have been used for both tumor imaging and macrophage repolarization to enhance immune clearance in glioblastoma (Israel et al. 2020). These examples illustrate how nanotechnology can be strategically optimized to enhance cancer immunotherapy through precise delivery and immune system engagement.

- **Factors Affecting Delivery of Cancer Antigens Through Nanoparticles**

Effectively transferring tumor antigens to lymph nodes is essential for producing a strong anticancer immune response since immune responses are mainly started in secondary lymphoid organs (Zhang et al. 2018b). Nanoparticles have emerged as promising delivery vehicles for this purpose, offering protection against enzymatic degradation and enabling targeted delivery to lymph nodes. Once delivered, nanoparticles encapsulating tumor antigens are efficiently internalized by antigen-presenting cells (APCs) (Tran et al. 2018). However, the efficiency of nanoparticle delivery to lymph nodes depends significantly on particle characteristics. Size is a critical factor: While big nanoparticles (>100 nm) may become stuck in the extracellular matrix (ECM), restricting their access to lymph nodes, tiny nanoparticles (<5 nm) may leak from blood vessels during circulation. The best nanoparticle sizes are between 5 and 100 nm because they stay in circulation and are efficiently delivered to lymph nodes by lymphatic channels (Reddy et al. 2007; Irvine et al. 2015; Zhu et al. 2017). For instance, about half of the 20–45 nm poly(propylene sulfide) nanoparticles were internalized by dendritic cells and APCs, and they were transported to lymph nodes and absorbed into lymphatic vessels for up to 120 h after administration (Reddy et al. 2006). Comparably, 25 nm-sized NPs were more effectively transported to draining lymph nodes through the lymphatic system and interstitial fluid than 100 nm nanoparticles, of which only 10% made it to their intended location. By altering nanoparticles with ligands, such as mannose, for specific lymph node targeting, active targeting can further improve delivery (Reddy et al. 2007; Yoon et al. 2018). Particle shape also influences lymph node delivery. While traditional nanoparticles are typically spherical, recent advancements have introduced a diverse range of shapes, including rods, prisms, cubes, stars, and disks. Higher aspect ratio non-spherical particles typically have better penetration into solid tissues and tumors, longer circulation times, and improved margination effects (Gentile et al. 2008; Toy et al. 2011; Carboni et al. 2014). Surface charge has a significant impact on cellular internalization and immune activation. Though they may have decreased tissue permeability as a result of immobilization in the negatively charged extracellular matrix, positively charged NPs often trigger higher immune responses than neutral or negatively charged ones. Although cationic NPs are more easily absorbed by dendritic cells at the injection site, their positive charge can cause hemolysis or platelet aggregation, which can complicate lymphatic transport and result in altered uptake kinetics and premature antigen release (Goodman et al. 2004; Foged et al. 2005; Manolova et al. 2008; Dewitte et al. 2014). Additionally, specific biomaterials, such as polymers with hydrophobic domains, exhibit intrinsic adjuvant activity, activating immune cells without the need for additional stimuli. For example, amphiphilic poly-γ-glutamic acid nanoparticles exhibit higher biological responses in vivo and improved antigen absorption and dendritic cell activation in vitro, which are correlated with their side chains' increased hydrophobicity (Da Silva et al. 2009; Park and Babensee 2012; Shima et al. 2013).

- **Delivery of Adjuvant Through NPs**

Adjuvants are molecules that boost the immunogenicity of tumor antigens, which often lack sufficient immune activation when presented alone. These compounds imitate molecular patterns linked to pathogens that pattern recognition receptors (PRRs) can identify. 3-O-desacyl-4′-monophosphoryl lipid A (MPLA), lipopolysaccharide (LPS), CpG oligodeoxynucleotides (CpG ODNs), polyinosinic: polycytidylic acid (poly I:C), and stimulator of IFN genes (STING) agonists are examples of common adjuvants in cancer immunotherapy. When internalized by APCs alongside tumor antigens, adjuvants enhance anticancer immune responses. To improve delivery, NPs-based strategies have been developed to efficiently transport adjuvants into APCs. By forming electrostatic complexes with positively charged NPs or polymers, negatively charged adjuvants such as CpG ODNs, poly I:C, and STING agonists facilitate their uptake into APCs through endocytosis and enhance anti-tumor immunity (Chen et al. 2014; Hanson et al. 2015; Luo et al. 2017). For example, PLGA nanoparticles loaded with CpG ODNs and tumor lysate antigens have effectively activated cytotoxic and helper T-cell responses, leading to tumor suppression in TRAMP mice (Mueller et al. 2012). Likewise, a study revealed that dual delivery nanoparticle vaccines co-delivering antigens and poly I:C adjuvants into the cytosol enhanced antigen cross-presentation and dendritic cell activation, significantly boosting CD8 + T-cell responses and inhibiting tumor growth in mouse models, offering a promising strategy for effective cancer immunotherapy (Carson et al. 2022). Furthermore, lipid-based nanoparticles carrying a STING agonist (cGAMP) were used to target pancreatic cancer, resulting in localized immune activation and tumor suppression (Shaji et al. 2024). These studies highlight the versatility of NP platforms for efficient adjuvant delivery and immune potentiation in cancer immunotherapy.

NPs-mediated delivery of adjuvants and antigens to the APC cytoplasm is critical for eliciting robust antigen-specific T-cell responses (Wilson et al. 2013). Immunomodulatory nanoliposomes, called tumosomes, with a size of 100 nm, have been designed to co-deliver cancer antigens and adjuvants (Noh et al. 2017). To further improve therapeutic results, tumosomes can also be used in conjunction with other therapies, including gene therapy, chemotherapy, or cell-based therapies (such as T cells or natural killer cells) (Park et al. 2018).

- **Modulation of the Immune-Suppressive Tumor Microenvironment using NPs**

Through the creation of an immunosuppressive tumor microenvironment (TME), tumors can promote the spread and metastasis of cancer. As a result, modulating the tumor microenvironment (TME) has become a key strategy in cancer immunotherapy (Jeanbart and Swartz 2015). Anti-tumor effector T cells are inhibited by regulatory T cells (Tregs), a subset of immunosuppressive T cells. Tregs limit immunological responses inside the TME and reduce anti-tumor activity in cancer, despite being crucial for immune tolerance maintenance and autoimmunity prevention. Suppressing or eliminating Tregs can help stimulate anti-tumor immunity (Ou et al. 2018). One commonly used approach involves checkpoint blockade therapy (anti-CTLA-4) to control Treg activity. Notably, iron oxide nanoparticle-mediated photothermal therapy (PTT), when combined with anti-CTLA-4 therapy, significantly depleted tumor Tregs in 4T1 breast tumors, offering a potent strategy to boost immunotherapy (Chen et al. 2020). Additionally, engineered NPs targeting Tregs have been developed to remove them from the TME selectively. Recent studies demonstrated that PEG-modified single-walled carbon nanotubes (PEG-SWCNTs) armed with GITR-targeting ligands can selectively target and internalize into intratumoral Tregs in B16 melanoma models. This selective delivery exploits both passive tumor targeting and active receptor-mediated uptake, offering a novel strategy for precise immune modulation in cancer therapy (Sacchetti et al. 2013). Tumor-associated macrophages (TAMs) are another type of immune cell frequently found in high numbers within the TME. In addition to secreting inflammatory mediators such as IL-12, IL-1β, TNF-α, and IL-6 that inhibit anti-tumor immune responses, they also produce immunoregulatory cytokines, including IL-10 and TGF-β. Eliminating TAMs from the TME is considered critical for improving the effectiveness of cancer immunotherapy. Recently, surface-modified NPs designed to target and destroy TAMs have shown promise in enhancing anticancer treatment outcomes. For instance, the multifunctional nanoparticle LyP-SA/AgNP@Dox was created to target immunosuppressive M2-type tumor-associated macrophages (TAMs) as well as triple-negative breast cancer (TNBC) cells. The nanoparticles induced ROS-mediated apoptosis in vitro and showed strong tumor targeting and therapeutic efficacy in vivo, offering a dual-targeted immunotherapy strategy (Liu et al. 2024a, b). Further, mannose-decorated PLGA nanoparticles (Man-pD-PLGA-NP@R848) were designed to simultaneously target and reprogram tumor-associated macrophages (TAMs) and tumor-infiltrating dendritic cells (TIDCs), converting an immunosuppressive "cold" tumor into an immune-active "hot" tumor. This dual-targeting strategy enhanced T-cell infiltration and inhibited tumor growth in B16F10 mouse models, offering a promising approach for cancer immunotherapy (Dang et al. 2024).

Transforming growth factor-beta (TGF-β), a cytokine often overexpressed in cancers such as breast, liver, and lung, interferes with the activation, maturation, and differentiation of immune cells. Inhibiting the TGF-β signaling pathway within the TME can help trigger immune responses against tumors. Recent studies have demonstrated that lipid nanoparticle-encapsulated TGF-β inhibitors can be delivered to the TME to stimulate both innate and adaptive immunity, suppress tumor growth, and improve survival rates in mice with metastatic melanoma (Park and Babensee 2012). Myeloid-derived suppressor cells (MDSCs), commonly found in the TMEs of breast, lung, gastrointestinal, and liver cancers, contribute to immune suppression by releasing factors such as IL-10, ARG1, NOS2, and IDO, which activate Tregs and inhibit other immune cells. Reducing the presence of MDSCs can substantially enhance the effectiveness of cancer immunotherapy (Hu et al. 2025). Immunomodulatory agents can be delivered to the TME by nanoparticle-based delivery systems either actively, using particular ligands, or passively, through the enhanced permeability and retention (EPR) effect. This localized delivery can amplify anti-tumor immune responses while minimizing systemic side effects. Nanotechnology-based strategies demonstrate the potential to overcome current limitations in cancer immunotherapy (Deivayanai et al. 2024).

## 11.9 Conclusion and Future Perspectives

The goal of cancer nanomedicine is to overcome the shortcomings of conventional cancer detection and therapy. Despite its potential and growing interest, clinical translation remains a challenge, leading some to question its full realization. The future of cancer nanomedicine is promising, with research focusing on personalized and precision medicine. Advancements in artificial intelligence and machine learning are enabling the optimization of nanomedicine applications, thereby ensuring personalized treatments for individual patients. Another important area of interest to reduce long-term toxicity is the development of biodegradable and biocompatible nanomaterials. Combination therapies that integrate nanotechnology with conventional treatments are expected to enhance treatment efficacy and overcome drug resistance. Despite the potential of nanomedicine, regulatory challenges and high production costs remain significant hurdles. Addressing these barriers through collaborative efforts among researchers, policymakers, and industry stakeholders will be crucial for the successful clinical translation of nanomedicine-based therapies.

Cancer nanomedicine has the potential to transform oncology as technology advancements continue, providing hope for more potent, tailored, and less harmful cancer treatments. Numerous cancer nanomedicines have exhibited significant efficacy in preclinical studies, enhancing therapeutic outcomes, extending survival, and minimizing adverse effects. However, their clinical translation remains a major hurdle. Only a small number of candidates have received regulatory approval for therapeutic usage, even though some have progressed to clinical trials.

## References


Aboushoushah SFO (2025) Iron oxide nanoparticles enhancing magnetic resonance imaging: a review of the latest advancements. J Sci: Adv Mater Devices 10(2):100875

Adair JE, Johnston SK, Mrugala MM, Beard BC, Guyman LA, Baldock AL, Kiem HP (2014) Gene therapy enhances chemotherapy tolerance and efficacy in glioblastoma patients. J Clin Investig 124(9):4082–4092

Adhikari HS, Yadav PN (2018) Anticancer activity of chitosan, chitosan derivatives, and their mechanism of action. Int J Biomater 2018(1):2952085

Alemu D, Getachew E, Mondal AK (2023) Study on the physicochemical properties of chitosan and their applications in the biomedical sector. Int J Polym Sci 2023(1):5025341

Alemzadeh E, Izadpanah K, Ahmadi F (2017) Generation of recombinant protein shells of Johnson grass chlorotic stripe mosaic virus in tobacco plants and their use as drug carrier. J Virol Methods 248:148–153

Alexander-Bryant AA, Vanden Berg-Foels WS, Wen X (2013) Bioengineering strategies for designing targeted cancer therapies. Adv Cancer Res 118:1–59

Aliabadi M, Shagholani H, Gerayeli A (2018) Planar polymer-graphene oxide nanohybrid as a 5-fluorouacil carrier in pH-responsive controlled release. J Drug Deliv Sci Technol 43:103–106

Allen TM (1998) Liposomal drug formulations: rationale for development and what we can expect for the future. Drugs 56(5):747–756

Allen TM, Martin FJ (2004) Advantages of liposomal delivery systems for anthracyclines. In: Seminars in oncology, vol 31. WB Saunders, pp 5–15

Alshehri S, Imam SS, Rizwanullah M, Akhter S, Mahdi W, Kazi M, et al (2020) Progress of cancer nanotechnology as diagnostics, therapeutics, and theranostics nanomedicine: preclinical promise and translational challenges. Pharmaceutics 13(1):24. https://doi.org/10.3390/pharmaceutics13010024

Alshetaili AS (2021) Gefitinib loaded PLGA and chitosan coated PLGA nanoparticles with magnified cytotoxicity against A549 lung cancer cell lines. Saudi J Biol Sci 28(9):5065–5073

Al-Thani AN, Jan AG, Abbas M, Geetha M, Sadasivuni KK (2024) Nanoparticles in cancer theragnostic and drug delivery: a comprehensive review. Life Sci 352:122899

Amjad MT, Chidharla A, Kasi A (2025) Cancer chemotherapy. [Updated 2023 Feb 27]. In: StatPearls [Internet]. Treasure Island (FL): StatPearls Publishing; 2025 Jan-. https://www.ncbi.nlm.nih.gov/books/NBK564367/

Amreddy N, Muralidharan R, Babu A, Mehta M, Johnson EV, Zhao YD, Ramesh R (2015) Tumor-targeted and pH-controlled delivery of doxorubicin using gold nanorods for lung cancer therapy. Int J Nanomedicine 6773–6788

Arif Y, Hasan N, Gowda BJ, Gupta G, Alsayari A, Wahab S, Kesharwani P (2024) Advancements in dendrimer-based drug delivery for combinatorial cancer therapy. J Drug Deliv Sci Technol 97:105755

Asmatulu R, Zalich MA, Claus RO, Riffle JS (2005) Synthesis, characterization and targeting of biodegradable magnetic nanocomposite particles by external magnetic fields. J Magn Magn Mater 292:108–119

Astruc D, Boisselier E, Ornelas C (2010) Dendrimers designed for functions: from physical, photophysical, and supramolecular properties to applications in sensing, catalysis, molecular electronics, photonics, and nanomedicine. Chem Rev 110(4):1857–1959

Attia MF, Anton N, Wallyn J, Omran Z, Vandamme TF (2019) An overview of active and passive targeting strategies to improve the nanocarriers efficiency to tumour sites. J Pharm Pharmacol 71(8):1185–1198
Bagheri B, Surwase SS, Lee SS, Park H, Rad ZF, Trevaskis NL, Kim YC (2022) Carbon-based nanostructures for cancer therapy and drug delivery applications. J Mater Chem B 10(48):9944–9967
Balkrishna A, Rohela A, Kumar A, Mishra S, Arya V, Kala V, Thakur N, Thakur N, Kumari A, Khan N (2023) Elucidating the role of plant extracts mediated gold nanoparticles as smart antimicrobials: two-way attack. J Nanomater 2023(1):4085090
Balkrishna A, Sinha S, Kumar A, Arya V, Kumar D, Dhansekhran M (2024) Nanoparticles as drug delivery systems: advances and challenges. In: Nanotechnology: agriculture, environment and health. Singapore: Springer Nature Singapore, pp 245–288
Barua S, Mitragotri S (2014) Challenges associated with penetration of nanoparticles across cell and tissue barriers: a review of current status and future prospects. Nano Today 9(2):223–243. https://doi.org/10.1016/j.nantod.2014.04.008
Bauer TA, Horvat NK, Marques O, Chocarro S, Mertens C, Colucci S et al (2021) Core cross-linked polymeric micelles for specific iron delivery: inducing sterile inflammation in macrophages. Adv Healthc Mat 10(19):e2100385. https://doi.org/10.1002/adhm.202100385
Beg S, Rizwan M, Sheikh AM, Hasnain MS, Anwer K, Kohli K (2011) Advancement in carbon nanotubes: basics, biomedical applications and toxicity. J Pharm Pharmacol 63(2):141–163
Bekyarova E, Ni Y, Malarkey EB, Montana V, McWilliams JL, Haddon RC, Parpura V (2005) Applications of carbon nanotubes in biotechnology and biomedicine. J Biomed Nanotechnol 1(1):3–17
Bhagyaraj S M, Oluwafemi OS (2018) Nanotechnology: the science of the invisible, in synthesis of inorganic nanomaterials. Elsevier, p 1–18. https://linkinghub.elsevier.com/retrieve/pii/B9780081019757000014
Biabanikhankahdani R, Alitheen NBM, Ho KL, Tan WS (2016) pH-responsive virus-like nanoparticles with enhanced tumour-targeting ligands for cancer drug delivery. Sci Rep 6(1):37891
Bielecki PA, Lorkowski ME, Becicka WM, Atukorale PU, Moon TJ, Zhang Y, Karathanasis E (2021) Immunostimulatory silica nanoparticle boosts innate immunity in brain tumors. Nanoscale Horiz 6(2):156–167
Bobo D, Robinson KJ, Islam J, Thurecht KJ, Corrie SR (2016) Nanoparticle-based medicines: a review of FDA-approved materials and clinical trials to date. Pharm Res 33(10):2373–2387
Bodunde OP, Ikumapayi OM, Akinlabi ET, Oladapo BI, Adeoye AOM, Fatoba SO (2021) A futuristic insight into a "nano-doctor": a clinical review on medical diagnosis and devices using nanotechnology. Mater Today Proc 44:1144–1153
Bogdan J, Pławińska-Czarnak J, Zarzyńska J (2017) Nanoparticles of titanium and zinc oxides as novel agents in tumor treatment: a review. Nanoscale Res Lett 12(1):225
Boisseau P, Loubaton B (2011) Nanomedicine, nanotechnology in medicine. C R Phys 12(7):620–636
Bozzuto G, Molinari A (2015) Liposomes as nanomedical devices. Int J Nanomedicine 975–999
Bray F, Laversanne M, Sung H, Ferlay J, Siegel RL, Soerjomataram I, Jemal A (2024) Global cancer statistics 2022: GLOBOCAN estimates of incidence and mortality worldwide for 36 cancers in 185 countries. CA: Cancer J Clin 74(3):229–263. https://doi.org/10.3322/caac.21834
Brown SD, Nativo P, Smith JA, Stirling D, Edwards PR, Venugopal B, Wheate NJ (2010) Gold nanoparticles for the improved anticancer drug delivery of the active component of oxaliplatin. J Am Chem Soc 132(13):4678–4684
Bruckman MA, Jiang K, Simpson EJ, Randolph LN, Luyt LG, Yu X, Steinmetz NF (2014) Dual-modal magnetic resonance and fluorescence imaging of atherosclerotic plaques in vivo using VCAM-1 targeted tobacco mosaic virus. Nano Lett 14(3):1551–1558
Bruckman MA, Czapar AE, VanMeter A, Randolph LN, Steinmetz NF (2016) Tobacco mosaic virus-based protein nanoparticles and nanorods for chemotherapy delivery targeting breast cancer. J Control Release 231:103–113

Cao J, Huang D, Peppas NA (2020) Advanced engineered nanoparticulate platforms to address key biological barriers for delivering chemotherapeutic agents to target sites. Adv Drug Deliv Rev 167:170–188

Carboni E, Tschudi K, Nam J, Lu X, Ma AW (2014) Particle margination and its implications on intravenous anticancer drug delivery. AAPS PharmSciTech 15(3):762–771

Carson CS, Becker KW, Garland KM, Pagendarm HM, Stone PT, Arora K, Wilson JT (2022) A nanovaccine for enhancing cellular immunity via cytosolic co-delivery of antigen and polyIC RNA. J Control Release 345:354–370

Chakraborty S, Rahman T (2012) The difficulties in cancer treatment. ecancermedicalscience, 6:ed16

Chamundeeswari M, Jeslin J, Verma ML (2019) Nanocarriers for drug delivery applications. Environ Chem Lett 17(2):849–865

Chang SH, Lee HJ, Park S, Kim Y, Jeong B (2018) Fast degradable polycaprolactone for drug delivery. Biomacromol 19(6):2302–2307

Chaudhari HS, Popat RR, Adhao VS, Shrikhande VN (2016) Dendrimers: novel carriers for drug delivery. J Appl Pharm Res 4(1):01–19

Chen T, Fu J (2012) PH-responsive nanovalves based on hollow mesoporous silica spheres for controlled release of corrosion inhibitor. Nanotechnology 23(23):235605

Chen XS, Casini G, Harrison SC, Garcea RL (2001) Papillomavirus capsid protein expression in Escherichia coli: purification and assembly of HPV11 and HPV16 L1. J Mol Biol 307(1):173–182

Chen JF, Ding HM, Wang JX, Shao L (2004) Preparation and characterization of porous hollow silica nanoparticles for drug delivery application. Biomaterials 25(4):723–727

Chen C, Han D, Cai C, Tang X (2010) An overview of liposome lyophilization and its future potential. J Control Release 142(3):299–311

Chen X, Zaro JL, Shen WC (2013) Fusion protein linkers: property, design and functionality. Adv Drug Deliv Rev 65(10):1357–1369

Chen S, Zhang H, Shi X, Wu H, Hanagata N (2014) Microfluidic generation of chitosan/CpG oligodeoxynucleotide nanoparticles with enhanced cellular uptake and immunostimulatory properties. Lab Chip 14(11):1842–1849

Chen G, Svirskis D, Lu W, Ying M, Huang Y, Wen J (2018) N-trimethyl chitosan nanoparticles and CSKSSDYQC peptide: N-trimethyl chitosan conjugates enhance the oral bioavailability of gemcitabine to treat breast cancer. J Control Release 277:142–153

Chen Z, Wang Z, Gu Z (2019) Bioinspired and biomimetic nanomedicines. Acc Chem Res 52(5):1255–1264

Chen H, Luan X, Paholak HJ, Burnett JP, Stevers NO, Sansanaphongpricha K, Sun D (2020) Depleting tumor-associated Tregs via nanoparticle-mediated hyperthermia to enhance anti-CTLA-4 immunotherapy. Nanomedicine 15(1):77–92

Chen J, Hu S, Sun M, Shi J, Zhang H, Yu H, Yang Z (2024) Recent advances and clinical translation of liposomal delivery systems in cancer therapy. Eur J Pharm Sci 193:106688

Clemente N., Ferrara B, Gigliotti CL, Boggio E, Capucchio MT, Biasibetti E, Dianzani C (2018) Solid lipid nanoparticles carrying temozolomide for melanoma treatment. Preliminary in vitro and in vivo studies. Int J Mol Sci 19(2):255

Coey T (2024) Nanomedicine in oncology: a critical review on epidemiology, health impacts, challenges, and future outlook. Prog Drug Discov Biomed Sci 7(1)

Crintea A, Motofelea AC, Șovrea AS, Constantin AM, Crivii CB, Carpa R, Duțu AG (2023) Dendrimers: advancements and potential applications in cancer diagnosis and treatment—an overview. Pharmaceutics 15(5):1406

Cui T, Liang JJ, Chen H, Geng DD, Jiao L, Yang JY, Ding Y (2017) Performance of doxorubicin-conjugated gold nanoparticles: regulation of drug location. ACS Appl Mater Interfaces 9(10):8569–8580

Da Silva CA, Chalouni C, Williams A, Hartl D, Lee CG, Elias JA (2009) Chitin is a size-dependent regulator of macrophage TNF and IL-10 production. J Immunol 182(6):3573–3582

Dadfar SM, Roemhild K, Drude NI, von Stillfried S, Knüchel R, Kiessling F, Lammers T (2019) Iron oxide nanoparticles: diagnostic, therapeutic and theranostic applications. Adv Drug Deliv Rev 138:302–325

Dang BTN, Duwa R, Lee S, Kwon TK, Chang JH, Jeong JH, Yook S (2024) Targeting tumor-associated macrophages with mannosylated nanotherapeutics delivering TLR7/8 agonist enhances cancer immunotherapy. J Control Release 372:587–608

Davis ME, Chen Z, Shin DM (2008) Nanoparticle therapeutics: an emerging treatment modality for cancer. Nat Rev Drug Discov 7(9):771–782

Davodabadi F, Sajjadi SF, Sarhadi M, Mirghasemi S, Hezaveh MN, Khosravi S, Ghavami S (2023) Cancer chemotherapy resistance: mechanisms and recent breakthrough in targeted drug delivery. Eur J Pharmacol 958:176013

Degertekin M, Serruys PW, Foley DP, Tanabe K, Regar E, Vos J, Popma JJ (2002) Persistent inhibition of neointimal hyperplasia after sirolimus-eluting stent implantation: long-term (up to 2 years) clinical, angiographic, and intravascular ultrasound follow-up. Circulation 106(13):1610–1613

Deivayanai VC, Thamarai P, Karishma S, Saravanan A, Yaashikaa PR, Vickram AS, Shruthi S (2024) Advances in nanoparticle-mediated cancer therapeutics: current research and future perspectives. Cancer Pathog Ther 3(4):293

Devarakonda B, Hill RA, de Villiers MM (2004) The effect of PAMAM dendrimer generation size and surface functional group on the aqueous solubility of nifedipine. Int J Pharm 284(1–2):133–140

Dewitte H, Verbeke R, Breckpot K, De Smedt SC, Lentacker I (2014) Nanoparticle design to induce tumor immunity and challenge the suppressive tumor microenvironment. Nano Today 9(6):743–758

Drummond DC, Meyer O, Hong K, Kirpotin DB, Papahadjopoulos D (1999) Optimizing liposomes for delivery of chemotherapeutic agents to solid tumors. Pharmacol Rev 51(4):691–743

Duncan B, Kim C, Rotello VM (2010) Gold nanoparticle platforms as drug and biomacromolecule delivery systems. J Control Release 148(1):122–127

Edis Z, Wang J, Waqas MK, Ijaz M, Ijaz M (2021) Nanocarriers-mediated drug delivery systems for anticancer agents: an overview and perspectives. Int J Nanomedicine 1313–1330

Escalante PI, Quinones LA, Contreras HR (2021) Epithelial-mesenchymal transition and MicroRNAs in colorectal cancer chemoresistance to FOLFOX. Pharmaceutics 13(1):75

Esfahani MKM, Alavi SE, Cabot PJ, Islam N, Izake EL (2022) Application of mesoporous silica nanoparticles in cancer therapy and delivery of repurposed anthelmintics for cancer therapy. Pharmaceutics 14(8):1579

Fan Q, Chen Z, Wang C, Liu Z (2018) Toward biomaterials for enhancing immune checkpoint blockade therapy. Adv Func Mater 28(37):1802540

Fan D, Cao Y, Cao M, Wang Y, Cao Y, Gong T (2023) Nanomedicine in cancer therapy. Signal Transduct Target Ther 8(1):293

Fang J (2022) EPR effect-based tumor targeted nanomedicine: a promising approach for controlling cancer. J Pers Med 12(1):95

Far BF, Naimi-Jamal MR, Daneshgar H, Rabiee N (2023) Co-delivery of doxorubicin/sorafenib by DNA-decorated green ZIF-67-based nanocarriers for chemotherapy and hepatocellular carcinoma treatment. Environ Res 225:115589

Fatfat Z, Fatfat M, Gali-Muhtasib H (2022) Micelles as potential drug delivery systems for colorectal cancer treatment. World J Gastroenterol 28(25):2867

Ferreira DDS, Lopes SCDA, Franco MS, Oliveira MC (2013) PH-sensitive liposomes for drug delivery in cancer treatment. Ther Deliv 4(9):1099–1123

Fischer M, Vögtle F (1999) Dendrimers: from design to application—a progress report. Angew Chem Int Ed 38(7):884–905

Foged C, Brodin B, Frokjaer S, Sundblad A (2005) Particle size and surface charge affect particle uptake by human dendritic cells in an in vitro model. Int J Pharm 298(2):315–322

Forssen EA, Tökès ZA (1981) Use of anionic liposomes for the reduction of chronic doxorubicin-induced cardiotoxicity. Proc Natl Acad Sci 78(3):1873–1877

Fulton MD, Najahi-Missaoui W (2023) Liposomes in cancer therapy: how did we start and where are we now. Int J Mol Sci 24(7):6615

Gao A, Hu XL, Saeed M, Chen BF, Li YP, Yu HJ (2019) Overview of recent advances in liposomal nanoparticle-based cancer immunotherapy. Acta Pharmacol Sin 40(9):1129–1137

Gavas S, Quazi S, Karpiński TM (2021) Nanoparticles for cancer therapy: current progress and challenges. Nanoscale Res Lett 16(1):173

Gentile F, Chiappini C, Fine D, Bhavane RC, Peluccio MS, Cheng MMC, Decuzzi P (2008) The effect of shape on the margination dynamics of non-neutrally buoyant particles in two-dimensional shear flows. J Biomech 41(10):2312–2318

Germain M, Caputo F, Metcalfe S, Tosi G, Spring K, Åslund AK, Schmid R (2020) Delivering the power of nanomedicine to patients today. J Control Release 326:164–171

German-Cortés J, Vilar-Hernández M, Rafael D, Abasolo I, Andrade F (2023) Solid lipid nanoparticles: multitasking nano-carriers for cancer treatment. Pharmaceutics 15(3):831

Gharpure KM, Wu SY, Li C, Lopez-Berestein G, Sood AK (2015) Nanotechnology: future of oncotherapy. Clin Cancer Res 21(14):3121–3130

Ghasemiyeh P, Mohammadi-Samani S (2018) Solid lipid nanoparticles and nanostructured lipid carriers as novel drug delivery systems: applications, advantages and disadvantages. Res Pharm Sci 13(4):288–303

Gholami A, Emadi F, Nazem M, Aghayi R, Khalvati B, Amini A, Ghasemi Y (2020) Expression of key apoptotic genes in hepatocellular carcinoma cell line treated with etoposide-loaded graphene oxide. J Drug Deliv Sci Technol 57:101725

Ghosh B, Biswas S (2021) Polymeric micelles in cancer therapy: state of the art. J Control Release 332:127–147

Gi HJ, Chen SN, Hwang JS, Tien C, Kuo MT (1992) Studies of formation and interface of oil-water microemulsion. Chin J Phys 30(5):665–678

Goddard ZR, Marín MJ, Russell DA, Searcey M (2020) Active targeting of gold nanoparticles as cancer therapeutics. Chem Soc Rev 49(23):8774–8789

Gong Z, Cheng C, Sun C, Cheng X (2025) Harnessing engineered extracellular vesicles for enhanced therapeutic efficacy: advancements in cancer immunotherapy. J Exp Clin Cancer Res 44(1):138

Goodman CM, McCusker CD, Yilmaz T, Rotello VM (2004) Toxicity of gold nanoparticles functionalized with cationic and anionic side chains. Bioconjug Chem 15(4):897–900

Gottesman MM, Fojo T, Bates SE (2002) Multidrug resistance in cancer: role of ATP–dependent transporters. Nat Rev Cancer 2(1):48–58

Grewal AK, Salar RK (2024) Chitosan nanoparticle delivery systems: an effective approach to enhancing efficacy and safety of anticancer drugs. Nano TransMed 3:100040

Guo K (2023) Application of polymeric micelles for cancer treatment: a review. Highlights Sci Eng Technol 45:160–169

Gurunathan S, Kang MH, Qasim M, Kim JH (2018) Nanoparticle-mediated combination therapy: two-in-one approach for cancer. Int J Mol Sci 19(10):3264

Han R, Tang K, Hou Y, Yu J, Wang C, Wang Y (2020) Ultralow-intensity near infrared light synchronously activated collaborative chemo/photothermal/photodynamic therapy. Biomater Sci 8(2):607–618

Hanson MC, Crespo MP, Abraham W, Moynihan KD, Szeto GL, Chen SH, Irvine DJ (2015) Nanoparticulate STING agonists are potent lymph node–targeted vaccine adjuvants. J Clin Investig 125(6):2532–2546

Hare JI, Neijzen RW, Anantha M, Dos Santos N, Harasym N, Webb MS, Waterhouse DN (2013) Treatment of colorectal cancer using a combination of liposomal irinotecan (Irinophore C™) and 5-fluorouracil. PLoS ONE 8(4):e62349

He Y, Liang S, Long M, Xu H (2017) Mesoporous silica nanoparticles as potential carriers for enhanced drug solubility of paclitaxel. Mater Sci Eng, C 78:12–17

Hersh DS, Wadajkar AS, Roberts N, Perez JG, Connolly NP, Frenkel V, Winkles JA, Woodworth GF, Kim AJ (2016) Evolving drug delivery strategies to overcome the blood brain barrier. Curr Pharm des 22(9):1177–1193. https://doi.org/10.2174/1381612822666151221150733

Hirsch LR, Stafford RJ, Bankson JA, Sershen SR, Rivera B, Price RE, West JL (2003) Nanoshell-mediated near-infrared thermal therapy of tumors under magnetic resonance guidance. Proc Natl Acad Sci 100(23):13549–13554

Ho TC, Chang CC, Chan HP, Chung TW, Shu CW, Chuang KP, Tyan YC (2022) Hydrogels: properties and applications in biomedicine. Molecules 27(9):2902

Hobbs SK, Monsky WL, Yuan F, Roberts WG, Griffith L, Torchilin VP, Jain RK (1998) Regulation of transport pathways in tumor vessels: role of tumor type and microenvironment. Proc Natl Acad Sci 95(8):4607–4612

Hoffman AS (2008) The origins and evolution of "controlled" drug delivery systems. J Control Release 132(3):153–163

Hofheinz RD, Gnad-Vogt SU, Beyer U, Hochhaus A (2005) Liposomal encapsulated anti-cancer drugs. Anticancer Drugs 16(7):691–707

Holder JE, Ferguson C, Oliveira E, Lodeiro C, Trim CM, Byrne LJ, Wilson CM (2023) The use of nanoparticles for targeted drug delivery in non-small cell lung cancer. Front Oncol 13:1154318

Holohan C, Van Schaeybroeck S, Longley DB, Johnston PG (2013) Cancer drug resistance: an evolving paradigm. Nat Rev Cancer 13(10):714–726

Hu Q, Sun W, Wang C, Gu Z (2016) Recent advances of cocktail chemotherapy by combination drug delivery systems. Adv Drug Deliv Rev 98:19–34

Hu T, Zhai J, Yang Z, Peng J, Wang C, Liu X, Wang H (2025) Myeloid-derived suppressor cells in cancer: mechanistic insights and targeted therapeutic innovations. MedComm 6(6):e70231

Huang X, El-Sayed MA (2010) Gold nanoparticles: optical properties and implementations in cancer diagnosis and photothermal therapy. J Adv Res 1(1):13–28

Huang P, Song H, Zhang Y, Liu J, Zhang J, Wang W, Kong D (2016) Bridging the gap between macroscale drug delivery systems and nanomedicines: a nanoparticle-assembled thermosensitive hydrogel for peritumoral chemotherapy. ACS Appl Mater Interfaces 8(43):29323–29333

Huang J, Yang B, Peng Y, Huang J, Wong SHD, Bian L et al (2021a) Nanomedicine-boosting tumor immunogenicity for enhanced immunotherapy. Adv Funct Mat 31(21):2011171. https://doi.org/10.1002/adfm.202011171

Huang L, Zhao S, Fang F, Xu T, Lan M, Zhang J (2021b) Advances and perspectives in carrier-free nanodrugs for cancer chemo-monotherapy and combination therapy. Biomaterials 268:120557. https://doi.org/10.1016/j.biomaterials.2020.120557

Huang P, Lian D, Ma H, Gao N, Zhao L, Luan P, Zeng X (2021c) New advances in gated materials of mesoporous silica for drug controlled release. Chin Chem Lett 32(12):3696–3704

Huang X, Yin Z, Wu S, Qi X, He Q, Zhang Q, Zhang H (2011) Graphene-based materials: synthesis, characterization, properties, and applications. Small 7(14):1876–1902

Iram S, Zahera M, Khan S, Khan I, Syed A, Ansary AA, Khan MS (2017) Gold nanoconjugates reinforce the potency of conjugated cisplatin and doxorubicin. Colloids Surf, B 160:254–264

Irvine DJ, Hanson MC, Rakhra K, Tokatlian T (2015) Synthetic nanoparticles for vaccines and immunotherapy. Chem Rev 115(19):11109–11146

Israel LL, Galstyan A, Holler E, Ljubimova JY (2020) Magnetic iron oxide nanoparticles for imaging, targeting and treatment of primary and metastatic tumors of the brain. J Control Release 320:45–62

Jabir NR, Tabrez S, Ashraf GM, Shakil S, Damanhouri GA, Kamal MA (2012) Nanotechnology-based approaches in anticancer research. Int J Nanomedicine 4391–4408

Jafernik K, Ładniak A, Blicharska E, Czarnek K, Ekiert H, Wiącek AE, Szopa A (2023) Chitosan-based nanoparticles as effective drug delivery systems—a review. Molecules 28(4):1963

Jang WD, Selim KK, Lee CH, Kang IK (2009) Bioinspired application of dendrimers: from bio-mimicry to biomedical applications. Prog Polym Sci 34(1):1–23

Javadi MM, Soleimani N, Zandi A (2025) Enhancing breast Cancer immunotherapy using gold nanoparticles carrying tumor antigens. Sci Rep 15(1):16747

Jeanbart L, Swartz MA (2015) Engineering opportunities in cancer immunotherapy. Proc Natl Acad Sci 112(47):14467–14472

Ji DK, Ménard-Moyon C, Bianco A (2019) Physically-triggered nanosystems based on two-dimensional materials for cancer theranostics. Adv Drug Deliv Rev 138:211–232

Jia Y, Jiang Y, He Y, Zhang W, Zou J, Magar KT, Boucetta H, Teng C, He W (2023) Approved nanomedicine against diseases. Pharmaceutics 15(3):774. https://doi.org/10.3390/pharmaceutics15030774

Jiang Y, Wang Y, Li Q, Yu C, Chu W (2020) Natural polymer-based stimuli-responsive hydrogels. Curr Med Chem 27(16):2631–2657

Jin W, Dong C, Yang D, Zhang R, Jiang T, Wu D (2020) Nanocarriers of combination tumor physical stimuli-responsive therapies. Curr Drug Deliv 17(7):577–587. https://doi.org/10.2174/1567201817666200525004225

Kam NWS, O'Connell M, Wisdom JA, Dai H (2005) Carbon nanotubes as multifunctional biological transporters and near-infrared agents for selective cancer cell destruction. Proc Natl Acad Sci 102(33):11600–11605

Kappel C, Seidl C, Medina-Montano C, Schinnerer M, Alberg I, Leps C et al (2021) Density of conjugated antibody determines the extent of Fc receptor dependent capture of nanoparticles by liver sinusoidal endothelial cells. ACS Nano 15(9):15191–15209. https://doi.org/10.1021/acsnano.1c05713

Karki N, Tiwari H, Pal M, Chaurasia A, Bal R, Joshi P, Sahoo NG (2018) Functionalized graphene oxides for drug loading, release and delivery of poorly water soluble anticancer drug: a comparative study. Colloids Surf, B 169:265–272

Karthika C, Sureshkumar R, Zehravi M, Akter R, Ali F, Ramproshad S, Mondal B, Tagde P, Ahmed Z, Khan FS, Rahman MH, Cavalu S (2022) Multidrug resistance of cancer cells and the vital role of P-glycoprotein. Life (Basel, Switzerland) 12(6):897. https://doi.org/10.3390/life12060897

Karunanidhi P, Ekambaram P, Abdul A, Sathali H, Priyanka K (2012) Solid lipid nanoparticles: a review. Sci Revs Chem Commun 2:80–102

Karve S, Werner ME, Sukumar R, Cummings ND, Copp JA, Wang EC, Wang AZ (2012) Revival of the abandoned therapeutic wortmannin by nanoparticle drug delivery. Proc Natl Acad Sci 109(21):8230–8235

Kass LE, Nguyen J (2022) Nanocarrier-hydrogel composite delivery systems for precision drug release. Wiley Interdiscip Rev: Nanomedicine Nanobiotechnology 14(2):e1756

Kedar U, Phutane P, Shidhaye S, Kadam V (2010) Advances in polymeric micelles for drug delivery and tumor targeting. Nanomedicine: Nanotechnol, Biol Med 6(6):714–729

Kemp JA, Kwon YJ (2021) Cancer nanotechnology: current status and perspectives. Nano Converg 8(1):34

Khan AK, Rashid R, Murtaza G, Zahra AJTR (2014) Gold nanoparticles: synthesis and applications in drug delivery. Trop J Pharm Res 13(7):1169–1177

Kim S, Shi Y, Kim JY, Park K, Cheng JX (2010) Overcoming the barriers in micellar drug delivery: loading efficiency, in vivo stability, and micelle–cell interaction. Expert Opin Drug Deliv 7(1):49–62

Kim KR, Lee AS, Kim SM, Heo HR, Kim CS (2023) Virus-like nanoparticles as a theranostic platform for cancer. Front Bioeng Biotechnol 10:1106767

Kitchens KM, El-Sayed ME, Ghandehari H (2005) Transepithelial and endothelial transport of poly (amidoamine) dendrimers. Adv Drug Deliv Rev 57(15):2163–2176

Klemm D, Cranston ED, Fischer D, Gama M, Kedzior SA, Kralisch D, Rauchfuß F (2018) Nanocellulose as a natural source for groundbreaking applications in materials science: today's state. Mater Today 21(7):720–748

Kobayashi H, Kawamoto S, Jo SK, Bryant HL, Brechbiel MW, Star RA (2003) Macromolecular MRI contrast agents with small dendrimers: pharmacokinetic differences between sizes and cores. Bioconjug Chem 14(2):388–394

Kranz LM, Diken M, Haas H, Kreiter S, Loquai C, Reuter KC, Sahin U (2016) Systemic RNA delivery to dendritic cells exploits antiviral defence for cancer immunotherapy. Nature 534(7607):396–401
Kshirsagar NA, Pandya SK, Kirodian BG, Sanath S (2005) Liposomal drug delivery system from laboratory to clinic. J Postgrad Med 51(Suppl 1):S5–S15
Kulhari H, Kulhari DP, Prajapati SK, Chauhan AS (2013) Pharmacokinetic and pharmacodynamic studies of poly (amidoamine) dendrimer based simvastatin oral formulations for the treatment of hypercholesterolemia. Mol Pharm 10(7):2528–2533
Kulhari H, Pooja D, Shrivastava S, Kuncha M, Naidu VGM, Bansal V, Adams DJ (2016) Trastuzumab-grafted PAMAM dendrimers for the selective delivery of anticancer drugs to HER2-positive breast cancer. Sci Rep 6(1):23179
Kumar C, Kohli S, Chiliveru S, Bapsy PP, Jain M, Suresh Attili VS, Sharan B (2017) A retrospective analysis comparing APCEDEN® dendritic cell immunotherapy with best supportive care in refractory cancer. Immunotherapy 9(11):889–897
Kumar C, Bagga J, Chiliveru S, Kohli S, Bharadwaj A, Jain M, Sharan B (2019) Substantial remission of prostate adenocarcinoma with dendritic cell therapy APCEDEN® in combination with chemotherapy. Futur Sci OA 5(10):FSO435
Kurul F, Turkmen H, Cetin AE, Topkaya SN (2025) Nanomedicine: How nanomaterials are transforming drug delivery, bio-imaging, and diagnosis. Next Nanotechnol 7:100129
Kushwaha SKS, Ghoshal S, Rai AK, Singh S (2013) Carbon nanotubes as a novel drug delivery system for anticancer therapy: a review. Braz J Pharm Sci 49:629–643
Lakshmi AP, Kumar MA, Hao S, Yannan Y, Manasi J, Jun Z, Chengzhong Y (2017) Asymmetric silica nanoparticles with tunable head–tail structures enhance hemocompatibility and maturation of immune cells
Lamprecht C, Liashkovich I, Neves V, Danzberger J, Heister E, Rangl M, Ebner A (2009) AFM imaging of functionalized carbon nanotubes on biological membranes. Nanotechnology 20(43):434001
Lancet JE, Cortes JE, Hogge DE, Tallman M, Kovacsovics T, Damon LE, Feldman EJ (2010) Phase 2B randomized study of CPX-351 vs. cytarabine (CYT)+ daunorubicin (DNR)(7+ 3 regimen) in newly diagnosed AML patients aged 60–75. Blood 116(21):655
Langer R, Peppas NA (2003) Advances in biomaterials, drug delivery, and bionanotechnology. AIChE J 49(12):2990–3006
Lassalle V, Ferreira ML (2007) PLA nano-and microparticles for drug delivery: an overview of the methods of preparation. Macromol Biosci 7(6):767–783
Lavan DA, McGuire T, Langer R (2003) Small-scale systems for in vivo drug delivery. Nat Biotechnol 21(10):1184–1191
Lee S, Margolin K (2011) Cytokines in cancer immunotherapy. Cancers 3(4):3856–3893
Lee KY, Wang Y, Nie S (2015) In vitro study of a pH-sensitive multifunctional doxorubicin–gold nanoparticle system: therapeutic effect and surface enhanced Raman scattering. RSC Adv 5(81):65651–65659
Leiva MC, Ortiz R, Contreras-Cáceres R, Perazzoli G, Mayevych I, López-Romero JM, Prados J (2017) Tripalmitin nanoparticle formulations significantly enhance paclitaxel antitumor activity against breast and lung cancer cells in vitro. Sci Rep 7(1):13506
Li Y, Chen H, Prins C (2016) Adaptive large neighborhood search for the pickup and delivery problem with time windows, profits, and reserved requests. Eur J Oper Res 252(1):27–38
Li M, Luo Z, Zhao Y (2018) Recent advancements in 2D nanomaterials for cancer therapy. Sci China Chem 61(10):1214–1226
Li Z, Song N, Yang YW (2019) Stimuli-responsive drug-delivery systems based on supramolecular nanovalves. Matter 1(2):345–368
Li Z, Yu Y, Zeng W, Ding F, Zhang D, Cheng W, Gao N (2022) Mussel-inspired ligand clicking and ion coordination on 2D black phosphorus for cancer multimodal imaging and therapy. Small 18(26):2201803

Liistro F, Stankovic G, Di Mario C, Takagi T, Chieffo A, Moshiri S, Colombo A (2002) First clinical experience with a paclitaxel derivate–eluting polymer stent system implantation for in-stent restenosis: immediate and long-term clinical and angiographic outcome. Circulation 105(16):1883–1886

Ling X, Wang H, Huang S, Xia F, Dresselhaus MS (2015) The renaissance of black phosphorus. Proc Natl Acad Sci 112(15):4523–4530

Liu W, Dong A, Wang B, Zhang H (2021) Current advances in black phosphorus-based drug delivery systems for cancer therapy. Adv Sci 8(5):2003033

Liu B, Zhou H, Tan L, Siu KTH, Guan XY (2024a) Exploring treatment options in cancer: tumor treatment strategies. Signal Transduct Target Ther 9(1):175

Liu Y, Zhang D, Zhang Z, Liang X, Yang X, Ding N, Li C (2024b) Multifunctional nanoparticles inhibit tumor and tumor-associated macrophages for triple-negative breast cancer therapy. J Colloid Interface Sci 657:598–610

Liu Y, Zhang Y, Li H, Hu TY (2025) Recent advances in the bench-to-bedside translation of cancer nanomedicines. Acta Pharmaceutica Sinica B 15(1):97–122

Loira-Pastoriza C, Todoroff J, Vanbever R (2014) Delivery strategies for sustained drug release in the lungs. Adv Drug Deliv Rev 75:81–91

Lu J, Liong M, Zink JI, Tamanoi F (2007) Mesoporous silica nanoparticles as a delivery system for hydrophobic anticancer drugs. Small 3(8):1341–1346

Luo M, Wang H, Wang Z, Cai H, Lu Z, Li Y, Gao J (2017) A STING-activating nanovaccine for cancer immunotherapy. Nat Nanotechnol 12(7):648–654

Ma L, Kohli M, Smith A (2013) Nanoparticles for combination drug therapy. ACS Nano 7(11):9518–9525. https://doi.org/10.1021/nn405674m

Maeda H, Wu J, Sawa T, Matsumura Y, Hori K (2000) Tumor vascular permeability and the EPR effect in macromolecular therapeutics: a review. J Control Release 65(1–2):271–284

Maiti D, Tong X, Mou X, Yang K (2019) Carbon-based nanomaterials for biomedical applications: a recent study. Front Pharmacol 9:1401

Malam Y, Loizidou M, Seifalian AM (2009) Liposomes and nanoparticles: nanosized vehicles for drug delivery in cancer. Trends Pharmacol Sci 30(11):592–599

Malekimusavi H, Ghaemi A, Masoudi G, Chogan F, Rashedi H, Yazdian F, Faal Hamedani N (2019) Graphene oxide-l-arginine nanogel: a pH-sensitive fluorouracil nanocarrier. Biotechnol Appl Biochem 66(5):772–780

Manolova V, Flace A, Bauer M, Schwarz K, Saudan P, Bachmann MF (2008) Nanoparticles target distinct dendritic cell populations according to their size. Eur J Immunol 38(5):1404–1413

Mason TG, Wilking JN, Meleson K, Chang CB, Graves SM (2006) Nanoemulsions: formation, structure, and physical properties. J Phys: Condens Matter 18(41):R635

Mastelic-Gavillet B, Balint K, Boudousquie C, Gannon PO, Kandalaft LE (2019) Personalized dendritic cell vaccines—recent breakthroughs and encouraging clinical results. Front Immunol 10:766

Matsumura Y, Maeda H (1986) A new concept for macromolecular therapeutics in cancer chemotherapy: mechanism of tumoritropic accumulation of proteins and the antitumor agent smancs. Cancer Res 46(12_Part_1):6387–6392

Mekuye B, Abera B (2023) Nanomaterials: an overview of synthesis, classification, characterization, and applications. Nano Select 4(8):486–501

Merisko-Liversidge EM, Liversidge GG (2008) Drug nanoparticles: formulating poorly water-soluble compounds. Toxicol Pathol 36(1):43–48. https://doi.org/10.1177/0192623307310946

Miele E, Spinelli GP, Miele E, Tomao F, Tomao S (2009) Albumin-bound formulation of paclitaxel (Abraxane® ABI-007) in the treatment of breast cancer. Int J Nanomedicine 99–105

Min J, Jung H, Shin HH, Cho G, Cho H, Kang S (2013) Implementation of p22 viral capsids as intravascular magnetic resonance T 1 contrast conjugates via site-selective attachment of Gd (III)-chelating agents. Biomacromol 14(7):2332–2339

Mishra V, Bansal KK, Verma A, Yadav N, Thakur S, Sudhakar K, Rosenholm JM (2018) Solid lipid nanoparticles: emerging colloidal nano drug delivery systems. Pharmaceutics 10(4):191

Missaoui WN, Arnold RD, Cummings BS (2018) Toxicological status of nanoparticles: what we know and what we don't know. Chem Biol Interact 295:1–12
Mitran RA, Berger D, Munteanu C, Matei C (2015) Evaluation of different mesoporous silica supports for energy storage in shape-stabilized phase change materials with dual thermal responses. J Phys Chem C 119(27):15177–15184
Momekova D, Rangelov S, Yanev S, Nikolova E, Konstantinov S, Romberg B, Lambov N (2007) Long-circulating, pH-sensitive liposomes sterically stabilized by copolymers bearing short blocks of lipid-mimetic units. Eur J Pharm Sci 32(4–5):308–317
Moon JJ, Huang B, Irvine DJ (2012) Engineering nano-and microparticles to tune immunity. Adv Mater 24(28):3724–3746
Mozafari MR (2005) Liposomes: an overview of manufacturing techniques. Cell Mol Biol Lett 10(4):711
Mueller M, Reichardt W, Koerner J, Groettrup M (2012) Coencapsulation of tumor lysate and CpG-ODN in PLGA-microspheres enables successful immunotherapy of prostate carcinoma in TRAMP mice. J Control Release 162(1):159–166
Mukwaya G, Forssen EA, Schmidt P, Ross M (1998) DaunoXome®(Liposomal Daunorubicin) for first-line treatment of advanced, HIV-related Kaposi's Sarcoma. In: Long circulating liposomes: old drugs, new therapeutics. Berlin, Heidelberg: Springer Berlin Heidelberg, pp 147–163
Muthu MS, Feng S-S (2009) Pharmaceutical stability aspects of nanomedicines. Nanomedicine 4(8):857–860. https://doi.org/10.2217/nnm.09.75
Muthu MS, Leong DT, Mei L, Feng SS (2014) Nanotheranostics—application and further development of nanomedicine strategies for advanced theranostics. Theranostics 4(6):660–677. https://doi.org/10.7150/thno.8698
National Cancer Institute (NCI) (2024) Cancer Statistics. https://www.cancer.gov/about-cancer/understanding/statistics#:~:text=Cancer%20is%20among%20the%20leading,ranks%20low%20on%20these%20measures (Accessed on 25 Feb 2025)
Noh YW, Kim SY, Kim JE, Kim S, Ryu J, Kim I, Lim YT (2017) Multifaceted immunomodulatory nanoliposomes: reshaping tumors into vaccines for enhanced cancer immunotherapy. Adv Func Mater 27(8):1605398
Okada H, Doken Y, Ogawa Y, Toguchi H (1994) Preparation of three-month depot injectable microspheres of leuprorelin acetate using biodegradable polymers. Pharm Res 11:1143–1147
Ou W, Thapa RK, Jiang L, Soe ZC, Gautam M, Chang JH, Kim JO (2018) Regulatory T cell-targeted hybrid nanoparticles combined with immuno-checkpoint blockage for cancer immunotherapy. J Control Release 281:84–96
Padera TP, Stoll BR, Tooredman JB, Capen D, Tomaso ED, Jain RK (2004) Pathology: cancer cells compress intratumour vessels. Nature 427(6976):695–695
Pan Y, Neuss S, Leifert A, Fischler M, Wen F, Simon U, Jahnen-Dechent W (2007) Size-dependent cytotoxicity of gold nanoparticles. Small 3(11):1941–1949
Pande V, Kothawade S, Kuskar S, Bole S, Chakole D (2023) Fabrication of mesoporous silica nanoparticles and its applications in drug delivery. In: Nanofabrication techniques-principles, processes and applications. IntechOpen
Pang HH, Chen PY, Wei KC, Huang CW, Shiue YL, Huang CY, Yang HW (2019) Convection-enhanced delivery of a virus-like nanotherapeutic agent with dual-modal imaging for besiegement and eradication of brain tumors. Theranostics 9(6):1752
Papahadjopoulos D, Nir S, Ohki S (1972) Permeability properties of phospholipid membranes: effect of cholesterol and temperature. Biochim Biophys Acta (BBA)-Biomembr 266(3);561–583
Pareek A, Kumar D, Pareek A, Gupta MM (2025) Advancing cancer therapy with quantum dots and other nanostructures: a review of drug delivery innovations, applications, and challenges. Cancers 17(5):878
Parekh HS (2007) The advance of dendrimers-a versatile targeting platform for gene/drug delivery. Curr Pharm des 13(27):2837–2850
Park J, Babensee JE (2012) Differential functional effects of biomaterials on dendritic cell maturation. Acta Biomater 8(10):3606–3617

Park JW, Hong K, Kirpotin DB, Papahadjopoulos D, Benz CC (1997) Immunoliposomes for cancer treatment. Adv Pharmacol 40:399–435
Park W, Heo YJ, Han DK (2018) New opportunities for nanoparticles in cancer immunotherapy. Biomater Res 22(1):24
Patel NR, Pattni BS, Abouzeid AH, Torchilin VP (2013) Nanopreparations to overcome multidrug resistance in cancer. Adv Drug Deliv Rev 65(13–14):1748–1762
Patra JK, Das G, Fraceto LF, Campos EVR, Rodriguez-Torres MDP, Acosta-Torres LS, Shin HS (2018) Nano based drug delivery systems: recent developments and future prospects. J Nanobiotechnology 16(1):71
Pelicano H, Martin DS, Xu RA, Huang P (2006) Glycolysis inhibition for anticancer treatment. Oncogene 25(34):4633–4646
Phi LTH, Sari IN, Yang YG, Lee SH, Jun N, Kim KS, Lee YK, Kwon HY (2018) Cancer stem cells (CSCs) in drug resistance and their therapeutic implications in cancer treatment. Stem Cells Int 2018:5416923. https://doi.org/10.1155/2018/5416923
Pooja D, Reddy TS, Kulhari H, Kadari A, Adams DJ, Bansal V, Sistla R (2020) N-acetyl-d-glucosamine-conjugated PAMAM dendrimers as dual receptor-targeting nanocarriers for anticancer drug delivery. Eur J Pharm Biopharm 154:377–386
Pucci C, Martinelli C, Ciofani G (2019) Innovative approaches for cancer treatment: Current perspectives and new challenges. ecancermedicalscience 13:961
Purohit G, Sakthivel T, Florence AT (2001) Interaction of cationic partial dendrimers with charged and neutral liposomes. Int J Pharm 214(1–2):71–76
Quader S, Kataoka K (2017) Nanomaterial-enabled cancer therapy. Mol Ther 25(7):1501–1513. https://doi.org/10.1016/j.ymthe.2017.04.026
Ranganathan R, Madanmohan S, Kesavan A, Baskar G, Krishnamoorthy YR, Santosham R, Venkatraman G (2012) Nanomedicine: towards development of patient-friendly drug-delivery systems for oncological applications. Int J Nanomedicine 1043–1060
Rastegari E, Hsiao YJ, Lai WY, Lai YH, Yang TC, Chen SJ, Chien Y (2021) An update on mesoporous silica nanoparticle applications in nanomedicine. Pharmaceutics 13(7):1067
Reddy ST, Rehor A, Schmoekel HG, Hubbell JA, Swartz MA (2006) In vivo targeting of dendritic cells in lymph nodes with poly (propylene sulfide) nanoparticles. J Control Release 112(1):26–34
Reddy ST, Van Der Vlies AJ, Simeoni E, Angeli V, Randolph GJ, O'Neil CP, Hubbell JA (2007) Exploiting lymphatic transport and complement activation in nanoparticle vaccines. Nat Biotechnol 25(10):1159–1164
Rezvantalab S, Drude NI, Moraveji MK, Güvener N, Koons EK, Shi Y, Kiessling F (2018) PLGA-based nanoparticles in cancer treatment. Front Pharmacol 9:1260
Rianasari I, de Jong MP, Huskens J, van der Wiel WG (2013) Covalent coupling of nanoparticles with low-density functional ligands to surfaces via click chemistry. Int J Mol Sci 14(2):3705–3717. https://doi.org/10.3390/ijms14023705
Riley RS, June CH, Langer R, Mitchell MJ (2019) Delivery technologies for cancer immunotherapy. Nat Rev Drug Discov 18(3):175–196
Rodríguez F, Caruana P, De la Fuente N, Español P, Gámez M, Balart J, Llurba E, Rovira R, Ruiz R, Martín-Lorente C, Corchero JL, Céspedes MV (2022) Nano-based approved pharmaceuticals for cancer treatment: present and future challenges. Biomolecules 12(6):784. https://doi.org/10.3390/biom12060784
Sábio RM, Meneguin AB, Ribeiro TC, Silva RR, Chorilli M (2019) New insights towards mesoporous silica nanoparticles as a technological platform for chemotherapeutic drugs delivery. Int J Pharm 564:379–409
Sacchetti C, Rapini N, Magrini A, Cirelli E, Bellucci S, Mattei M, Bottini M (2013) In vivo targeting of intratumor regulatory T cells using PEG-modified single-walled carbon nanotubes. Bioconjug Chem 24(6):852–858
Saha RN, Vasanthakumar S, Bende G, Snehalatha M (2010) Nanoparticulate drug delivery systems for cancer chemotherapy. Mol Membr Biol 27(7):215–231

Sahoo NG, Bao H, Pan Y, Pal M, Kakran M, Cheng HKF, Tan LP (2011) Functionalized carbon nanomaterials as nanocarriers for loading and delivery of a poorly water-soluble anticancer drug: a comparative study. Chem Commun 47(18):5235–5237
Salahuddin N, Elbarbary AA, Alkabes HA (2017) Antibacterial and antitumor activities of 3-amino-phenyl-4 (3 H)-quinazolinone/polypyrrole chitosan core shell nanoparticles. Polym Bull 74(5):1775–1790
Santi M, Maccari G, Mereghetti P, Voliani V, Rocchiccioli S, Ucciferri N, Signore G (2017) Rational design of a transferrin-binding peptide sequence tailored to targeted nanoparticle internalization. Bioconjug Chem 28(2):471–480
Satapathy MK, Yen TL, Jan JS, Tang RD, Wang JY, Taliyan R, Yang CH (2021) Solid lipid nanoparticles (SLNs): an advanced drug delivery system targeting brain through BBB. Pharmaceutics 13(8):1183
Mahanta SS, Kumar AK, Maiti SP (2018) Controlled drug delivery vehicles for cancer treatment and their performance. Signal Transduct Target Ther 3(1):7. https://doi.org/10.1038/s41392-017-0004-3.
Sercombe L, Veerati T, Moheimani F, Wu SY, Sood AK, Hua S (2015) Advances and challenges of liposome assisted drug delivery. Front Pharmacol 6:286
Shaji SG, Patel P, Mamani UF, Guo Y, Koirala S, Lin CY, Cheng K (2024) Delivery of a STING agonist using lipid nanoparticles inhibits pancreatic cancer growth. Int J Nanomedicine 8769–8778
Shanmuganathan R, Edison TNJI, LewisOscar F, Kumar P, Shanmugam S, Pugazhendhi A (2019) Chitosan nanopolymers: an overview of drug delivery against cancer. Int J Biol Macromol 130:727–736
Shao K, Singha S, Clemente-Casares X, Tsai S, Yang Y, Santamaria P (2015) Nanoparticle-based immunotherapy for cancer. ACS Nano 9(1):16–30
Sharma VK, Liu X, Oyarzún DA, Abdel-Azeem AM, Atanasov AG, Hesham AEL, Singh BN (2022) Microbial polysaccharides: an emerging family of natural biomaterials for cancer therapy and diagnostics. In: Seminars in cancer biology, vol 86. Academic Press, pp 706–731
Sharma V, Mukhopadhyay CD (2024) Exosome as drug delivery system: current advancements. Extracell Vesicle 3:100032
Shen X, Pan D, Gong Q, Gu Z, Luo K (2024) Enhancing drug penetration in solid tumors via nanomedicine: evaluation models, strategies and perspectives. Bioact Mater 32:445–472
Sherje AP, Jadhav M, Dravyakar BR, Kadam D (2018) Dendrimers: a versatile nanocarrier for drug delivery and targeting. Int J Pharm 548(1):707–720
Shima F, Akagi T, Uto T, Akashi M (2013) Manipulating the antigen-specific immune response by the hydrophobicity of amphiphilic poly (γ-glutamic acid) nanoparticles. Biomaterials 34(37):9709–9716
Shrestha B, Tang L, Romero G (2019) Nanoparticles-mediated combination therapies for cancer treatment. Adv Ther 2(11):1900076
Shrestha B, Wang L, Brey EM, Uribe GR, Tang L (2021) Smart nanoparticles for chemo-based combinational therapy. Pharmaceutics 13(6):853
Siemer S, Bauer TA, Scholz P, Breder C, Fenaroli F, Harms G et al (2021) Targeting cancer chemotherapy resistance by precision medicine-driven nanoparticle-formulated cisplatin. ACS Nano 15(11):18541–18556. https://doi.org/10.1021/acsnano.1c08632
Silva CO, Pinho JO, Lopes JM, Almeida AJ, Gaspar MM, Reis C (2019) Current trends in cancer nanotheranostics: metallic, polymeric, and lipid-based systems. Pharmaceutics 11(1):22
Singh K, Singhal S, Pahwa S, Sethi VA, Sharma S, Singh P, Sagadevan S (2024) Nanomedicine and drug delivery: a comprehensive review of applications and challenges. Nano-Struct Nano-Objects 40:101403
Sun LQ, Li MJ, Sun K, Yu SH, Wang RS, Xie HM (2012) Electrochemical activity of black phosphorus as an anode material for lithium-ion batteries. J Phys Chem C 116(28):14772–14779

Sung H, Ferlay J, Siegel RL, Laversanne M, Soerjomataram I, Jemal A, Bray F (2021) Global cancer statistics 2020: GLOBOCAN estimates of incidence and mortality worldwide for 36 cancers in 185 countries. CA: Cancer J Clin 71(3):209–249. https://doi.org/10.3322/caac.21660

Susewind M, Schilmann AM, Heim J, Henkel A, Link T, Fischer K, Tremel W (2015) Silica-coated Au@ ZnO Janus particles and their stability in epithelial cells. J Mater Chem B 3(9):1813–1822

Tan B, Huang L, Wu Y, Liao J (2021) Advances and trends of hydrogel therapy platform in localized tumor treatment: a review. J Biomed Mater Res, Part A 109(4):404–425

Teow HM, Zhou Z, Najlah M, Yusof SR, Abbott NJ, D'Emanuele A (2013) Delivery of paclitaxel across cellular barriers using a dendrimer-based nanocarrier. Int J Pharm 441(1–2):701–711

Thakur N, Kumar A, Thakur N (2023) *Tinospora cordifolia* and polyvinylpyrrolidone encapsulated dual doped Ni-Cu $TiO_2$ emerging nanocatalyst for the removal of organic dyes from wastewater and its free radical assay activity. Hybrid Advances 4:100086

Thakur N, Arya V, Kumar A, Thakur N (2024) Catalytic efficiency of encapsulated coupled doped Zn-Ni $TiO_2$ nanocatalyst for the removal of organic pollutants and free radical assay. Int J Environ Anal Chem 104(20):9754–9771

Timmermans SB, Ramezani A, Montalvo T, Nguyen M, van der Schoot P, van Hest JC, Zandi R (2022) The dynamics of viruslike capsid assembly and disassembly. J Am Chem Soc 144(28):12608–12612

Toy R, Hayden E, Shoup C, Baskaran H, Karathanasis E (2011) Effect of particle size, density and shape on margination of nanoparticles in microcirculation. Nanotechnology 22(11):115101

Tran TH, Tran TTP, Nguyen HT, Dai Phung C, Jeong JH, Stenzel MH, Kim JO (2018) Nanoparticles for dendritic cell-based immunotherapy. Int J Pharm 542(1–2):253–265

Trifonova EA, Nikitin NA, Arkhipenko MV, Donchenko EK, Atabekov JG, Karpova OV (2017) Comparative study of thermal remodeling of viruses with icosahedral and helical symmetry. Mosc Univ Biol Sci Bull 72(4):179–183

Urruticoechea A, Alemany R, Balart J, Villanueva A, Vinals F, Capella G (2010) Recent advances in cancer therapy: an overview. Curr Pharm des 16(1):3–10

Vahed SZ, Salehi R, Davaran S, Sharifi S (2017) Liposome-based drug co-delivery systems in cancer cells. Mater Sci Eng, C 71:1327–1341

Van Der Poll DG, Kieler-Ferguson HM, Floyd WC, Guillaudeu SJ, Jerger K, Szoka FC, Fréchet JM (2010) Design, synthesis, and biological evaluation of a robust, biodegradable dendrimer. Bioconjug Chem 21(4):764–773

Wang X, Cheng L (2019) Multifunctional two-dimensional nanocomposites for photothermal-based combined cancer therapy. Nanoscale 11(34):15685–15708

Wang F, Wang YC, Dou S, Xiong MH, Sun TM, Wang J (2011) Doxorubicin-tethered responsive gold nanoparticles facilitate intracellular drug delivery for overcoming multidrug resistance in cancer cells. ACS Nano 5(5):3679–3692

Wang X, Cai X, Hu J, Shao N, Wang F, Zhang Q, Cheng Y (2013) Glutathione-triggered "off–on" release of anticancer drugs from dendrimer-encapsulated gold nanoparticles. J Am Chem Soc 135(26):9805–9810

Wang G, Jia T, Xu X, Chang L, Zhang R, Fu Y, Li J (2016) Novel miR-122 delivery system based on MS2 virus like particle surface displaying cell-penetrating peptide TAT for hepatocellular carcinoma. Oncotarget 7(37):59402

Wang M, Meng Y, Zhu H, Hu Y, Xu CP, Chao X, Pan C (2021) Green synthesized gold nanoparticles using Viola betonicifolia leaves extract: characterization, antimicrobial, antioxidant, and cytobiocompatible activities. Int J Nanomedicine 7319–7337

Weber S, Zimmer A, Pardeike J (2014) Solid lipid nanoparticles (SLN) and nanostructured lipid carriers (NLC) for pulmonary application: a review of the state of the art. Eur J Pharm Biopharm 86(1):7–22

Wei L, Li G, Lu T, Wei Y, Nong Z, Wei M, Li X (2021) Functionalized graphene oxide as drug delivery systems for platinum anticancer drugs. J Pharm Sci 110(11):3631–3638

Wiener EC, Auteri FP, Chen JW, Brechbiel MW, Gansow OA, Schneider DS, Lauterbur PC (1996) Molecular dynamics of ion– chelate complexes attached to dendrimers. J Am Chem Soc 118(33):7774–7782
Wilhelm S, Tavares AJ, Dai Q, Ohta S, Audet J, Dvorak HF, Chan WC (2016) Analysis of nanoparticle delivery to tumours. Nat Rev Mater 1(5):1–12
Wilson JT, Keller S, Manganiello MJ, Cheng C, Lee CC, Opara C, Stayton PS (2013) pH-Responsive nanoparticle vaccines for dual-delivery of antigens and immunostimulatory oligonucleotides. ACS Nano 7(5):3912–3925
Wu J (2021) The enhanced permeability and retention (EPR) effect: the significance of the concept and methods to enhance its application. J Pers Med 11(8):771. https://doi.org/10.3390/jpm11080771
Xie Z, Wang D, Fan T, Xing C, Li Z, Tao W, Zhang H (2018) Black phosphorus analogue tin sulfide nanosheets: synthesis and application as near-infrared photothermal agents and drug delivery platforms for cancer therapy. J Mater Chem B 6(29):4747–4755
Yang P, Gai S, Lin J (2012) Functionalized mesoporous silica materials for controlled drug delivery. Chem Soc Rev 41(9):3679–3698
Yang G, Liu Z, Li Y, Hou Y, Fei X, Su C, Guo Z (2017) Facile synthesis of black phosphorus–Au nanocomposites for enhanced photothermal cancer therapy and surface-enhanced Raman scattering analysis. Biomater Sci 5(10):2048–2055
Yanikoglu R, Karakas CY, Ciftci F, Insel MA, Karavelioglu Z, Varol R, Ustundag CB (2024) Development of graphene oxide-based anticancer drug combination functionalized with folic acid as nanocarrier for targeted delivery of methotrexate. Pharmaceutics 16(6):837
Yetisgin AA, Cetinel S, Zuvin M, Kosar A, Kutlu O (2020) Therapeutic nanoparticles and their targeted delivery applications. Mol (Basel, Switz) 25(9):2193. https://doi.org/10.3390/molecules25092193
Yih TC, Al-Fandi M (2006) Engineered nanoparticles as precise drug delivery systems. J Cell Biochem 97(6):1184–1190
Yiyun C, Tongwen X (2005) Dendrimers as potential drug carriers. Part I. Solubilization of non-steroidal anti-inflammatory drugs in the presence of polyamidoamine dendrimers. Eur J Med Chem 40(11):1188–1192
Yokoyama M (2011) Clinical applications of polymeric micelle carrier systems in chemotherapy and image diagnosis of solid tumors. J Exp Clin Med 3(4):151–158
Yoon HY, Selvan ST, Yang Y, Kim MJ, Yi DK, Kwon IC, Kim K (2018) Engineering nanoparticle strategies for effective cancer immunotherapy. Biomaterials 178:597–607
Yuan H, Guo H, Luan X, He M, Li F, Burnett J, Sun D (2020) Albumin nanoparticle of paclitaxel (Abraxane) decreases while taxol increases breast cancer stem cells in treatment of triple negative breast cancer. Mol Pharm 17(7):2275–2286
Zang X, Zhao X, Hu H, Qiao M, Deng Y, Chen D (2017) Nanoparticles for tumor immunotherapy. Eur J Pharm Biopharm 115:243–256
Zhang Y, Liu Y, Liu J, Guo P, Heng L (2017) Super water absorbency OMMT/PAA hydrogel materials with excellent mechanical properties. RSC Adv 7(24):14504–14510
Zhang H, De Smedt SC, Remaut K (2018a) Fluorescence correlation spectroscopy to find the critical balance between extracellular association and intracellular dissociation of mRNA complexes. Acta Biomater 75:358–370. https://doi.org/10.1016/j.actbio.2018.05.016
Zhang H, De Smedt SC, Remaut K (2018b) Fluorescence correlation spectroscopy to find the critical balance between extracellular association and intracellular dissociation of mRNA complexes. Acta Biomater 75:358–370
Zhang D, Liu L, Wang J, Zhang H, Zhang Z, Xing G, Liu M (2022) Drug-loaded PEG-PLGA nanoparticles for cancer treatment. Front Pharmacol 13:990505
Zhang L, Xia J, Zhao Q, Liu L, Zhang Z (2010) Functional graphene oxide as a nanocarrier for controlled loading and targeted delivery of mixed anticancer drugs. Small 6(4):537–544

Zhao P, Li L, Zhou S, Qiu L, Qian Z, Liu X, Zhang H (2018) TPGS functionalized mesoporous silica nanoparticles for anticancer drug delivery to overcome multidrug resistance. Mater Sci Eng, C 84:108–117

Zhao Q, Cheng N, Sun X, Yan L, Li W (2023) The application of nanomedicine in clinical settings. Front Bioeng Biotechnol 11:1219054

Zhou W, Cui H, Ying L, Yu XF (2018) Enhanced cytosolic delivery and release of CRISPR/Cas9 by black phosphorus nanosheets for genome editing. Angew Chem 130(32):10425–10429

Zhu G, Zhang F, Ni Q, Niu G, Chen X (2017) Efficient nanovaccine delivery in cancer immunotherapy. ACS Nano 11(3):2387–2392

# Chapter 12
# Nanoecotoxicology: Illusion or Reality

**Acharya Balkrishna, Nidhi Sharma, Deepika Srivastava, Amar Jeet, Hemant Sharma, Amita Kumari, Vedpriya Arya, Sonia Malik, and Ashwani Kumar**

## 12.1 Introduction

The rapid development and widespread use of engineered nanoparticles (ENPs) in various sectors, such as medicine, electronics, and environmental remediation, have led to significant advances in science and technology. However, their introduction into the environment raises serious concerns about their potential ecological and environmental impacts. ENPs are particles with a size range of 1–100 nm in diameter.

A. Balkrishna · N. Sharma · D. Srivastava · A. Jeet · H. Sharma · A. Kumari · V. Arya · A. Kumar (✉)
Patanjali Herbal Research Department, Patanjali Research Foundation, Haridwar, India
e-mail: dr.ashwanikumar@patanjali.res.in

A. Balkrishna
e-mail: acharya.balkrishnapri@patanjali.res.in

N. Sharma
e-mail: nidhi.sharma@patanjali.res.in

D. Srivastava
e-mail: deepika.srivastava@patanjali.res.in

A. Jeet
e-mail: amarjeet@patanjali.res.in

H. Sharma
e-mail: hemant.sharma@patanjali.res.in

A. Kumari
e-mail: amita.kumari@patanjali.res.in

V. Arya
e-mail: vedpriya.arya@patanjali.res.in

S. Malik
Department of Biotechnology, Baba Farid College, Bathinda, India
e-mail: soniamalik@babafaridgroup.edu.in



A. Balkrishna et al. (eds.), *Nanotechnology in Environmental Science and Healthcare*,
https://doi.org/10.1007/978-981-95-6714-0_12

These nanoparticles (NPs) are manufactured through controlled processes that result in low impurity levels and consistent morphology, such as spheres, tubes, and plates (Sukhanova et al. 2018). Due to their small size and large surface area, NPs possess unique physical and chemical properties that make them highly reactive and enable them to interact with biological systems in ways that larger particles cannot (Medina et al. 2007; Sharma et al. 2025; Kumar et al. 2025). In recent years, ENPs have been incorporated into various products, including cosmetics, sunscreens, toothpaste, food additives, sporting goods, stain-resistant clothing, and construction materials (Buzea et al. 2007; Ribeiro et al. 2017). As these products make their way into the environment, intentionally or unintentionally, concerns have emerged about the adverse effects of ENPs on living organisms, including humans (Gottschalk and Nowack 2011; Nowack et al. 2012). This has led to the development of nanoecotoxicology, which focuses on understanding how these ENPs affect ecosystems, organisms, and biodiversity at various levels (Corsi et al. 2022). The goal of nanoecotoxicology is to assess the risk posed by NPs to the environment and living organisms and to develop strategies for minimizing or mitigating these risks.

Over the past few years, numerous research studies have been conducted to better understand the toxicity of ENPs in the environment (Matranga and Corsi 2012; Salieri et al. 2015; Talreja and Kumar 2018; Zhang et al. 2022). Due to their small size and unique properties, such as enhanced magnetic characteristics and uneven electron distribution, NPs exhibit different behaviors than larger particles, which in turn influence their toxicity. For example, the large surface area of NPs allows for more significant interaction with target sites in organisms, which can contribute to their toxicity, including damage to organs such as the liver (Albanese et al. 2012; Ma et al. 2016). Additionally, NPs are more likely to be internalized by cells, which can lead to cytotoxicity (Manzanares and Ceña 2020; Means et al. 2022). NPs can easily penetrate biological barriers and cell membranes, causing cellular damage upon entry into living organisms (Tran et al. 2023). Furthermore, many NPs are designed to be stable and resistant to degradation, enabling them to persist in the environment for significantly longer than larger particles (Liu et al. 2010; Levard et al. 2012). This persistence raises concerns about the long-term exposure risk posed to organisms and ecosystems. In addition to their inherent toxicity, NPs can undergo environmental transformations that alter their properties and, consequently, their toxicity (Lei et al. 2018). They can also interact with other environmental pollutants, such as heavy metals or organic chemicals, potentially forming more toxic compounds (Deng et al. 2017). Additionally, NPs may aggregate or agglomerate in the environment, which can influence their bioavailability and toxicity (Lei et al. 2018).

The released NPs also pose a risk to marine organisms. Studies have shown that they can increase cytotoxicity and oxidative stress in marine microalgae and copepod (Gong et al. 2019), and certain ENPs have been linked to organ failure in fish (Lekamge et al. 2020; López-Barrera et al. 2021). Once released, NPs can enter an organism's cell through diffusion, endocytosis, or adhesion (Kim et al. 2006; Lin and Xing 2007). In plants, algae, and fungi, ENPs can penetrate the cell wall through tiny pores (5–20 nm), which may enlarge upon interaction with the particles, facilitating their uptake (Navarro et al. 2008). Once inside the cell, NPs can interact with organelles and disrupt essential metabolic processes. Subsequently, large leaf plants are more likely to interact with airborne NPs that enter through stomata, wounds, or cuts, whereas roots take up NPs from the soil or water (Navarro et al. 2008). In the case of titanium oxide NPs ($TiO_2$-NPs), these particles can enter through the respiratory route, translocating to the brain and other organs such as the kidneys and liver (Shakeel et al. 2017). High concentrations of $TiO_2$-NPs, in particular, have been shown to inhibit the growth of green algae, such as *Desmodesmus subspicatus* (Hund-Rinke and Simon 2006). Other types of ENPs, including fullerenes and single-walled carbon nanotubes (SWCNTs), have also been reported to induce neurotoxicity in fish (Oberdorster et al. 2006; Smith et al. 2007).

Additionally, mechanistic studies of ENPs have been shown to generate reactive oxygen species (ROS) in organisms, leading to damage to cell membranes, DNA, and other cellular structures. The production of ROS increases membrane permeability, making cells more susceptible to osmotic stress. For instance, ENPs like $TiO_2$, when exposed to UV light, produce ROS that can cause DNA damage (Shah et al. 2017). Other ENPs, such as C60 fullerenes and silver NPs, have been found to break DNA strands when exposed to light (Ávalos et al. 2015; Afanasieva et al. 2015). Certain ENPs, including quantum dots (QDs), can release toxic metals that accumulate in cells, causing further cellular damage (Klaine et al. 2008). For example, silver ions ($Ag^+$) can disrupt enzyme function, deplete antioxidants, and interfere with DNA replication (Feng et al. 2000; Hussain et al. 2005), whereas cerium oxide ($CeO_2$) NPs are known to damage cell membranes and affect energy production (Thill et al. 2006).

Overall, the unique properties of ENPs, while offering numerous technological benefits, also pose significant challenges to environmental and human health. The need for continued research into their toxicity, environmental fate, and strategies for risk mitigation remains critical to fully understanding their impact on ecosystems and living organisms. Therefore, this chapter aims to explore the effect of NPs on humans and the environment.

## 12.2 Type of NPs

NPs represent a diverse class of materials with unique properties that make them suitable for various applications, particularly in fields such as medicine, materials science, and environmental technology. The classification of NPs can be approached from multiple perspectives, including their composition, structure, and intended application. Classification of NPs encompasses a wide range of materials, each with distinct properties and applications. From liposomes and polymeric NPs to metallic and carbon-based NPs, the versatility of NPs continues to drive innovation across various scientific fields. Based on their composition, NPs can be categorized into three types: organic, inorganic, and carbon-based NPs (Fig. 12.1).

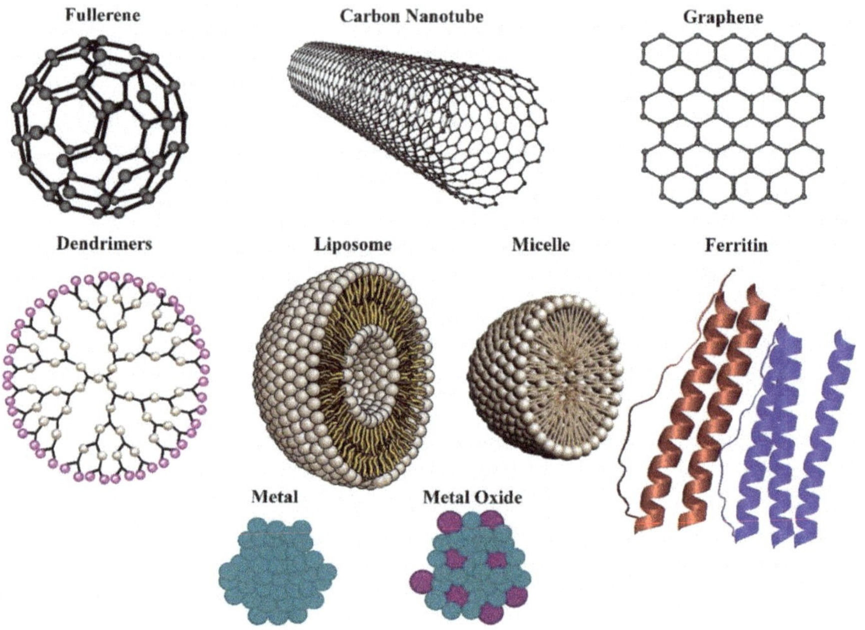


**Fig. 12.1** Types of nanoparticles

### 12.2.1 *Organic NPs*

Organic NPs, are composed of organic compounds such as lipids, polymers, carbohydrates, proteins, and other organic materials (Bhardwaj and Saxena 2017). They include structures like micelles, dendrimers, liposomes, nanogels, and polymeric NPs (Alshammari et al. 2023). Some organic NPs, such as micelles and liposomes, are hollow spheres that are non-toxic and biodegradable. These NPs are sensitive to environmental factors like heat and light and can naturally degrade (nanocapsule) (Rao and Geckeler 2011; Ealia and Saravanakumar 2017). Their ease of breakdown and non-covalent interactions make them more labile and easier to eliminate from the body. Organic NPs are commonly employed in biomedical fields, particularly in targeted drug delivery and cancer therapy (Gujrati et al. 2014; Ealia and Saravanakumar 2017).

### 12.2.2 *Carbon-Based NPs*

Carbon-based nanomaterials (CBNs) have recently garnered significant attention due to their diverse allotropes, including carbon nanotubes (CNTs), graphene oxide

(GO), graphene quantum dots (GQDs), and fullerene (Maiti et al. 2019). These materials exhibit remarkable properties and are widely utilized in various applications, including biosensing, drug delivery, tissue engineering, and cancer therapy (Hong et al. 2015; Bhattacharya et al. 2016). CNTs are hollow cylindrical structures made of graphitic sheets and are classified into single-walled (SWCNTs) and multi-walled carbon nanotubes (MWCNTs) (Odom et al. 1998; Eatemadi et al. 2014). Graphene, a single layer of carbon atoms arranged in a hexagonal lattice, has exceptional electrical, mechanical, and chemical properties, making it highly suitable for bio-applications. GQDs are small graphene sheets that exhibit enhanced fluorescence, making them ideal for imaging and cancer therapy (Maiti et al. 2019).

### *12.2.3 Inorganic NPs*

Inorganic NPs are composed of inorganic materials, such as metals (gold, silver, platinum), metal oxides (titanium oxide, zinc oxide), ceramics, or semiconductors (silica, DTs) (Nasirzadeh et al. 2016; Balkrishna et al. 2023). These NPs offer unique physical and chemical properties, including high surface area, catalytic activity, and optical properties, making them useful in drug delivery, imaging, and electronics (Ealia and Saravanakumar 2017; Balkrishna et al. 2024). Inorganic NPs are often hydrophilic, nontoxic, and biocompatible with living organisms (Alshammari et al. 2023).

## 12.3 Pathways and Fate of NPs in the Environment

ENPs are widely used in various consumer products such as household items, textiles, personal care products, agriculture, and biomedical devices, owing to their distinct physical, chemical, optical, and electrical properties (Lien et al. 2017; Donia and Carbone 2019). However, the extensive use of these materials results in their release into the environment at various stages of their life cycle- during production, use, and disposal (Keller and Lazareva 2014; Gottschalk et al. 2015). For instance, $TiO_2$-NPs and zinc oxide NPs are commonly found in products like sunscreens, paints, and food additives (Chen and Mao 2007). AgNPs are used in various applications, including antibacterial products, food packaging, and textiles (Marin et al. 2015). $TiO_2$ in tattoos is particularly common for creating white colors or enhancing darker shades. It tends to participate in the deeper layer of the skin, much like other pigments (Donia and Carbone 2019).

ENPs such as silver (Ag), $TiO_2$, and ZnO are often released into the environment through consumer products like textiles and paints. Carbon-based ENPs, in particular, may adsorb organic and inorganic contaminants, thereby altering their environmental behavior and reactivity (Canesi et al. 2015; Sigmund et al. 2018). Whether intentional or accidental, releasing these materials raises concerns about their potential risks to human health and ecosystems (Abbas et al. 2019; Sendra et al. 2019). Multiple studies have indicated that metal oxide NPs accumulate in plants due to the increased use of NP-based fertilizers and pesticides. This accumulation can have various impacts on plant development, including changes in germination, photosynthesis, antioxidant response, and other physiological and biochemical functions (Cox et al. 2017; Du et al. 2017). The emission of NPs may also occur directly into the environment or indirectly through systems such as WWTPs or landfills, where they can undergo transformations, including coating or aggregation (Kaegi et al. 2010, 2011; Zuin et al. 2013; Al-Kattan et al. 2015). NPs are also widely utilized in biomedical applications. For instance, titanium alloys have been commonly used in surgical implants (plastic surgery) and biomedical devices because of their excellent biocompatibility, mechanical strength, and corrosion resistance. However, the growing use of titanium-based products has increased human exposure, leading to cases of hypersensitivity to this metal (Fage et al. 2016).

The increasing application of nanomaterials also necessitates understanding the potential routes of exposure and associated risks for humans. NPs can enter the body primarily through the skin, lungs, or gastrointestinal tract, posing risks of adverse health effects (Hoet et al. 2004; Warheit et al. 2004; Oberdörster et al. 2005; Davoren et al. 2007; Li et al. 2007). Biomedical applications may also enter via intravenous, intradermal, or peritoneal routes (Stern and McNeil 2008). Factors such as particle size, charge, surface area, and shape influence the degree of exposure, with particles smaller than 30 nm having a high surface-to-volume ratio, making them more reactive and potentially toxic (Auffan et al. 2008). The toxicity of NPs is further influenced by their resistance in the body and their ability to be cleared, which depends on the immune response and other physiological factors (Card et al. 2008). The environmental distribution of NPs is depicted in Fig. 12.2.

## 12.4 Mechanisms of Nanotoxicity

NPs commonly used in industries like medicine, electronics, and cosmetics, exhibit unique properties that make them highly functional. However, their environmental persistence and interactions with biological systems raise concerns. These particles can accumulate in ecosystems and interact with biological membranes, potentially leading to toxic outcomes. The mechanisms of nanotoxicity include oxidative stress, membrane damage, protein misfolding, genotoxicity, and inflammatory responses. When NPs enter biological systems, they can generate (ROS, leading to oxidative damage in DNA, lipids, and proteins. These particles can also penetrate cell membranes and interact with cellular organelles, disrupting normal cellular functions and causing cytotoxic effects. Understanding the specific mechanisms underlying nanotoxicity is essential for mitigating risks and promoting the safe application of NPs. Here are some key mechanisms by which NPs may exert toxic effects.

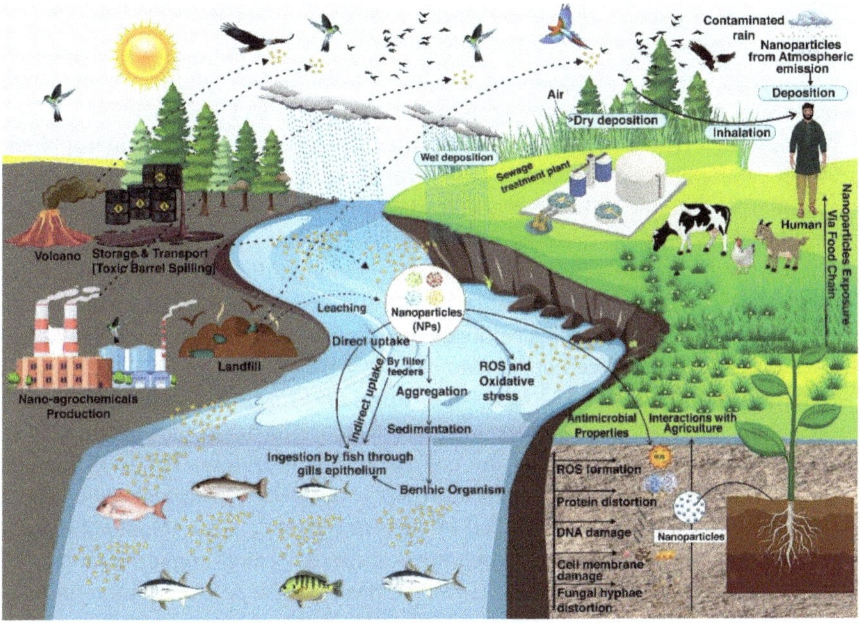


**Fig. 12.2** An overview of the environmental distribution of nanoparticles. Reproduced from Srivastava et al. (2024) with permission from Springer Nature (License no.: 5959300466617)

## *12.4.1 Interaction with Biological Membranes*

NPs can disrupt cell permeability and cause membrane damage through multiple mechanisms upon interacting with biological membranes. These effects depend on the type of NPs, their size, shape, surface charge, and functionalization, as well as the specific characteristics of the target cell membrane. Here are some key mechanisms, which describe how NPs can interact with and disrupt cell membranes:

- **Direct Physical Interaction and Membrane Penetration**: NPs, particularly those with sharp or elongated shapes (rods, wires, or needles), can cause mechanical punctures or physical stress on the cell membrane. When NPs come into contact with the membrane, they penetrate the lipid bilayer, creating pores or small holes. This disrupts the membrane's integrity, leading to cellular content leakage, ion imbalance, and potential cell death (Roiter et al. 2008).
- **Interaction with Membrane Proteins**: NPs can also interact with membrane proteins by physically binding to them or inducing conformational changes that compromise their function. These interactions can cause the proteins to lose their functionality, disrupting membrane signaling, transport, and structural integrity (Verma and Stellacci 2009; Shang et al. 2014).
- **Endocytosis and Cytotoxicity**: NPs can be internalized by cells through endocytosis, where they may accumulate in intracellular vesicles. In some cases, the

particles can puncture or destabilize the vesicle or cell membrane upon release, leading to leakage of intracellular contents. Some NPs are designed to escape from endosomes into the cytoplasm, where the NPs can lead to cytotoxicity mediated by several mechanisms including oxidative stress (Oh and Park 2014).

- **Nanoparticle-Induced Membrane Permeabilization**: Some NPs, such as lipid-coated or amphiphilic ones have both hydrophobic and hydrophilic regions, which can interact with the cell membrane, leading to permeabilization. This causes the membrane to become more permeable to ions, small molecules, and proteins, ultimately disturbing cellular homeostasis. Amphiphilic or charged NPs enter into the lipid bilayer, causing the formation of transient pores or channels. These pores can lead to leakage of cellular contents and disruption of ion gradients, eventually resulting in cell death (Mu et al. 2014).
- **Reactive Oxygen Species (ROS) Generation and Lipid Peroxidation**: Certain NPs, especially metal-based ones (silver, copper, or iron oxide), can generate ROS either on the surface or within the cell upon internalization. The ROS include the superoxide anion ($O_2^{\cdot -}$), hydrogen peroxide ($H_2O_2$), and hydroxyl radical ($HO^{\cdot}$). ROS are normally produced by cells at certain levels to maintain regular metabolism and homeostasis, which are considered critical signaling molecules in cell proliferation and survival. However, they may be produced through interactions with exogenous sources such as NPs. If this event produces excessive ROS that the cellular antioxidant defense system (enzymatic antioxidants such as glutathione (GSH) peroxidases) cannot handle, oxidative stress is triggered. This may lead to the destruction of organelles and bio-molecules, including triggering membrane damage, lipid peroxidation, DNA damage, protein damage, apoptosis, necrosis, and inflammatory response (Awashra and Młynarz 2023).

These mechanisms often act together, and the extent of membrane damage depends on the NPs' properties and the type of cell membrane. Understanding these interactions helps design safer NPs for drug delivery, diagnostics, and environmental sciences applications.

### *12.4.2 Genotoxic and Mutagenic Effects*

The small size and high reactivity of NPs raise concerns about their potential impact on biological systems, particularly at the cellular and genetic levels. NPs, particularly metal-based types such as silver, zinc oxide, and titanium dioxide, can penetrate cellular barriers and reach the nucleus, where they may interact directly with DNA, leading to oxidative stress capable of damaging DNA and other cellular components. This genotoxic effect may result in single-strand and double-strand DNA breaks, and if these are not adequately repaired, they can lead to mutations that disrupt normal cellular functions or activate oncogenes, potentially increasing cancer risk or causing genetic disorders. Additionally, some NPs interfere with the cell's DNA repair mechanisms, inhibiting key proteins and allowing mutations to accumulate, which can cause cellular dysfunction, apoptosis, or uncontrolled cell division (Kisin et al. 2007; Ahamed et al. 2008; Egbuna et al. 2021).

Research into biodegradable NPs and the regulation of nanoparticle use in agriculture, medicine, and industry can help minimize these impacts. Long-term studies and environmental monitoring can also aid in understanding the potential risks and controlling nanoparticle release into sensitive ecosystems.

### *12.4.3 Bioaccumulation and Biomagnification*

NPs, owing to their tiny size and unique chemical properties, have the potential to accumulate in organisms and bioaccumulate up the food chain. This possibility raises environmental and health concerns as NPs are increasingly used in various industries, including medicine, agriculture, cosmetics, and electronics. NPs accumulation and trophic transfer may occur, leading to potential risks such as toxicity in organisms at various levels of the food chain.

- **Bioaccumulation within Individual Organisms**: NPs can quickly enter cells through passive diffusion or endocytosis due to their small size. Once inside, they may not be readily metabolized or excreted, leading to accumulation within tissues and organs, particularly in the liver, kidneys, brain, and spleen, where they may remain for long periods and induce chronic toxicity (Oberdörster et al. 2005; Nel 2006; Uddin et al. 2020).
- **Movement Up the Food Chain (Trophic Transfer)**: As larger predators consume smaller organisms with accumulated NPs, these particles can move up the food chain. For instance, if NPs are absorbed by phytoplankton, which are then eaten by zooplankton, these particles may eventually reach fish and higher-level consumers. As NPs move up the food chain, their concentration may increase in larger predators (a process called biomagnification). This poses potential health risks to top-level predators, including humans, who consume seafood or meat containing these NPs (Handy and Shaw 2007; Ferry et al. 2009; Baker et al. 2014).

The bioaccumulation of NPs can disrupt ecosystem dynamics by negatively affecting primary producers (such as algae), foundational organisms (like earthworms), and decomposers. In aquatic environments, NPs settling in water bodies may be ingested by fish, amphibians, and microorganisms. At the top of the food chain, humans face potential risks from NPs exposure through consuming contaminated food. Ingested NPs can accumulate in human tissues, posing risks of chronic diseases and organ damage depending on the type of NPs and level of exposure (Nel 2006; Baker et al. 2014; Huang et al. 2021). Research has shown that certain NPs can induce oxidative stress, DNA damage, and cell membrane disruption, resulting in reduced fertility, impaired growth, and increased mortality among aquatic organisms. In conclusion, while NPs offer significant benefits, their potential for accumulation in organisms and trophic transfer highlights the need for careful monitoring and

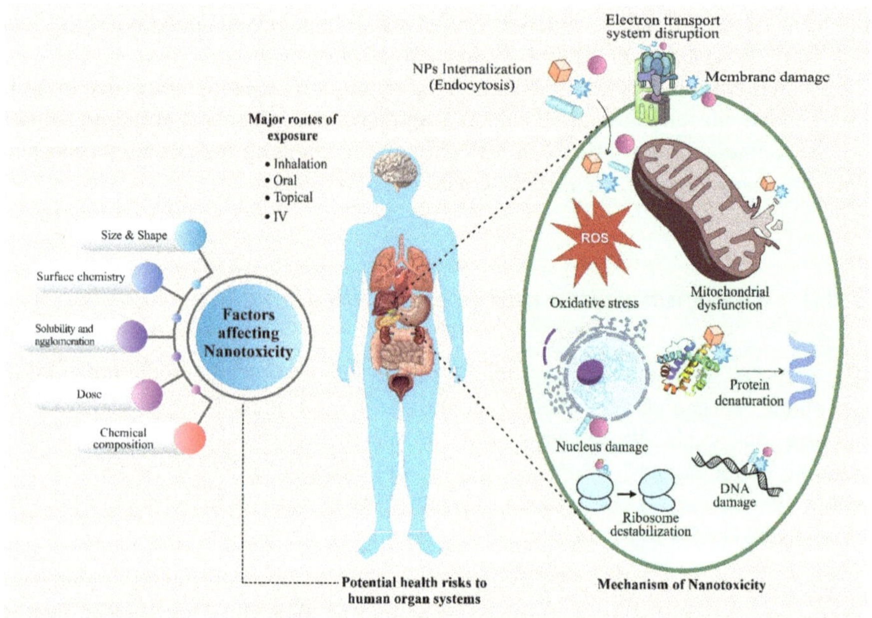


**Fig. 12.3** Mechanism of nanotoxicity

regulation to prevent unintended ecological and health consequences. The factors affecting NPs' toxicity, such as size, shape, surface area, composition, properties, dose, and mechanism of nanotoxicity, are highlighted in Fig. 12.3.

## 12.5 Nanotoxicology in Different Environments

The emergence of nanotechnology has brought significant advancements in various fields, but it has also raised concerns regarding the potential environmental impacts of NPs in different ecosystems.

### *12.5.1 Nanotoxicity in Aquatic Ecosystems*

NPs pose significant risks to aquatic ecosystems, affecting plants, algae, invertebrates, and fishes (Schirmer 2014). Accumulating in water, NPs transfer to higher trophic levels via algae and filter feeders, forming toxic sediments harmful to benthic organisms. Fish absorb NPs through gills and epithelium, while invertebrates target gut and hepatopancreas tissues. Temperature affects NPs toxicity, with higher temperatures creating smaller, more harmful aggregates (Tiple et al. 2020). Various studies demonstrate the toxicological impact of NPs on aquatic species, emphasizing histological, biochemical, and physiological alterations. Ag-NPs (0.09 mg/l) caused severe damage to gills, intestine, muscles, and liver in common carp (*Cyprinus carpio*) after 20 days (Kakakhel et al. 2021). In another study $SiO_2$NPs exposure (20–100 mg/l over three weeks) in Nile tilapia (*Oreochromis niloticus*) resulted in hepato-renal damage (Abdel-Latif et al. 2021). Further, ZnO-NPs were found to induce oxidative stress and histopathological changes in the liver, kidneys, and gills of goldfish (*Carassius auratus*) at 1.5 mg/L after 14 days (Ghafarifarsani et al. 2023). CdS-NPs caused epithelial and cellular damage in gills, including the loss of epithelial, mucous, and pillar cells with the collapse of micro-ridges and microbridges of *Channa punctatus* (Verma et al. 2020), while $TiO_2$-NPs triggered genotoxicity in gills and liver of Mozambique tilapia (*Oreochromis mossambicus*) (Shahzad et al. 2022).

Algae are foundational to aquatic ecosystems, and their health significantly influences the overall ecosystem dynamics. CuO-NPs were reported to cause DNA and oxidative damage to *Microcystis aeruginosa* at concentrations up to 2 mg/l, with an $EC_{50}$ of 0.47 mg/l after 72 h (Angela et al. 2014). In *Skeletonema costatum*, exposure to $TiO_2$-NPs concentrations up to 30 mg/l inhibited growth and increased malondialdehyde (MDA) content, indicating oxidative stress. The $EC_{50}$ value for this study was determined to be 7.37 mg/l after 72 h (Li et al. 2015a, b). Similarly, Cadmium telluride quantum dots (CdTe QDs) at concentration as low as 0.01 mg/l induced oxidative stress in *Chlorella pyrenoidosa*, reducing chlorophyll and protein levels while increasing superoxide dismutase (SOD) activity and MDA content, highlighting oxidative damage with an $EC_{50}$ of 0.015 mg/l after 96 h demonstrating the high sensitivity of algae to these NPs (Xiao et al. 2016). Further, Carbon nanotubes (CNTs) were reported to inhibit *Chlorella vulgaris* growth at concentrations of 1–50 mg/l, with $EC_{50}$ values of 1.8 mg/l for dispersed and 24 mg/l for agglomerated suspensions (Schwab et al. 2011).

The aquatic invertebrates, molluscs and worms are key indicators of aquatic ecosystem health. The DNA damage caused by CuO-NPs was evident in marine bivalves such as *Scrobicularia plana* and ragworms (*Hediste diversicolor*). Comet assay studies revealed genotoxic effects in these organisms, especially with higher concentrations of NPs (Buffet et al. 2013). In another study, AgNPs (16 and 32 μg/l) caused significant dose-dependent toxicity in *Hediste diversicolor* (Cong et al. 2011). Similarly, Mussels (*Mytilus galloprovincialis*) were exposed to 10 mg/l of AgNPs and ionic silver. Both forms of silver demonstrated bioaccumulation potential, leading to toxic effects (Gomes et al. 2014). Further, the other sea dweller Sea urchins (*Paracentrotus lividus*) exposed to AgNPs exhibited developmental defects such as development, asymmetry, and behavioral changes at a concentration of 0.3 mg/l. These defects highlighted the developmental toxicity of NPs in marine invertebrates (Šiller et al. 2013). In coral species such as *Seriatopora caliendrum*, exposure to ZnO-NPs at 50–200 μg/l concentrations caused membrane damage. This was evident through alterations in lyso-glycerophosphocholines (lyso-GPCs) and fatty acid profiles (DPA and DHA-GPCs) while decreasing arachidonic acid-possessing GPCs (involved in inflammation). Additionally, there was an increase in polyunsaturated plasmanylcholines, likely a response to membrane damage. These findings underline the potential risks NPs pose to coral health, even at relatively low concentrations (Tang et al. 2017).

Studies have also explored the combined effects of different NPs on aquatic organisms. In common carp (*Cyprinus carpio*), co-exposure to Ag-NPs and $TiO_2$-NPs enhanced toxicity, as reflected in the reduced $LC_{50}$ values during 96-h acute toxicity tests. Chronic exposure for 10 days also showed aggravated oxidative stress and gill histopathology compared to exposure to individual NPs. Such findings emphasize the importance of evaluating combined nanoparticle toxicity, as it may be more severe than their individual NPs (Haghighat et al. 2021). The toxicological effects of NPs on fish, algae, and invertebrates highlight their potential to disrupt aquatic ecosystems. The alterations in histology, growth, and biochemical markers in these organisms indicate that they can induce oxidative stress, DNA damage, and tissue degeneration. These impacts may cascade through the food web, affecting ecosystem balance and biodiversity. For instance, as primary producers, algae play a crucial role in nutrient cycling. The inhibition of algal growth and photosynthetic activity due to NPs could limit food availability for higher trophic levels. Similarly, histopathological damage in fish and invertebrates can compromise their survival, reproduction, and ecological roles.

### 12.5.2 Nanotoxicity in Soil Ecosystems

The increasing prevalence of NPs in the environment, especially in agricultural systems, has raised significant concerns about their impact on soil ecosystems. Continuous deposition, low biodegradability, and longer persistence of metal NPs in soils adversely impact the population of soil-beneficial bacteria and fungi and soil-dwelling organisms like earthworms. Research indicates that the accumulation of NPs in treated soils could endanger microbial species and other invertebrates, disrupting soil health and ecosystem balance (Tourinho et al. 2012; Lead et al. 2018; Sanzari et al. 2019).

The structural composition of soil microbial communities is significantly affected by NPs, with cellular toxicity influenced by factors such as the size, shape, chemical composition, concentration, and duration of exposure. Studies reveal that specific plant growth-promoting rhizobacteria, such as *Bacillus thuringiensis* and *Pseudomonas mosselii*, exhibit varying degrees of tolerance CuO, $TiO_2$, and $Al_2O_3$ NPs. Still, they are sensitive to Ag and ZnO at even lower doses (Ahmed et al. 2020). Adverse effects on key microbial activities, such as ammonification, respiration, and enzyme production, have also been documented, particularly with ZnO and Ag-NPs (Shahrokh et al. 2014; Shen et al. 2015). $TiO_2$ and ZnO-NPs have been shown to disrupt symbiotic relationships between *Rhizobium* and legumes, which is crucial for nitrogen fixation. These NPs cause morphological changes in rhizobial cells, which hinder their ability to recognize host plants, reducing nitrogen fixation efficiency by up to 54% (Fan et al. 2014; Moll et al. 2016). Furthermore, transcriptomic studies on bacteria like *Desulfovibrio vulgaris* exposed to CuO-NPs reveal impaired energy conversion, iron availability, and cell motility due to oxidative stress induced by ROS (Chen et al. 2019).

NPs also impose destructive effects on unicellular, mycorrhizal, and cellulose/lignin-degrading fungi (Ameen et al. 2021). Arbuscular mycorrhizal fungi (AMF), essential for nutrient uptake in plants, are adversely affected by various NPs. AgNPs have been found to reduce AMF colonization in AMF-Tomato and rice at low concentrations, while ZnO and $TiO_2$-NPs significantly inhibited AMF symbiosis in maize and rice ecosystems (Li et al. 2015a, b; Wang et al. 2016; Noori et al. 2017). These disruptions underscore the broader ecological impact of NPs on plant–microbe interactions, potentially affecting nutrient cycles and soil fertility. NPs also pose risks to soil invertebrates, such as earthworms, which are critical for soil health. Studies show that Ag, Cu, and Au NPs can affect earthworm growth, reproduction, and genetic expression (Unrine et al. 2010). Similarly, ZnO-NPs negatively impacted earthworm productivity and were absorbed more readily than traditional forms of zinc, suggesting higher bioavailability and toxicity (García-Gomez et al. 2014).

NPs directly affect plant roots by generating ROS, leading to oxidative stress, lipid peroxidation, and cellular damage. They alter membrane fluidity, nutrient uptake, and root growth, disrupting plant development. Ag-NPs, for example, penetrate root cell walls at higher concentrations, causing oxidative damage and DNA fragmentation (Patlolla et al. 2012; Mirzajani et al. 2013). Studies on crops such as rice (*Oryza sativa*), cucumber (*Cucumis sativus*), and coriander (*Coriandrum sativum*) reveal significant reductions in biomass, photosynthetic pigments, and root length due to exposure to CuO and ZnO-NPs (Mosa et al. 2018; AlQuraidi et al. 2019). Transcriptomic analyses in *Arabidopsis thaliana* suggest that MO-NPs suppress genes related to root morphogenesis, pathogen defense, and salicylic acid signaling, indicating broader impacts on plant defense systems and development (García-Sánchez et al. 2021).

High concentrations of NPs in soil can lead to reduced crop productivity by disrupting photosynthesis, nutrient uptake, and energy conversion processes. For instance, $CeO_2$ NPs were found to inhibit photosynthesis in soybean (*Glycine max*) and degrade grain quality in wheat (*Triticum aestivum*) (Du et al. 2015; Li et al. 2020). NiO NPs were also observed to reduce biomass and antioxidant activity in *Allium* species, demonstrating their genotoxic potential (Manna and Bandyopadhyay 2017). These studies reveal that the deposition of NPs in soil ecosystems poses significant challenges for soil microorganisms, plant–microbe interactions, and overall soil health. From disrupting microbial communities and nitrogen fixation processes to impairing root development and crop productivity, the adverse effects of NPs highlight the need for comprehensive risk assessments and sustainable management practices.

### 12.5.3 Toxicity of Airborne NPs

The rapid development and utilization of NPs have revolutionized numerous industries. Still, their increasing presence in the environment, particularly in the air, raises significant concerns regarding their ecological and human health impacts (Tran et al. 2023). While NPs offer many advantages, such as increased product performance and functionality, their unintended release into the air during manufacturing, usage, and disposal presents serious environmental and health concerns. Airborne NPs include many substances, with metal-based NPs such as Ag, $TiO_2$, and ZnO being particularly common (Khan et al. 2021). For instance, AgNPs are used for their antimicrobial properties but have been detected in air concentrations ranging from $6 \times 10^{-8}$ to over $6.9 \times 10^{-4}$ mg/l air, posing potential toxicological consequences (Ham et al. 2012; Lee et al. 2012, 2013). These emissions, although small, accumulate over time and exacerbate air pollution, posing risks to human health and ecosystems. On the other hand, anthropogenic carbon-based NPs also significantly impact air pollution and global warming by absorbing solar radiation and posing inhalation risks through particulate matter with a 2.5-μm diameter exposure (Marelle et al. 2021; Thangavel et al. 2022). Carbon nanotubes amplify toxicity by generating ROS via interactions with OH radicals and ozone (Saleem et al. 2022).

Exposure to nano-$La_2O_3$ caused dose-dependent adverse effects on honeybee survival, body weight, and pollen consumption, alongside dysbiosis in gut microbiota, with an increase in pathogenic bacteria like *Serratia* and *Frischella.* These findings highlight the potential ecological and health risks of nano-$La_2O_3$ on honeybees (Liu et al. 2021). Additionally, NPs such as iron oxide have been found to impair the development of plants like *Tillandsia usneoides*, an epiphytic species sensitive to environmental changes. These findings highlight the potential of airborne NPs to disrupt ecological processes, affecting plant and animal health and the broader environmental balance (Falsini et al. 2024). Such disruptions can have cascading effects on ecosystems, ultimately compromising their ability to sustain biodiversity and ecosystem services.

Case studies on airborne NPs reveal diverse health risks, emphasizing the complex pathways through which these ultrafine particles affect human health. Inhalation of NPs such as ZnO and $CeO_2$ has been linked to respiratory toxicity, including asthma, lung inflammation, and fibrosis. These effects stem from oxidative stress caused by ROS, which damages cellular structures and triggers inflammation (Ohlwein et al. 2019). Beyond respiratory issues, NPs like nano-silver have been shown to cause systemic effects, including lung inflammation, oxidative stress, thrombosis, genetic toxicity, and cardiomyocyte apoptosis (Ferdous et al. 2019). Both in vivo and in vitro studies demonstrate the ability of NPs to penetrate the alveolar–capillary barrier, entering the bloodstream and affecting systemic organs (Du et al. 2017; He et al. 2020). Comparative research shows that AgNPs induce more significant gastrointestinal toxicity, with liver lymphocytic infiltration and elevated serum markers, compared to other particles like $SiO_2$ and IONPs (Yun et al. 2015). Furthermore, the combined exposure to $Fe_2O_3$NPs and AgNPs in rats exacerbates oxidative

DNA damage, lipid peroxidation, and cytokine production in the lungs compared to individual NP exposures (Yousef et al. 2019).

Specific NPs like $CoFe_2O_4$ and n-$SiO_2$ have also been studied for their endocrine-disrupting potential. n-$SiO_2$ enhances bioaccumulation of tetrabromobisphenol A (TBBPA), disrupting thyroid function in zebrafish larvae (Zhu et al. 2021). Similarly, $CoFe_2O_4$ NPs have shown thyroid-related risks, necessitating further research into their endocrine impacts (Ahmad et al. 2016). Moreover, polyethyleneimine-coated IONPs activate toll-like receptor 4 signaling, leading to ROS production, heightened macrophage activation, and pro-inflammatory effects (Mulens-Arias et al. 2015). Studies also highlight the neurotoxicity of NPs, with particles like FeO, $TiO_2$, and $SiO_2$ crossing the blood–brain barrier and contributing to neurodegenerative diseases. Mechanisms include oxidative stress, cytokine release, DNA damage, and p53 pathway activation, leading to conditions such as Alzheimer's and Parkinson's (Saleem et al. 2022). These findings underline the need for stricter monitoring and mitigation strategies to address the health risks posed by airborne NPs. NPs, widely used across industries, present serious health and ecological concerns due to their ability to penetrate biological systems and cause toxicity. Studies link them to respiratory, systemic, and neurological damage in humans and disruptions in animal and plant ecosystems. Addressing these risks requires stringent regulations, sustainable practices, and further research into their long-term impacts.

NPs pose significant risks to aquatic, soil and air ecosystems. Their effects, ranging from genotoxicity to oxidative stress and bioaccumulation, call for rigorous environmental monitoring and regulation of NPs release into these ecosystems. Future research should focus on long-term ecotoxicological studies, considering chronic exposure scenarios and the interactions between different NPs. Additionally, developing eco-friendly NPs and strategies to mitigate their environmental impact will be crucial for sustainable nanotechnology applications. An overview of the toxicity of airborne NPs and nanoecotoxicity in soil and aquatic ecosystems is presented in Fig. 12.4.

## 12.6 Regulatory Perspectives and Ethical Concerns

The integration of ENPs into consumer products and industrial applications raises significant concerns regarding their potential environmental impacts, leading to an urgent need for effective regulation. Nanoecotoxicology, the study of the ecotoxicological effects of nanomaterials on environmental systems has emerged as a critical field, yet the regulatory landscape and ethical challenges surrounding these materials remain complex and evolving. The rapid development and commercialization of nanomaterials have prompted regulatory bodies around the world to adapt existing frameworks to manage their potential risks. In the United States, the Environmental Protection Agency (EPA) oversees the regulation of chemicals under the Toxic Substances Control Act (TSCA), which mandates that new chemical substances—including nanomaterials be reviewed for safety before they are manufactured or

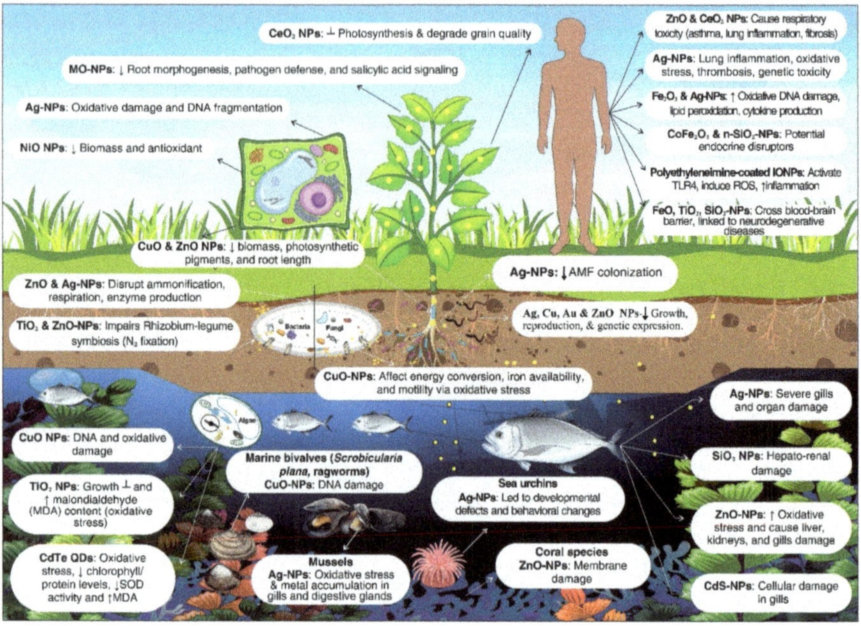


**Fig. 12.4** Nanoecotoxicity in soil, aquatic, and airborne settings

imported. Under this regulation, companies must submit a Pre-Manufacture Notice (PMN), which is assessed for potential environmental and health risks (USEPA 2015). If nanomaterials pose a risk, further assessments are required to evaluate their environmental impact and safety, especially given their novel properties compared to conventional chemicals.

In the European Union, nanomaterials are addressed under the Registration, Evaluation, Authorization, and Restriction of Chemicals (REACH) regulation. REACH requires the gathering of comprehensive data on a substance's environmental behavior, including its Predicted Environmental Concentration (PEC) and Predicted No-Effect Concentration (PNEC), as well as an assessment of whether the substance is persistent, bioaccumulative or toxic (PBT) (European Commission 2008; ECHA 2015, 2016). This process, which has been adapted for conventional chemicals, is now considered broadly applicable to nanomaterials. However, the distinct physicochemical properties of nanomaterials, such as their size, surface area, and potential to aggregate or agglomerate, pose challenges to traditional risk assessment models (Aitken et al. 2011; Hartmann et al. 2017; Hariyadi et al. 2020).

Apart from these regulatory frameworks, various global entities have recognized the potential of nanomaterials for drug-related applications and are working to harmonize guidelines for their safe use. In Europe, the European Chemicals Agency (ECHA), in collaboration with the European Food Safety Authority (EFSA) and the European Medicines Agency (EMA), initiated the first phase of the European Union Observatory for Nanomaterials (EUON) in 2017. This initiative aims to increase awareness and transparency regarding the use of nanomaterials in consumer products, including those used in food, cosmetics, and medical devices. Similarly, in South America, the Brazilian Health Surveillance Agency (ANVISA) emphasizes drugs and medical devices that incorporate nanomaterials, highlighting the need for stringent safety evaluations in these sectors (Marques et al. 2019; Allan et al. 2021). Regulatory bodies such as the United States Food and Drug Administration (USFDA), the European Medicines Agency (EMA), the Therapeutic Goods Administration (TGA), and the Medicines and Healthcare Products Regulatory Agency (MHRA) are collaborating to harmonize information on nano-specific products and establish uniform policies and guidelines (Ali et al. 2023).

Despite significant efforts, challenges remain in adapting regulatory frameworks to address the unique characteristics of nanomaterials, particularly in ecotoxicity testing. Unlike conventional chemicals, nanomaterials often demonstrate unpredictable behaviors in test systems due to their complex interactions within aqueous environments. Many nanomaterials are either insoluble or only partially soluble, persisting as particles or colloidal suspensions that may transform, such as agglomeration, aggregation, or dissolution into ions or chemical species (Skjolding et al. 2016). Regulatory challenges are further compounded by the highly variable properties of NPs, which can differ based on size, shape, surface coating, and chemical composition. These variations significantly influence their toxicity and environmental behavior, complicating the establishment of standardized safety guidelines (Kühnel and Nickel 2014; Khan et al. 2017; Ali et al. 2023).

One critical issue in nanomaterial ecotoxicity testing is that nanomaterials may release ions or particles into the environment, which can have distinct toxicological profiles depending on their size, coating, and surface charge (Brinch et al. 2016). Standard ecotoxicity tests, designed for soluble chemicals, often fail to account for these complexities, making it difficult to evaluate the full range of potential environmental impacts of nanomaterials. As a result, regulators such as the Organisation for Economic Co-operation and Development (OECD) have recognized the need for nano-specific testing guidelines (Hansen et al. 2017). While these guidelines are currently under development, there is still a lack of precise, standardized approaches for how nanomaterials should be handled in ecotoxicity assessments.

Regulators and industry leaders must adopt a precautionary approach to nanomaterial development and use to address these concerns. This approach should include comprehensive risk assessments, transparent reporting, and ongoing monitoring of nanomaterials' environmental impacts. Furthermore, ethical considerations should guide regulatory decisions, ensuring that nanomaterials are introduced into the market only when sufficient evidence supports their safety for human health and the environment.

## 12.7 The Future of Nanoecotoxicology: Illusion or Reality?

The field of nanotoxicology is evolving rapidly, with advancements in analytical and experimental techniques enhancing the assessment of nanomaterials' environmental impacts. High-throughput screening (HTS) methods are emerging as a cornerstone in nanotoxicology, enabling the rapid evaluation of numerous nanomaterials across diverse biological endpoints. HTS platforms integrate robotic automation and miniaturized assays, allowing for efficient testing of NPs' toxicity profiles under varying environmental conditions (Kühnel and Nickel 2014; Domingues et al. 2022). These methods are particularly valuable in screening for potential toxicological impacts while conserving resources and minimizing animal testing. Complementing HTS, in vitro models are gaining traction as alternatives to traditional ecotoxicological assessments. These models leverage cell cultures and organotypic systems to simulate the interactions between nanomaterials and biological systems. Recent innovations include developing co-culture systems replicating complex ecological interactions and providing insights into NPs' bioavailability, uptake, and transformation (Skjolding et al. 2016; Huang et al. 2021). Additionally, advancements in imaging and spectroscopy techniques, such as transmission electron microscopy (TEM) and single-particle inductively coupled plasma mass spectrometry (spICP-MS), have enhanced the characterization of nanomaterials in test systems. These tools allow for precise measurement of nanoparticle size, aggregation state, and dissolution behavior, addressing some challenges posed by their dynamic transformations (Pettitt and Lead 2013; Patil et al. 2022).

Despite these advancements, significant challenges remain in bridging the gap between laboratory findings and real-world environmental scenarios. Testing methods must adapt to account for the heterogeneity of natural ecosystems, where variables such as pH, organic matter content, and microbial communities influence nanoparticle behavior. Future research should prioritize the development of environmentally relevant test systems that integrate complex, multispecies interactions, providing a more holistic understanding of nanomaterial impacts.

### *12.7.1 Risk–Benefit Analysis*

Integrating nanotechnology into diverse sectors, including medicine, agriculture, and energy, underscores its immense potential to address global challenges. However, the ecological risks associated with nanomaterials necessitate a balanced appraisal of their benefits and drawbacks. The risk–benefit analysis framework offers a structured approach to evaluate whether its advantages outweigh the environmental costs of nanotechnology. On the benefits side, nanotechnology has revolutionized numerous fields, contributing to cleaner energy solutions, advanced water treatment technologies, and precision medicine. For instance, NPs such as titanium dioxide ($TiO_2$) have demonstrated efficacy in photocatalytic degradation of pollutants, offering promising applications in environmental remediation. Similarly, AgNPs are widely used for their antimicrobial properties, reducing the spread of infectious diseases in healthcare settings (Notter et al. 2014; Murali et al. 2022; Huang et al. 2024). These advancements highlight the transformative potential of nanotechnology in improving human and environmental health.

Conversely, the ecological risks of nanomaterials cannot be overlooked. Studies have reported that NPs, such as ZnO and CuO, release ions that are toxic to aquatic organisms, raising concerns about their accumulation and persistence in natural habitats (Brinch et al. 2016; Jogaiah et al. 2021; Srivastava et al. 2024). Furthermore, the dynamic transformations of NPs, including aggregation, dissolution, and interaction with organic matter, complicate their risk assessment. The uncertainty surrounding long-term environmental impacts fuels skepticism about the sustainability of nanotechnology. A critical question arises: are the current concerns surrounding nanoecotoxicology substantiated, or are they amplified by the limitations of existing assessment methods? Emerging evidence suggests that nanomaterials pose potential risks but are context-dependent and vary significantly across ecosystems and application scenarios (Malysheva et al. 2015; Khan et al. 2017; Huang et al. 2024). Incorporating life cycle assessments into risk evaluations can provide a more comprehensive perspective, capturing the environmental implications of nanomaterials from production to disposal. This holistic approach may reveal opportunities to mitigate risks through better design and end-of-life management practices.

### *12.7.2 Moving Toward Sustainable Nanotechnology*

The path toward sustainable nanotechnology lies in integrating responsible design principles and proactive regulation aimed at minimizing ecological risks while maximizing societal benefits. Green chemistry approaches, for instance, advocate for developing nanomaterials with reduced toxicity and enhanced biodegradability. By tailoring nanoparticle properties, such as size, shape, and surface functionalization, researchers can design materials that degrade into non-toxic byproducts upon environmental release (Ali et al. 2023). Regulatory frameworks must also evolve to address the unique challenges posed by nanomaterials. Current guidelines, such as the OECD Test Guidelines, provide a foundation but require nano-specific adaptations to ensure their relevance and reliability. The development of nanoCRED, a framework for evaluating the reliability of nanomaterial ecotoxicity data, represents a step forward in aligning regulatory standards with scientific advancements (Moermond et al. 2016). Additionally, fostering collaboration between scientists, policymakers, and industry stakeholders can accelerate the translation of research findings into actionable regulations. Public engagement and education are equally vital in promoting sustainable nanotechnology. Increasing awareness about the benefits and risks of nanomaterials can empower consumers and industries to make informed choices, driving demand for safer, eco-friendly alternatives. Initiatives such as eco-labeling nanotechnology-enabled products can further incentivize adopting sustainable practices (Pokrajac et al. 2021; José and Shinde 2024).

## 12.8 Conclusion

In conclusion, nanoecotoxicology remains a complex and evolving field with both significant challenges and opportunities. Technological advancements in testing methods and a balanced risk–benefit analysis provide a foundation for addressing ecological concerns. By embracing sustainable design and proactive regulation, society can mitigate the risks of nanotechnology, transforming nanoecotoxicology from a pressing reality into a manageable issue. The future of nanoecotoxicology lies in its ability to balance innovation with sustainability, ensuring that the benefits of nanotechnology are realized without compromising environmental integrity.

## References

Abbas Q, Liu G, Yousaf B, Ali MU, Ullah H, Ahmed R (2019) Effects of biochar on uptake, acquisition and translocation of silver nanoparticles in rice (*Oryza sativa* L.) in relation to growth, photosynthetic traits and nutrients displacement. Environ Pollut 250:728–736. https://doi.org/10.1016/j.envpol.2019.04.083

Abdel-Latif HM, Shukry M, El Euony OI, Mohamed Soliman M, Noreldin AE, Ghetas HA, Dawood MAO, Khallaf MA (2021) Hazardous effects of $SiO_2$ nanoparticles on liver and kidney functions, histopathology characteristics, and transcriptomic responses in Nile Tilapia (*Oreochromis niloticus*) Juveniles. Biology 10(3):183. https://doi.org/10.3390/biology10030183

Afanasieva KS, Prylutska SV, Lozovik AV, Bogutska KI, Sivolob AV, Prylutskyy YI, Ritter U, Scharff P (2015) C60 fullerene prevents genotoxic effects of doxorubicin in human lymphocytes in vitro. Ukr Biochem J 87(1):91–98

Ahamed M, Karns M, Goodson M, Rowe J, Hussain SM, Schlager JJ, Hong Y (2008) DNA damage response to different surface chemistry of silver nanoparticles in mammalian cells. Toxicol Appl Pharmacol 233(3):404–410. https://doi.org/10.1016/j.taap.2008.09.015

Ahmad F, Liu X, Zhou Y, Yao H, Zhao F, Ling Z, Xu C (2016) Assessment of thyroid endocrine system impairment and oxidative stress mediated by cobalt ferrite ($CoFe_2O_4$) nanoparticles in zebrafish larvae. Environ Toxicol 31(12):2068–2080

Ahmed B, Ameen F, Rizvi A, Ali K, Sonbol H, Zaidi A, Khan MS, Musarrat J (2020) Destruction of cell topography, morphology, membrane, inhibition of respiration, biofilm formation, and bioactive molecule production by nanoparticles of Ag, ZnO, CuO, $TiO_2$, and $Al_2O_3$ toward beneficial soil bacteria. ACS Omega 5(14):7861–7876

Aitken RA, Bassan A, Friedrichs S, Hankin SM, Hansen SF, Holmqvist J, Peters SAK, Poland CA, Tran CL (2011) Specific advice on exposure assessment and hazard/risk characterisation for nanomaterials under REACH (RIP-oN 3)-final project report. European Commission. http://ec.europa.eu/environment/chemicals/nanotech/pdf/report_ripon3.pdf

Albanese A, Tang PS, Chan WC (2012) The effect of nanoparticle size, shape, and surface chemistry on biological systems. Annu Rev Biomed Eng 14(1):1–16

Ali F, Neha K, Parveen S (2023) Current regulatory landscape of nanomaterials and nanomedicines: a global perspective. J Drug Deliv Sci Technol 80:104118

Al-Kattan A, Wichser A, Vonbank R, Brunner S, Ulrich A, Zuin S, Arroyo Y, Golanski L, Nowack B (2015) Characterization of materials released into water from paint containing nano-$SiO_2$. Chemosphere 119:1314–1321

Allan J, Belz S, Hoeveler A, Hugas M, Okuda H, Patri A, Rauscher H, Silva P, Slikker W, Sokull-Kluettgen B, Tong W, Anklam E (2021) Regulatory landscape of nanotechnology and nanoplastics from a global perspective. Regul Toxicol Pharmacol 122:104885

AlQuraidi AO, Mosa KA, Ramamoorthy K (2019) Phytotoxic and genotoxic effects of copper nanoparticles in coriander (*Coriandrum sativum*—Apiaceae). Plants 8(1):1–12

Alshammari BH, Lashin MM, Mahmood MA, Al-Mubaddel FS, Ilyas N, Rahman N, Sohail M, Khan A, Abdullaev SS, Khan R (2023) Organic and inorganic nanomaterials: fabrication, properties and applications. RSC Adv 13(20):13735–13785

Ameen F, Alsamhary K, Alabdullatif JA, ALNadhari S (2021) A review on metal-based nanoparticles and their toxicity to beneficial soil bacteria and fungi. Ecotoxicol Environ Safety 213:112027

Angela I, Imbi K, Kaja K, Irina B, Villem A, Sandra S, Heiki V, Aleksandr K, Tiina T, Margit H (2014) Size-dependent toxicity of silver nanoparticles to bacteria, yeast, algae, crustaceans and mammalian cells in vitro. PLoS ONE 9(7):e102108

Auffan M, Achouak W, Rose J, Chane´ C, Waite DT, Masion A, Woicik J, Wiesner MR, Bottero JY (2008) Relation between the redox state of ironbased nanoparticles and their cytotoxicity towards *Escherichia coli*. Environ Sci Technol 42(17):6730–6735

Ávalos A, Haza AI, Morales P (2015) Manufactured silver nanoparticles of different sizes induced DNA strand breaks and oxidative DNA damage in hepatoma and leukaemia cells and in dermal and pulmonary fibroblasts. Folia Biol 61(1):33–42

Awashra M, Młynarz P (2023) The toxicity of nanoparticles and their interaction with cells: an *in vitro* metabolomic perspective. Nanoscale Adv 5(10):2674–2723. https://doi.org/10.1039/d2na00534d

Baker TJ, Tyler CR, Galloway TS (2014) Impacts of metal and metal oxide nanoparticles on marine organisms. Environ Pollut 186:257–271. https://doi.org/10.1016/j.envpol.2013.11.014

Balkrishna A, Rohela A, Kumar A, Mishra S, Arya V, Kala V, Thakur N, Thakur N, Kumari A, Khan N (2023) Elucidating the role of plant extracts mediated gold nanoparticles as smart antimicrobials: two-way attack. J Nanomater 2023(1):4085090

Balkrishna A, Sinha S, Kumar A, Arya V, Kumar D, Dhansekhran M (2024) Nanoparticles as drug delivery systems: advances and challenges. In: Nanotechnology: agriculture, environment and health. Springer Nature Singapore, Singapore, pp 245–288

Bhardwaj M, Saxena DC (2017) Preparation of organic and inorganic nanoparticles and their subsequent application in nanocomposites for food packaging systems: a review. Indian J Sci Technol 10(31):1–8

Bhattacharya K, Mukherjee SP, Gallud A, Burkert SC, Bistarelli S, Bellucci S, Bottini M, Star A, Fadeel B (2016) Biological interactions of carbon-based nanomaterials: from coronation to degradation. Nanomedicine 12(2):333–351. https://doi.org/10.1016/j.nano.2015.11.011

Brinch A, Hansen SF, Hartmann NB, Baun A (2016) EU regulation of nanobiocides: challenges in implementing the biocidal product regulation (BPR). Nanomaterials 6(2):1–16. https://doi.org/10.3390/nano6020033

Buffet PE, Richard M, Caupos F, Vergnoux A, Perrein-Ettajani H, Luna-Acosta A, Akcha F, Amiard JC, Amiard-Triquet C, Guibbolini M, Risso-de Faverney C, Thomas-Guyon H, Reip P, Dybowska A, Berhanu D, Valsami-Jones E, Mouneyrac C (2013) A mesocosm study of fate and effects of CuO nanoparticles on endobenthic species (*Scrobicularia plana, Hediste diversicolor*). Environ Sci Technol 47(3):1620–1628

Buzea C, Pacheco II, Robbie K (2007) Nanomaterials and nanoparticles: sources and toxicity. Biointerphases 2(4):MR17-MR71

Canesi L, Ciacci C, Balbi T (2015) Interactive effects of nanoparticles with other contaminants in aquatic organisms: friend or foe? Mar Environ Res 111:128–134. https://doi.org/10.1016/J.MARENVRES.2015.03.010

Card JW, Zeldin DC, Bonner JC, Nestmann ER (2008) Pulmonary applications and toxicity of engineered nanoparticles. Am J Physiol Lung Cell Mol Physiol 295(3):L400–L411

Chen X, Mao SS (2007) Nanostructured titanium dioxide materials: properties, preparation and applications. Chem Rev 107(7):2891–2959

Chen Z, Gao SH, Jin M, Sun S, Lu J, Yang P, Bond PL, Yuan Z, Guo J (2019) Physiological and transcriptomic analyses reveal CuO nanoparticle inhibition of anabolic and catabolic activities of sulfate-reducing bacterium. Environ Int 125:65–74

Cong Y, Banta GT, Selck H, Berhanu D, Valsami-Jones E, Forbes VE (2011) Toxic effects and bioaccumulation of nano-, micron-and ionic-Ag in the polychaete, Nereis Diversicolor. Aquat Toxicol 105(3–4):403–411

Corsi I, Desimone MF, Cazenave J (2022) Nanotoxicology: challenges and solutions to safeguard human health and the environment. Front Bioeng Biotechnol 10:1110246

Cox A, Venkatachalam P, Sahi S, Sharma N (2017) Reprint of: silver and titanium dioxide nanoparticle toxicity in plants: a review of current research. Plant Physiol Biochem 110:33–49

Davoren M, Herzog E, Casey A, Cottineau B, Chambers G, Byrne HJ, Lyng FM (2007) *In vitro* toxicity evaluation of single walled carbon nanotubes on human A549 lung cells. Toxicol in Vitro 21(3):438–448

Deng R, Lin D, Zhu L, Majumdar S, White JC, Gardea-Torresdey JL, Xing B (2017) Nanoparticle interactions with co-existing contaminants: joint toxicity, bioaccumulation and risk. Nanotoxicology 11(5):591–612

Domingues C, Santos A, Alvarez-Lorenzo C, Concheiro A, Jarak I, Veiga F, Barbosa I, Dourado M, Figueiras A (2022) Where is nano today and where is it headed? A review of nanomedicine and the dilemma of nanotoxicology. ACS Nano 16(7):9994–10041

Donia DT, Carbone M (2019) Fate of the nanoparticles in environmental cycles. Int J Environ Sci Technol 16:583–600

Du W, Gardea-Torresdey JL, Ji R, Yin Y, Zhu J, Peralta-Videa JR, Guo H (2015) Physiological and biochemical changes imposed by $CeO_2$ nanoparticles on wheat: a life cycle field study. Environ Sci Technol 49:11884–11893. https://doi.org/10.1021/acs.est.5b03055

Du W, Tan W, Peralta-Videa JR, Gardea-Torresdey JL, Ji R, Yin Y, Guo H (2017) Interaction of metal oxide nanoparticles with higher terrestrial plants: physiological and biochemical aspects. Plant Physiol Biochem 110:210–225

Ealia SAM, Saravanakumar MP (2017) A review on the classification, characterisation, synthesis of nanoparticles and their application. In: IOP conference series: materials science and engineering, vol 263(3), 032019. IOP Publishing

Eatemadi A, Daraee H, Karimkhanloo H, Kouhi M, Zarghami N, Akbarzadeh A, Abasi M, Hanifehpour Y, Joo SW (2014) Carbon nanotubes: properties, synthesis, purification, and medical applications. Nanoscale Res Lett 9(1):1–13. https://doi.org/10.1186/1556-276X-9-393

ECHA (European Chemicals Agency) (2015) Guidance on information requirements and chemical safety assessment chapter R.7a: endpoint specific guidance, Version 4.1

ECHA (European Chemicals Agency) (2016) Guidance on information requirements and chemical safety assessment chapter R.7b: endpoint specific guidance, Version 3.0

Egbuna C, Parmar VK, Jeevanandam J, Ezzat SM, Patrick-Iwuanyanwu KC, Adetunji CO, Khan J, Onyeike EN, Uche CZ, Akram M, Ibrahim MS, El M, Awuchi CG, Saravanan K, Tijjani H, Odoh UE, Messaoudi M, Ifemeje JC, Olisah MC, Ezeofor NJ, Ibeabuchi CG (2021) Toxicity of nanoparticles in biomedical application: nanotoxicology. J Toxicol 2021(1):1–21. https://doi.org/10.1155/2021/9954443

European Commission (EC) (2008) Regulation (EC) No 1272/2008 of the European parliament and of the council of 16 December 2008 on classification, labelling and packaging of substances and mixtures, amending and repealing directives 67/548/EEC and 1999/45/EC, and amending regulation (EC) No 1907/2006

Fage SW, Muris J, Jakobsen SS, Thyssen JP (2016) Titanium: a review on exposure, release, penetration, allergy, epidemiology, and clinical reactivity. Contact Dermatitis 74(6):323–345

Falsini S, Colzi I, Dainelli M, Parigi E, Salvatici MC, Papini A, Talbot D, Abou-Hassan A, Gonnelli C, Ristori S (2024) Impact of airborne iron oxide nanoparticles on *Tillandsia usneoides* as a model plant to assess pollution in heavy traffic areas. Chemosphere 355:141765. https://doi.org/10.1016/j.chemosphere.2024.141765

Fan R, Huang YC, Grusak MA, Huang CP, Sherrier DJ (2014) Effects of nano-$TiO_2$ on the agronomically-relevant Rhizobium-legume symbiosis. Sci Total Environ 466:503–512. https://doi.org/10.1016/j.scitotenv.2013.07.032

Feng QL, Wu J, Chen GQ, Cui FZ, Ki TN, Kim JO (2000) A mechanistic study of the antibacterial effect of silver ions on *Esherichia coli* and *Staphylococcus aureus*. J Biomed Mater Res 52(4):662–668

Ferdous Z, Al-Salam S, Greish YE, Ali BH, Nemmar A (2019) Pulmonary exposure to silver nanoparticles impairs cardiovascular homeostasis: effects of coating, dose and time. Toxicol Appl Pharmacol 367:36–50

Ferry JL, Craig PS, Hexel CR, Sisco PN, Frey R, Pennington PL, Fulton MH, Scott IG, Decho AW, Kashiwada S, Murphy CJ, Shaw TJ (2009) Transfer of gold nanoparticles from the water column to the estuarine food web. Nat Nanotechnol 4(7):441–444. https://doi.org/10.1038/nnano.2009.157

Garcia-Gomez C, Babin M, Obrador A, Alvarez JM, Fernandez MD (2014) Toxicity of ZnO nanoparticles, ZnO bulk, and $ZnCl_2$ on earthworms in a spiked natural soil and toxicological effects of leachates on aquatic organisms. Arch Environ Contam Toxicol 67:465–473. https://doi.org/10.1007/s00244-014-0025-7

García-Sánchez S, Gala M, Žoldák G (2021) Nanoimpact in plants: lessons from the transcriptome. Plants 10(4):1–17

Ghafarifarsani H, Hedayati SA, Yousefi M, Hoseinifar SH, Yarahmadi P, Mahmoudi SS, Van Doan H (2023) Toxic and bioaccumulative effects of zinc nanoparticle exposure to goldfish, *Carassius auratus* (Linnaeus, 1758). Drug Chem Toxicol 46(5):984–994

Gomes T, Pereira CG, Cardoso C, Sousa VS, Teixeira MR, Pinheiro JP, Bebianno MJ (2014) Effects of silver nanoparticles exposure in the mussel *Mytilus galloprovincialis*. Mar Environ Res 101:208–214

Gong N, Shao K, Che C, Sun Y (2019) Stability of nickel oxide nanoparticles and its influence on toxicity to marine algae *Chlorella vulgaris*. Mar Pollut Bull 149:110532

Gottschalk F, Nowack B (2011) The release of engineered nanomaterials to the environment. J Environ Monit 13(5):1145–1155

Gottschalk F, Lassen C, Kjoelholt J, Christensen F, Nowack B, Gottschalk F, Lassen C, Kjoelholt J, Christensen F, Nowack B (2015) Modeling flows and concentrations of nine engineered nanomaterials in the Danish environment. Int J Environ Res Public Health 12:5581–5602. https://doi.org/10.3390/ijerph120505581

Gujrati M, Malamas A, Shin T, Jin E, Sun Y, Lu ZR (2014) Multifunctional cationic lipid-based nanoparticles facilitate endosomal escape and reduction-triggered cytosolic siRNA release. Mol Pharm 11(8):2734–2744

Haghighat F, Kim Y, Sourinejad I, Yu IJ, Johari SA (2021) Titanium dioxide nanoparticles affect the toxicity of silver nanoparticles in common carp (*Cyprinus carpio*). Chemosphere 262:127805

Ham S, Yoon C, Lee E, Lee K, Park D, Chung E, Kim P, Lee B (2012) Task-based exposure assessment of nanoparticles in the workplace. J Nanopart Res 14:1–17

Handy RD, Shaw BJ (2007) Toxic effects of nanoparticles and nanomaterials: Implications for public health, risk assessment and the public perception of nanotechnology. Health Risk Soc 9(2):125–144. https://doi.org/10.1080/13698570701306807

Hansen SF, Hjorth R, Skjolding LM, Bowman DM, Maynard A, Baun A (2017) A critical analysis of the environmental dossiers in the OECD sponsorship programme for manufactured nanomaterials. Environ Sci Nano (4):282–291. http://dx.doi.org/ https://doi.org/10.1039/C6EN00465B

Hariyadi DM, Athiyah U, Pathak YV (2020) Nanomedicine: risk, safety, regulation, and public health. Nano Med Nano Saf: Recent Trends Clin EvidS 561–578. https://doi.org/10.1007/978-981-15-6255-6_21

Hartmann NB, Ågerstrand M, Lützhøft HCH, Baun A (2017) NanoCRED: a transparent framework to assess the regulatory adequacy of ecotoxicity data for nanomaterials-Relevance and reliability revisited. NanoImpact 6:81–89

He Z, Zhang Y, Feng N (2020) Cell membrane-coated nanosized active targeted drug delivery systems homing to tumor cells: a review. Mater Sci Eng C Mater Biol Appl 106:110298. https://doi.org/10.1016/j.msec.2019.110298

Hoet PH, Brüske-Hohlfeld I, Salata OV (2004) Nanoparticles-known and unknown health risks. J Nanobiotechnol 2:1–15

Hong G, Diao S, Antaris AL, Dai H (2015) Carbon nanomaterials for biological imaging and nanomedicinal therapy. Chem Rev 115:10816–10906. https://doi.org/10.1021/acs.chemrev.5b00008

Huang HJ, Lee YH, Hsu YH, Liao CT, Lin YF, Chiu HW (2021) Current strategies in assessment of nanotoxicity: alternatives to in vivo animal testing. Int J Mol Sci 22(8):1–14

Huang X, Auffan M, Eckelman MJ, Elimelech M, Kim JH, Rose J, Zuo K, Li Q, Alvarez PJJ (2024) Trends, risks and opportunities in environmental nanotechnology. Nat Rev Earth Environ 5(8):572–587

Hund-Rinke K, Simon M (2006) Ecotoxic effect of photocatalytic active nanoparticles $TiO_2$ on algae and daphnids. Environ Sci Pollut Res 13:225–232

Hussain SM, Hess KL, Gearhart JM, Geiss KT, Schlager JJ (2005) In vitro toxicity of nanoparticles in BRL 3A rat liver cells. Toxicol in Vitro 19(7):975–983

Jogaiah S, Paidi MK, Venugopal K, Geetha N, Mujtaba M, Udikeri SS, Govarthanan M (2021) Phytotoxicological effects of engineered nanoparticles: an emerging nanotoxicology. Sci Total Environ 801:149809

José BJA, Shinde MD (2024) Advancing science, technology, and engineering through sustainable nanotechnology. Asp Nanotechnol 6(1):103–109

Kaegi R, Sinnet B, Zuleeg S, Hagendorfer H, Mueller E, Vonbank R, Boller M, Burkhardt M (2010) Release of silver nanoparticles from outdoor facades. Environ Pollut 158(9):2900–2905

Kaegi R, Voegelin A, Sinnet B, Zuleeg S, Hagendorfer H, Burkhardt M, Siegrist H (2011) Behavior of metallic silver nanoparticles in a pilot wastewater treatment plant. Environ Sci Technol 45(9):3902–3908

Kakakhel MA, Wu F, Sajjad W, Zhang Q, Khan I, Ullah K, Wang W (2021) Long-term exposure to high-concentration silver nanoparticles induced toxicity, fatality, bioaccumulation, and histological alteration in fish (*Cyprinus carpio*). Environ Sci Europe 33:1–11. https://doi.org/10.21203/rs.3.rs-117675/v1

Keller AA, Lazareva A (2014) Predicted releases of engineered nanomaterials: from global to regional to local. Environ Sci Technol Lett 1(1):65–70

Khan FR, Syberg K, Palmqvist A (2017) Are standardized test guidelines adequate for assessing waterborne particulate contaminants? Environ Sci Technol 51(4):1948–1950

Khan M, Khan MSA, Borah KK, Goswami Y, Hakeem KR, Chakrabartty I (2021) The potential exposure and hazards of metal-based nanoparticles on plants and environment, with special emphasis on ZnO NPs, $TiO_2$ NPs, and AgNPs: a review. Environ Adv 6:100128

Kim JS, Yoon TJ, Yu KN, Kim BG, Park SJ, Kim HW, Lee KH, Park SB, Lee JK, Cho MH (2006) Toxicity and tissue distribution of magnetic nanoparticles in mice. Toxicol Sci 89:338–347

Kisin ER, Murray AR, Keane MJ, Shi XC, Schwegler-Berry D, Gorelik O, Arepalli S, Castranova V, Wallace WE, Kagan VE, Shvedova AA (2007) Single-walled carbon nanotubes: geno-and cytotoxic effects in lung fibroblast V79 cells. J Toxicol Environ Health A 70(24):2071–2079. https://doi.org/10.1080/15287390701601251

Klaine SJ, Alvarez PJJ, Batley GE, Fernandes TF, Handy RD, Lyon DY, Mahendra S, McLaughlin MJ, Lead JR (2008) Nanomaterials in the environment: behavior, fate, biovailability, and effects. Environ Toxicol Chem 27:1825–1851

Kühnel D, Nickel C (2014) The OECD expert meeting on ecotoxicology and environmental fate—towards the development of improved OECD guidelines for the testing of nanomaterials. Sci Total Environ 472:347–353
Kumar P, Arya V, Kumar A, Thakur N (2025) Polymer/phytochemical mediated eco-friendly synthesis of Cu/Zn doped hematite nanoparticles revealing biological properties and photocatalytic activity. Int J Mater Res 116:30–49
Lead JR, Batley GE, Alvarez PJ, Croteau MN, Handy RD, McLaughlin MJ, Judy JD, Schirmer K (2018) Nanomaterials in the environment: behavior, fate, bioavailability, and effects—an updated review. Environ Toxicol Chem 37(8):2029–2063
Lee JH, Ahn K, Kim SM, Jeon KS, Lee JS, Yu IJ (2012) Continuous 3-day exposure assessment of workplace manufacturing silver nanoparticles. J Nanopart Res 14:1–10
Lee JH, Sohn EK, Ahn JS, Ahn K, Kim KS, Lee JH, Lee TM, Yu IJ (2013) Exposure assessment of workers in printed electronics workplace. Inhalation Toxicol 25(8):426–434
Lei C, Sun Y, Tsang DC, Lin D (2018) Environmental transformations and ecological effects of iron-based nanoparticles. Environ Pollut 232:10–30
Lekamge S, Ball AS, Shukla R, Nugegoda D (2020) The toxicity of nanoparticles to organisms in freshwater. Rev Environ Contam Toxicol 248:1–80
Levard C, Hotze EM, Lowry GV, Brown GE Jr (2012) Environmental transformations of silver nanoparticles: impact on stability and toxicity. Environ Sci Technol 46(13):6900–6914
Li Z, Hulderman T, Salmen R, Chapman R, Leonard SS, Young SH, Shvedova A, Luster MI, Simeonova PP (2007) Cardiovascular effects of pulmonary exposure to single-wall carbon nanotubes. Environ Health Perspect 115(3):377–382
Li F, Liang Z, Zheng X, Zhao W, Wu M, Wang Z (2015a) Toxicity of nano-TiO2 on algae and the site of reactive oxygen species production. Aquat Toxicol 158:1–13
Li S, Liu XQ, Wang FY, Miao YF (2015b) Effects of ZnO nanoparticles, $ZnSO_4$ and arbuscular mycorrhizal fungus on the growth of Maize. Huan Jing Ke Xue 36(12):4615–4622
Li J, Mu Q, Du Y, Luo J, Liu Y, Li T (2020) Growth and photosynthetic inhibition of cerium oxide nanoparticles on soybean (*Glycine max*). Bull Environ Contam Toxicol 105:119–126
Lien C, Camille LA, Flahaut E (2017) Carbon nanotubes: Impacts and behaviour in the terrestrial ecosystem-a review. Carbon 123:767–785
Lin D, Xing B (2007) Phytotoxicity of nanoparticles: inhibition of seed germination and root growth. Environ Pollut 150:243–250
Liu JY, Sonshine DA, Shervani S, Hurt RH (2010) Controlled release of biologically active silver from nanosilver surfaces. ACS Nano 4(11):6903–6913
Liu YJ, Jing Z, Bai XT, Diao QY, Wang J, Wu YY, Zhao Q, Xia T, Xing B, Holden PA, Ge Y (2021) Nano-$La_2O_3$ induces honeybee (*Apis mellifera*) death and enriches for pathogens in honeybee gut bacterial communities. Front Microbiol 12:1–12
López-Barrera EA, Grotzner SR, Esquivel L, Voigt CL, Campos SX, de Oliveira Ribeiro CA (2021) Histopathological effects of silver nanoparticles in *Rhamdia quelen* after oral exposure. Ecotoxicol Environ Contam 16(1):83–89
Ma Y, Zhang Z, Xu Y, Ma M, Chen B, Wei L, Xiao L (2016) A bright carbon-dot-based fluorescent probe for selective and sensitive detection of mercury ions. Talanta 161:476–481
Maiti D, Tong X, Mou X, Yang K (2019) Carbon-based nanomaterials for biomedical applications: a recent study. Front Pharmacol 9:1–16
Malysheva A, Lombi E, Voelcker NH (2015) Bridging the divide between human and environmental nanotoxicology. Nat Nanotechnol 10(10):835–844
Manna I, Bandyopadhyay M (2017) Engineered nickel oxide nanoparticle causes substantial physicochemical perturbation in plants. Front Chem 5:1–16
Manzanares D, Ceña V (2020) Endocytosis: the nanoparticle and submicron nanocompounds gateway into the cell. Pharmaceutics 12(4):371
Marelle L, Thomas JL, Ahmed S, Tuite K, Stutz J, Dommergue A, Simpson WR, Frey MM, Baladima F (2021) Implementation and impacts of surface and blowing snow sources of Arctic bromine activation within WRF-Chem 4.1. 1. J Adv Model Earth Syst 13(8):e2020MS002391

Marin S, Mihail Vlasceanu G, Elena Tiplea R, Raluca Bucur I, Lemnaru M, Minodora Marin M, Mihai Grumezescu A (2015) Applications and toxicity of silver nanoparticles: a recent review. Curr Top Med Chem 15(16):1596–1604

Marques MR, Choo Q, Ashtikar M, Rocha TC, Bremer-Hoffmann S, Wacker MG (2019) Nanomedicines-tiny particles and big challenges. Adv Drug Deliv Rev 151:23–43

Matranga V, Corsi I (2012) Toxic effects of engineered nanoparticles in the marine environment: model organisms and molecular approaches. Mar Environ Res 76:32–40

Means N, Elechalawar CK, Chen WR, Bhattacharya R, Mukherjee P (2022) Revealing macropinocytosis using nanoparticles. Mol Aspects Med 83:1–30

Medina C, Santos-Martinez MJ, Radomski A, Corrigan OI, Radomski MW (2007) Nanoparticles: pharmacological and toxicological significance. Br J Pharmacol 150(5):552–558

Mirzajani F, Askari H, Hamzelou S, Farzaneh M, Ghassempour A (2013) Effect of silver nanoparticles on *Oryza sativa* L. and its rhizosphere bacteria. Ecotoxicol Environ Saf 88:48–54

Moermond C, Kase R, Korkaric M, Ågerstrand M (2016) CRED: criteria for reporting and evaluating ecotoxicity data. Environ Toxicol Chem 35(5):1297–1309

Moll J, Okupnik A, Gogos A, Knauer K, Bucheli TD, Van Der Heijden MG, Widmer F (2016) Effects of titanium dioxide nanoparticles on red clover and its rhizobial symbiont. PLoS ONE 11(5):e0155111

Mosa KA, El-Naggar M, Ramamoorthy K, Alawadhi H, Elnaggar A, Wartanian S, Ibrahim E, Hani H (2018) Copper nanoparticles induced genotoxicty, oxidative stress, and changes in superoxide dismutase (SOD) gene expression in cucumber (*Cucumis sativus*) plants. Front Plant Sci 9:1–13

Mu Q, Jiang G, Chen L, Zhou H, Fourches D, Tropsha A, Yan B (2014) Chemical basis of interactions between engineered nanoparticles and biological systems. Chem Rev 114(15):7740–7781

Mulens-Arias V, Rojas JM, Pérez-Yagüe S, Morales MP, Barber DF (2015) Polyethylenimine-coated SPIONs trigger macrophage activation through TLR-4 signaling and ROS production and modulate podosome dynamics. Biomaterials 52:494–506

Murali M, Gowtham HG, Singh SB, Shilpa N, Aiyaz M, Alomary MN, Alshamrani M, Salawi A, Almoshari Y, Ansari MA, Amruthesh KN (2022) Fate, bioaccumulation and toxicity of engineered nanomaterials in plants: current challenges and future prospects. Sci Total Environ 811:152249

Nasirzadeh K, Nazarian S, Gheibi Hayat SM (2016) Inorganic nanomaterials: a brief overview of the applications and developments in sensing and drug delivery. J Appl Biotechnol Rep 3(2):395–402

Navarro E, Baun A, Behra R, Hartmann NB, Filser J, Miao AJ, Quigg A, Santschi PH, Sigg L (2008) Environmental behaviour and ecotoxicity of engineered nanoparticles to algae, plants and fungi. Ecotoxicology 17:372–386

Nel A (2006) Toxic potential of materials at the nanolevel. Science 311(5761):622–627

Noori A, White JC, Newman LA (2017) Mycorrhizal fungi influence on silver uptake and membrane protein gene expression following silver nanoparticle exposure. J Nanopart Res 19:1–13

Notter DA, Mitrano DM, Nowack B (2014) Are nanosized or dissolved metals more toxic in the environment? A meta-analysis. Environ Toxicol Chem 33(12):2733–2739

Nowack B, Ranville JF, Diamond S, Gallego-Urrea JA, Metcalfe C, Rose J, Horne N, Koelmans AA, Klaine SJ (2012) Potential scenarios for nanomaterial release and subsequent alteration in the environment. Environ Toxicol Chem 31(1):50–59

Oberdörster G, Oberdörster E, Oberdörster J (2005) Nanotoxicology: an emerging discipline evolving from studies of ultrafine particles. Environ Health Perspect 113(7):823–839

Oberdorster E, Zhu SQ, Blickley TM, Clellan-Green P, Haasch ML (2006) Ecotoxicology of carbon-based engineered nanoparticles: effects of fullerene (C60) on aquatic organisms. Carbon 44:1112–1120

Odom TW, Huang JL, Kim P, Lieber MC (1998) Atomic structure and electronic properties of single-walled carbon nanotubes. Nature 391:62–64

Oh N, Park J-H (2014) Endocytosis and exocytosis of nanoparticles in mammalian cells. Int J Nanomed 9(Suppl 1):51–63

Ohlwein S, Kappeler R, Kutlar Joss M, Künzli N, Hoffmann B (2019) Health effects of ultrafine particles: a systematic literature review update of epidemiological evidence. Int J Public Health 64(4):547–559

Patil RM, Deshpande PP, Aalhate M, Gananadhamu S, Singh PK (2022) An update on sophisticated and advanced analytical tools for surface characterization of nanoparticles. Surf Interfaces 33:102165

Patlolla AK, Berry A, May L, Tchounwou PB (2012) Genotoxicity of silver nanoparticles in *Vicia faba*: a pilot study on the environmental monitoring of nanoparticles. Int J Environ Res Public Health 9(5):1649–1662

Pettitt ME, Lead JR (2013) Minimum physicochemical characterisation requirements for nanomaterial regulation. Environ Int 52:41–50

Pokrajac L, Abbas A, Chrzanowski W, Dias GM, Eggleton BJ, Maguire S, Maine E, Malloy T, Nathwani J, Nazar L, Sips A, Sone J, van den Berg A, Weiss PS, Mitra S (2021) Nanotechnology for a sustainable future: Addressing global challenges with the international network4sustainable nanotechnology. ACS Nano 15(12):18608–18623

Rao JP, Geckeler KE (2011) Polymer nanoparticles: preparation techniques and size-control parameters. Prog Polym Sci 36(7):887–913

Ribeiro AR, Leite PE, Falagan-Lotsch P, Benetti F, Micheletti C, Budtz HC, Jacobsen NR, Lisboa-Filho PN, Rocha LA, Kühnel D, Hristozov D, Granjeiro JM (2017) Challenges on the toxicological predictions of engineered nanoparticles. Nano Impact 8:59–72

Roiter Y, Ornatska M, Rammohan AR, Balakrishnan J, Heine DR, Minko S (2008) Interaction of nanoparticles with lipid membrane. Nano Lett 8(3):941–944

Saleem H, Zaidi SJ, Ismail AF, Goh PS (2022) Advances of nanomaterials for air pollution remediation and their impacts on the environment. Chemosphere 287(2):132083

Salieri B, Righi S, Pasteris A, Olsen SI (2015) Freshwater ecotoxicity characterisation factor for metal oxide nanoparticles: a case study on titanium dioxide nanoparticle. Sci Total Environ 505:494–502

Sanzari I, Leone A, Ambrosone A (2019) Nanotechnology in plant science: to make a long story short. Front Bioeng Biotechnol 7:1–12

Schirmer K (2014) Mechanisms of nanotoxicity. In: Frontiers of nanoscience, vol 7, pp 195–221. Elsevier

Schwab F, Bucheli TD, Lukhele LP, Magrez A, Nowack B, Sigg L, Knauer K (2011) Are carbon nanotube effects on green algae caused by shading and agglomeration? Environ Sci Technol 45(14):6136–6144

Sendra M, Moreno I, Blasco J (2019) Toxicity of metal and metal oxide engineered nanoparticles to phytoplankton. In: Ecotoxicology of nanoparticles in aquatic systems. Boca Raton, FL., CRC Press pp 1–37

Shah SNA, Shah Z, Hussain M, Khan M (2017) Hazardous effects of titanium dioxide nanoparticles in ecosystem. Bioinorg Chem Appl 2017(1):4101735

Shahrokh S, Hosseinkhani B, Emtiazi G (2014) The impact of nano-silver on bacterial aerobic nitrate reductase. J Bioprocess Biotechn 4(4):1–6

Shahzad K, Khan MN, Jabeen F, Chaudhry AS, Khan MKA, Ara C, Khan MS (2022) Study of some toxicological aspects of titanium dioxide nanoparticles through oxidative stress, genotoxicity, and histopathology in tilapia, *Oreochromis mossambicus*. Bionanoscience 12(4):1116–1124. https://doi.org/10.1007/s12668-022-01024-7

Shakeel M, Jabeen F, Shabbir S, Asghar MS, Khan MS, Chaudhry AS (2017) Toxicity of nano-titanium dioxide ($TiO_2$-NP) through various routes of exposure: a review. Biol Trace Elem Res 172(1):1–36

Shang L, Nienhaus K, Nienhaus GU (2014) Engineered nanoparticles interacting with cells: size matters. J Nanobiotechnol 12(1):1–11

Sharma S, Devi M, Kumar P, Thakur N, Kumar K, Jeet K, Sharma A, Kumar S, Kumar A, Thakur N (2025) Photocatalytic, antibacterial and antioxidant study of *Vitex negundo* mediated green synthesized nickel and neodymium doped zinc oxide nanoparticles. Toxicol Environ Chem 107:178–206

Shen Z, Chen Z, Hou Z, Li T, Lu X (2015) Ecotoxicological effect of zinc oxide nanoparticles on soil microorganisms. Front Environ Sci Eng 9:912–918

Sigmund G, Jiang C, Hofmann T, Chen W (2018) Environmental transformation of natural and engineered carbon nanoparticles and implications for the fate of organic contaminants. Environ Sci Nano 5:2500–2518. https://doi.org/10.1039/C8EN00676H

Šiller L, Lemloh ML, Piticharoenphun S, Mendis BG, Horrocks BR, Brümmer F, Medaković D (2013) Silver nanoparticle toxicity in sea urchin *Paracentrotus lividus*. Environ Pollut 178:498–502

Skjolding LM, Sørensen SN, Hartmann NB, Hjorth R, Hansen SF, Baun A (2016) A critical review of aquatic ecotoxicity testing of nanoparticles-the quest for disclosing nanoparticle effects. Angew Chem Int Ed 55:15224–15239

Smith CJ, Shaw BJ, Handy RD (2007) Toxicity of single walled carbon nanotubes to rainbow trout, (*Oncorhynchus mykiss*): respiratory toxicity, organ pathologies, and other physiological effects. Aquat Toxicol 82(2):94–109

Srivastava D, Sharma N, Arya V, Arya B, Kumar A, Amarowicz R (2024) Nanotoxicology: a threat to the environment and human health. In: Nanotechnology: agriculture, environment and health. Springer Nature Singapore, Singapore, pp 179–226

Stern ST, McNeil ES (2008) Nanotechnology safety concerns revisited. Toxicol Sci 101(1):4–21

Sukhanova A, Bozrova S, Sokolov P, Berestovoy M, Karaulov A, Nabiev I (2018) Dependence of nanoparticle toxicity on their physical and chemical properties. Nanoscale Res Lett 13:1–21. https://doi.org/10.1186/s11671-018-2457-x

Talreja N, Kumar D (2018) Engineered nanoparticles' toxicity: environmental aspects. Nanotechnol Environ Sci 737–758

Tang CH, Lin CY, Lee SH, Wang WH (2017) Membrane lipid profiles of coral responded to zinc oxide nanoparticle-induced perturbations on the cellular membrane. Aquat Toxicol 187:72–81

Thangavel P, Park D, Lee YC (2022) Recent insights into particulate matter (PM2. 5)-mediated toxicity in humans: an overview. Int J Environ Res Public Health 19(12):1–22

Thill A, Spalla O, Chauvat F, Rose J, Auffan M, Flank AM (2006) Cytotoxicity of $CeO_2$ nanoparticles for *Escherichia coli*: a physico-chemical insight of the cytotoxicity mechanism. Environ Sci Technol 40:6151–6156

Tiple AD, Badwaik VJ, Padwad SV, Chaudhary RG, Singh NB (2020) A review on nanotoxicology: aquatic environment and biological system. Mater Today Proceed 29:1246–1250

Tourinho PS, Van Gestel CA, Lofts S, Svendsen C, Soares AM, Loureiro S (2012) Metal-based nanoparticles in soil: fate, behavior, and effects on soil invertebrates. Environ Toxicol Chem 31(8):1679–1692

Tran TK, Nguyen MK, Lin C, Hoang TD, Nguyen TC, Lone AM, Khedulkar AP, Gaballah MS, Singh J, Chung WJ, Nguyen DD (2023) Review on fate, transport, toxicity and health risk of nanoparticles in natural ecosystems: emerging challenges in the modern age and solutions toward a sustainable environment. Sci Total Environ 912:169331

Uddin MdN, Desai F, Asmatulu E (2020) Engineered nanomaterials in the environment: bioaccumulation, biomagnification and biotransformation. Environ Chem Lett 18(4):1073–1083. https://doi.org/10.1007/s10311-019-00947-0

Unrine JM, Tsyusko OV, Hunyadi SE, Judy JD, Bertsch PM (2010) Effects of particle size on chemical speciation and bioavailability of copper to earthworms (*Eisenia fetida*) exposed to copper nanoparticles. J Environ Qual 39(6):1942–1953

US Environmental Protection Agency (USEPA) (2015) Regulatory actions under TSCA section 5. https://www.epa.gov/reviewing-new-chemicals-under-toxic-substances-control-act-tsca/actions-under-tsca-section-5

Verma A, Stellacci F (2009) Effect of surface properties on nanoparticle-cell interactions. Small 6(1):12–21. https://doi.org/10.1002/smll.200901158

Verma Y, Rani V, Rana SVS (2020) Assessment of cadmium sulphide nanoparticles toxicity in the gills of a fresh water fish. Environ Nanotechnol Monitor Manag 13:100280

Wang F, Liu X, Shi Z, Tong R, Adams CA, Shi X (2016) Arbuscular mycorrhizae alleviate negative effects of zinc oxide nanoparticle and zinc accumulation in maize plants-a soil microcosm experiment. Chemosphere 147:88–97

Warheit DB, Laurence BR, Reed KL, Roach DH, Reynolds GA, Webb TR (2004) Comparative pulmonary toxicity assessment of single-wall carbon nanotubes in rats. Toxicol Sci 77(1):117–125

Xiao A, Wang C, Chen J, Guo R, Yan Z, Chen J (2016) Carbon and metal quantum dots toxicity on the microalgae *Chlorella pyrenoidosa*. Ecotoxicol Environ Saf 133:211–217

Yousef MI, Abuzreda AA, Kamel MAN (2019) Cardiotoxicity and lung toxicity in male rats induced by long-term exposure to iron oxide and silver nanoparticles. Exp Ther Med 18(6):4329–4339

Yun JW, Kim SH, You JR, Kim WH, Jang JJ, Min SK, Kim HC, Chung JDHJ, Kang B-C, Che JH (2015) Comparative toxicity of silicon dioxide, silver and iron oxide nanoparticles after repeated oral administration to rats. J App Toxicol 35(6):681–693

Zhang J, Wang F, Yalamarty SSk, Filipczak N, Jin Y, Li X (2022) Nano silver-induced toxicity and associated mechanisms. Int J Nanomed 17:1851–1864

Zhu B, Han J, Lei L, Hua J, Zuo Y, Zhou B (2021) Effects of $SiO_2$ nanoparticles on the uptake of tetrabromobisphenol A and its impact on the thyroid endocrine system in zebrafish larvae. Ecotoxicol Environ Saf 209:111845

Zuin S, Gaiani M, Ferrari A, Golanski L (2013) Leaching of nanoparticles from experimental water-borne paints under laboratory test conditions. J Nanopart Res 16:1–17